Climate and Biodiversity in Infrastructure

Developed through the Zofnass Program at the Harvard Graduate School of Design, *Climate and Biodiversity in Infrastructure* proposes a high-level conceptual framework to holistically capture how integrated climate and biodiversity action is defined at the infrastructure project level. At its core, it proposes a set of criteria to jointly address two of the most significant, concurrent, and interconnected crises of our time: climate change and biodiversity loss. These criteria aim to mainstream biodiversity into climate, and vice versa, as a way to achieve multiple goals.

The criteria represent a framing and scoping approach for infrastructure project teams to identify and prioritize key issues and risks associated with climate change and biodiversity loss, outline the required level of ambition, identify options and opportunities, and inform decision-making.

Fields covered include landscape design, energy, water and wastewater, food, solid waste, transportation, and information. This book will be especially useful to decision-makers in countries with mandates for net zero carbon, such as Europe and the US, that are called to prioritize the right projects for funding to reach their goals. Moreover, the book will be particularly useful to students of the design professions and especially planning and engineering.

SPIRO N. POLLALIS is Professor of Design, Technology, and Management, Emeritus at the Harvard Design School, USA. He was the founder and Director of the Zofnass Program for the Sustainability of Infrastructure at Harvard (2007-2022), which led to the development of the Envision® Rating System and the establishment of the Institute of Sustainable Infrastructure. Envision is used worldwide for rating the sustainability of infrastructure and training professionals. Prof. Pollalis's research has focused on sustainability, and in his private practice, he has planned sustainable cities and consulted with governments and multilateral organizations.

EVGENIA HAGISTAVROU is an architect and landscape architect. She has been a core member of the Zofnass Program research team from 2011 to 2022. Her research work has focused on infrastructure sustainability assessment at both system and project levels, the analysis of Environmental, Social, and Governance (ESG) systems, and the integration of triple-bottom-line sustainability into lifecycle assessments.

Climate and Biodiversity in Infrastructure

DIRECTED BY
SPIRO N. POLLALIS

EDITED BY
SPIRO N. POLLALIS & EVGENIA HAGISTAVROU

Routledge
Taylor & Francis Group
NEW YORK AND LONDON

Designed cover image: Alexandra Papagianni,
Chrysostomos Tsimourdagkas

First published 2026
by Routledge
605 Third Avenue, New York, NY 10158

and by Routledge
4 Park Square, Milton Park, Abingdon, Oxon OX14 4RN

Routledge is an imprint of the Taylor & Francis Group, an informa business

ISBN: 9781032853048 (hbk)
ISBN: 9781032853031 (pbk)
ISBN: 9781003517573 (ebk)

DOI: 10.4324/9781003517573

Typeset in FF DIN OpenType TTF Std

Publisher's Note
This book has been prepared from camera-ready copy provided by the author.

CONTENTS

PREFACE vii
ACKNOWLEDGMENTS viii
BOOK CREDITS ix

PART A: SUSTAINABLE INFRASTRUCTURE AND THE CLIMATE-BIODIVERSITY CRISIS 2
Industry perspectives
Chapter 1: Urgency for Climate and Biodiversity Action 14
The twin climate and biodiversity crises as threats
Urgent need for addressing the action gap
Chapter 2: Need for Integration 30
Need for an integrated approach to biodiversity and climate crises
Biodiversity as an integral component of climate action

PART B: FRAMING CLIMATE AND BIODIVERSITY ACTION IN INFRASTRUCTURE PROJECTS 38
Industry perspectives
Chapter 3: General Framework 46
Scope
Methodology
Chapter 4: Climate Action Criteria 68
Key concepts and approaches shaping the climate action criteria
Climate change mitigation criteria
Climate change adaptation criteria
Mitigation and adaptation criteria overlaps
Chapter 5: Biodiversity Action Criteria 108
Key concepts and approaches shaping the biodiversity action criteria
Biodiversity action criteria
Chapter 6: Integrated Climate-Biodiversity Action Criteria 134
Merging the climate and biodiversity action criteria
Key strategies for integrated action

PART C: CONNECTING THE CRITERIA WITH SUSTAINABLE PROJECT STRATEGIES 158

Chapter 7: Methodology for Project Analysis 160
Chapter 8: Climate-Biodiversity Action in a Transit Project 164
Project overview
Analysis based on the integrated criteria
Key takeaways
Chapter 9: Climate-Biodiversity Action in a Wind Energy Project 214
Project overview
Analysis based on the integrated criteria
Key takeaways

BIBLIOGRAPHY AND SOURCES 262
IMAGE CREDITS 270
RESEARCH TEAM FOR THE DEVELOPMENT OF THE CRITERIA 272
CONTRIBUTORS 274
INDEX 276

PREFACE

This book is the result of the research effort of the Zofnass Program at the Harvard Graduate School of Design in the period 2020-2022 and the subsequent research by Prof. Pollalis's team. The book is triggered by the ongoing global discourse on the parallel climate change and biodiversity crises and the need for pairing net-zero climate targets with nature-positive targets. It calls infrastructure projects to respond to the urgency for addressing climate change mitigation, and adaptation needs, while mainstreaming biodiversity into project planning and decisions. To do so, it introduces high-priority criteria for framing climate and biodiversity action at the infrastructure project level.

The criteria bring forth the crucial issues that projects are called to address in the current context of urgency for climate and biodiversity action. They break down key concepts related to climate and biodiversity action as they appear in key publications and high-profile initiatives, e.g., national, sector-level, city-level, industry-level, and corporate, and translate them into infrastructure and project-specific language, to facilitate a better understanding of how these concepts can be applied in infrastructure projects.

Professor Spiro N. Pollalis
Zofnass Program Director
Professor of Design, Technology and Management at the Harvard Design School

ACKNOWLEDGMENTS

This research was made possible through the generous support of Paul Zofnass and Joan Zofnass and industry leaders, whose determination and commitment to sustainability led to the establishment and operations of the Zofnass Program at Harvard University in 2007. The Program's founding objective was to create a sustainability rating system for infrastructure projects, developed in close collaboration with the Harvard professional schools and with industry partners.

In 2007, Professor Spiro Pollalis founded the Zofnass Program for the Sustainability of Infrastructure, was appointed the first Director of the Program and assumed the task of formulating and implementing its vision. A team of researchers and students assembled to advance this work and the Criteria for Climate and Biodiversity Action is the direct or indirect outcome of a multi-year effort on sustainability of the many researchers who have contributed to the Zofnass Program over time. The Zofnass Program has also been supported by Harvard University faculty advisors from the Graduate School of Design, the Center for the Environment, the School of Public Health, the Business School, and the Law School, whose diverse expertise has been invaluable to the Program.

In parallel, the Zofnass Program established the Sustainable Industry Advisory Board (SIAB). SIAB members supported the Program financially and, equally important, with the commitment of the firms' leadership, sharing professional expertise and providing valuable feedback to the Program's research.

The Program organized quarterly, invitation-only workshops attended by its researchers, Harvard faculty advisors, SIAB members, and invited guests, including public authorities and private stakeholders involved in large-scale projects. These workshops fostered intensive dialogs on complex issues requiring the highest levels of expertise and an up-to-date understanding of industry practice. We attribute many of our research outcomes to the insights and exchanges made possible through these workshops.

The Zofnass Program maintained a close relation with the Institute for Sustainable Infrastructure (ISI, www.sustainableinfrastructure.org). Established in 2012 as a nonprofit organization with three founding members: the American Society of Civil Engineers (ASCE), the American Council of Engineering Companies (ACEC), and the American Public Works Association (APWA), ISI adopted our Zofnass rating system, giving it a new life after the research phase, and has transformed it into Envision®, a commercial-grade product for self-evaluating the sustainability of infrastructure projects, for rating them, and for training Envision® certified professionals (ENV SP). Our interactions with ISI have had an impact on our work.

The Climate and Biodiversity Action Criteria were presented, and discussed in workshops during the two-year research. Long-standing collaborators of the Zofnass Program provided valuable feedback on the research and contributed introductory essays sharing their valuable professional perspectives.

We extend our gratitude to Santa Clara Valley Transportation Company (VTA) and to Alliant Energy for granting permission to use their projects as case studies, and for providing access to essential project information.

Special thanks are due to Mrs. Ana Calnan, former Environmental Programs Manager at VTA; Mrs. Rachael Keish, Environmental Coordinator at VTA and CEO at Keish Environmental; Mrs. Christina Jaworski, current Environmental Programs Manager at VTA; Mrs. Lani Lee Ho, Environmental Planner at VTA; Mr. Alan Arnold, former Environmental Renewables Program Manager at Alliant Energy; Mr. William Skalitzky, Environmental Services Manager at Alliant Energy; Mr. Anthony Kane, President and CEO at ISI; and Ms. Judith Rodriguez, Research Manager at ISI.

BOOK CREDITS

The work presented in this book is the result of the research effort of the Zofnass Program at the Harvard Graduate School of Design in the period 2020-22, and was updated during 2024-25. The entire team seamlessly contributed to the methodology, graphics and texts.

Principal Investigator of the Zofnass Program
Director of the Climate and Biodiversity Research
Professor Spiro N. Pollalis

Lead Authors
Professor Spiro N. Pollalis
Evgenia Hagistavrou

Contributing Authors
Angela Kouveli
Eleonora Marinou
Olga Tzioti

Core Research Team
Evgenia Hagistavrou
Angela Kouveli
Dimosthenis Lappas
Eleonora Marinou
Olga Tzioti

Editors
Professor Spiro N. Pollalis
Evgenia Hagistavrou

Publication Coordination
Evgenia Hagistavrou

Design and Layout Coordination
Eleonora Marinou
Alexandra Papagianni
Chrysostomos Tsimourdagkas
Olga Tzioti

Climate and Biodiversity in Infrastructure

A

SUSTAINABLE INFRASTRUCTURE AND THE CLIMATE-BIODIVERSITY CRISIS

Part A sets the context for the book. Chapter 1 outlines the urgency of responding to the climate and biodiversity crises, framing the broader context in which infrastructure projects are expected to contribute. Chapter 2 highlights the need for coordinated action to address both crises together.

INDUSTRY PERSPECTIVES

LIVING WITHIN THE PLANETARY BOUNDARIES, ONE INFRASTRUCTURE AT A TIME

After over 20 years of debating, on March 4, 2024, the International Commission on Stratigraphy (ICS) voted against formally declaring the Anthropocene as a chronostratigraphic unit in the geological history of the Earth. Instead, the ICS has suggested that the Anthropocene can be considered as "an event, similar to the great transformative events in Earth history such as the Great Oxygenation (2.4–2.1 Ga), the Cambrian Explosion, or the Great Ordovician Biodiversification events."[1] Although disappointing for the proponents of the higher classification, this decision is a recognition that the magnitude of the changes brought about by humanity on Earth's natural processes has reached global proportions and will leave a permanent mark on the planet's geological history.

It is perhaps useful to remember some key data and information on the depth and breadth of just a few of the planetary changes due to human influence. The global cumulative greenhouse gas emissions generated by human activities from 1850 to 2022 amount to 2,575 Gtons CO_2, which represents 25% more than the amount of all the carbon stored in the vegetation at the global level (703 vs 560 Gtons of C).[2] This has caused an increase in the concentration of CO_2 in the atmosphere from 280 ppm in the pre-industrial levels to over 422 ppm in 2024,[3] a level not reached in the past 14 million years.[4] Globally, over one third of fish stocks are overexploited, with some fisheries that decreased by over 70% and others unable to recover even following the stop of fishing activities.[5] Of the total biomass of mammals on the planet, only 4% is represented by wild animals, while humans account for 34% and farmed animals like cows, pigs and sheep represent the remaining 62%. Almost half the Earth's habitable land is used for agriculture; of this, 80% is used for feeding livestock, even if they provide only 17% of the global calorie supply.[6] The total length of roads built on the planet is estimated to be between 20 and 32 million kilometers,[7] a distance that would take us halfway to Mars at its closest distance from Earth.

It is somewhat puzzling to think that changes of this magnitude are the result of the individual and collective work of billions of people over a period of a few thousand years, a mere blip in the planet's history. It is a multitude that, for the most part of human history, has looked at the Earth as a limitless place and has been completely unaware of its contribution to a series of global geophysical and biological events that could undermine the foundations on which the very civilization it built lies.

But the awareness is now, within the limits of science, complete, and with that comes the responsibility of considering in full the consequences of our individual and collective acts, particularly for those who are involved in planning, designing, and building infrastructures whose life can span the life of generations and whose environmental and social impacts can embrace nations, if not continents.

Perhaps the most powerful representation of the limits of the planet is that of "planetary boundaries,"[8] which define the limits of acceptable alteration to key Earth systems, developed by Johan Rockstrom and his colleagues at the Stockholm Resilience Centre. The first assessment of seven planetary boundaries in 2009 revealed that three (biosphere integrity, climate change, and biogeochemical flows of nitrogen) were already crossed. This already worrying picture worsened dramatically in the 2023 assessment, when six out of the nine boundaries assessed were crossed.[9] Besides the assessment of the individual boundaries, it is important to consider the interrelations among them, particularly between biodiversity and climate change.

The biosphere has a fundamental role in maintaining the global concentrations of greenhouse gas within certain boundaries, through the combination of metabolic and ecological processes that regulate the distribution of carbon among the different compartments (soil, ocean, atmosphere, biomass). However, the ability of the biosphere to perform this global ecosystem service is severely impaired by the effects of higher temperatures, catastrophic events like hurricanes and wildfires, and erratic changes in the precipitation patterns. Perhaps the most notable example of this relationship is the Amazon basin, historically considered a carbon sink of global relevance. According to recent studies, a combination of increased temperatures, decreased precipitations, and deforestation activities has flipped at least 20% of the Amazon forest from a carbon sink to a carbon source.[10] Similarly, the combination of increased forest fires, widespread tree mortality due to insects thriving in higher temperatures, and accelerated decomposition of organic material—again favored by higher temperatures—have turned Canadian boreal forests into a carbon source.[11] Infrastructure development is intimately connected to our ability to avoid crossing the planetary boundaries that are still in the safe area and to have a chance to bring back within safe boundaries those that are currently in a dangerous area. Among the most notable examples we can mention: stopping the development of infrastructures that increase the extraction and use of fossil fuels and increasing those that promote energy efficiency and sustainable energy production; the widespread adoption of nature-based solutions instead of grey infrastructures; and the planning and design of compact urban forms and smart public and private service planning to minimize the need for private transport.

In the past two decades, many (more or less genuine and successful) attempts have been made to define what makes an infrastructure "sustainable," and the best and more genuine attempts have tried to incorporate the best scientific and empirical knowledge about the effects of infrastructures on the ecological and human environment. One of the limits of these efforts is the consideration of the effects in a sectoral way, one environmental or social component at a time, with limited consideration for the secondary and cumulative effects, and for the feedback loops that are inherent in ecological and geochemical processes.

These limits are well known to those who practice environmental and social impact assessment of infrastructure projects, as they are embedded in the way the environmental legislation is written in most of the world, and in the way scientific and engineering education systems still work, not promoting holistic thinking and interdisciplinarity.

Trying to bring together in a coherent framework the multiple interrelations among infrastructures, climate change and biodiversity, as described in this volume, goes in the right direction as it offers practitioners a solid guide to dramatically improve their planning, design, and assessment work. The value of this book is particularly high since the past 20 years have seen the development of many different international standards and frameworks, with little effort to try to bring them under a single frame of reference that can be put into action at an infrastructure project level.

Thoroughly applying the framework will certainly improve the work of infrastructures planners, designers and assessors; however, it will not in itself address the enormous challenge facing humanity if those applying the standards do not fully exert their influence over the decision-making process. The mantra of the work that led to the Envision protocol by the Institute for Sustainable Infrastructure and the Zofnass Program at Harvard has been that one should not only do an infrastructure right, but rather focus on doing the right infrastructure.

This apparently simple dichotomy reveals the fact that there is only so much one can do to maximize the benefits and minimize the negative impacts of an infrastructure that has been decided through a flawed political process or that has been poorly planned and wrongly sited. In these cases, it takes courage and integrity to challenge the fundamentals of a "wrong" project, but it is exactly what practitioners should do to the best of their possibilities, and sometimes to the point of withdrawing from the truly irredeemable projects. Engineers, planners, and environmental scientists are not required to take the Hippocratic Oath, but in the face of the catastrophic consequences for humankind of uncontrolled climate change and widespread collapse of the biosphere, they should feel obligated to act with the same rigor doctors use to save a single human life.

Roberto Mezzalama
Global ESIA Subject Matter Expert at WSP Italia

NOTES

1. Håvard Kilhavn et al., "Comment to the ICS Anthropocene Decision: From Stratigraphy to Storytelling," *Quaternary Environments and Humans* 2, no. 6 (October 28, 2024): 100036, https://doi.org/10.1016/j.qeh.2024.100036.
2. P Friedlingstein et al, "Global Carbon Budget 2023," *Earth System Science Data* 15, no. 12 (2023): 5301-5369, https://doi.org/10.5194/essd-15-5301-2023.
3. NOAA Global Monitoring Laboratory, "Trends in Atmospheric Carbon Dioxide," *NOAA Global Monitoring Laboratory*, accessed October 12, 2024, https://gml.noaa.gov/ccgg/trends/.
4. The Cenozoic CO_2 Proxy Integration Project (CenCO_2PIP) Consortium, "Toward a Cenozoic History of Atmospheric CO_2," *Science* 382, no. 6676 (December 8, 2023): eadi5177, https://doi.org/10.1126/science.adi5177.
5. Hannah Ritchie and Max Roser, "Fish and Overfishing," *Our World in Data*, July 2021. https://ourworldindata.org/fish-and-overfishing.
6. Hannah Ritchie and Max Roser, "Global Land Use for Food Production," *Our World in Data*, November 2019, https://ourworldindata.org/global-land-for-agriculture.
7. Johan R Meijer et al., "Global Patterns of Current and Future Road Infrastructure," *Environmental Research Letters* 13, no. 6 (April 11, 2018): 064006, https://doi.org/10.1088/1748-9326/aabd42.
8. Nine planetary boundaries are currently defined and measured: biosphere integrity, climate change, ocean acidification, freshwater use, atmospheric aerosol loading, the introduction of novel entities, biogeochemical flows, land-system change, and stratospheric ozone depletion.
9. Katherine Richardson et al., "Earth Beyond Six of Nine Planetary Boundaries," *Science Advances* 9, no. 37 (September 13, 2023): eadh2458, https://doi.org/10.1126/sciadv.adh2458.
10. Luciana V. Gatti et al., "Amazonia as a Carbon Source Linked to Deforestation and Climate Change," *Nature* 595, no. 7867 (July 14, 2021): 388–93, https://doi.org/10.1038/s41586-021-03629-6.
11. Natural Resources Canada, "Forest Carbon," Government of Canada, accessed October 12, 2024. https://natural-resources.canada.ca/climate-change/climate-change-impacts-forests/forest-carbon/13085.

OVERCOMING BARRIERS IN SUSTAINABLE INFRASTRUCTURE: INTEGRATING NATURE'S BENEFITS IN THE UPSTREAM POLICY AND ENABLING ENVIRONMENT

Nature-based solutions (NbS) for climate adaptation are increasingly being applied by engineering firms around the world, with thousands of projects each year harnessing combinations of built infrastructure and restored natural vegetation. These approaches, however, remain far too limited in scale and overly reliant on restoration. To address the urgency of climate change and biodiversity loss, the values, services, and performance functions of intact nature must become integral to infrastructure planning and development.

Around the world, weather and climate extremes continue to break historical records as temperatures keep climbing with increasing CO_2 emissions. In 2024, global average temperatures officially passed 1.5°C above pre-industrial levels for the first time, and 2025 is on pace to be the second- or third-warmest year in recorded history.[1] A year of more extreme hazards provides the evidence: in Pakistan, flooding killed over 1,000 people and displaced 2 million; in central Texas, more than 20 inches of rainfall—at rates exceeding 5 inches an hour—killed hundreds; and in Los Angeles, historically devastating and fast-moving fires destroyed entire neighborhoods, causing an estimated $65 billion in damages.[2,3,4] At the same time, climate change is exacerbating biodiversity loss driven by habitat fragmentation, infrastructure development, pollution, and deforestation. A record 17 million acres of tropical primary forest were lost in 2024—the highest in two decades—with five times more primary forest burned than in 2023, partly due to climate change.[5] Globally, there has been a 73% decline in the average size of monitored wildlife populations, with losses as high as 95% in Latin America and the Caribbean.[6]

There has therefore never been a greater need for integrated solutions that halt and reverse nature loss while slowing climate change through economic

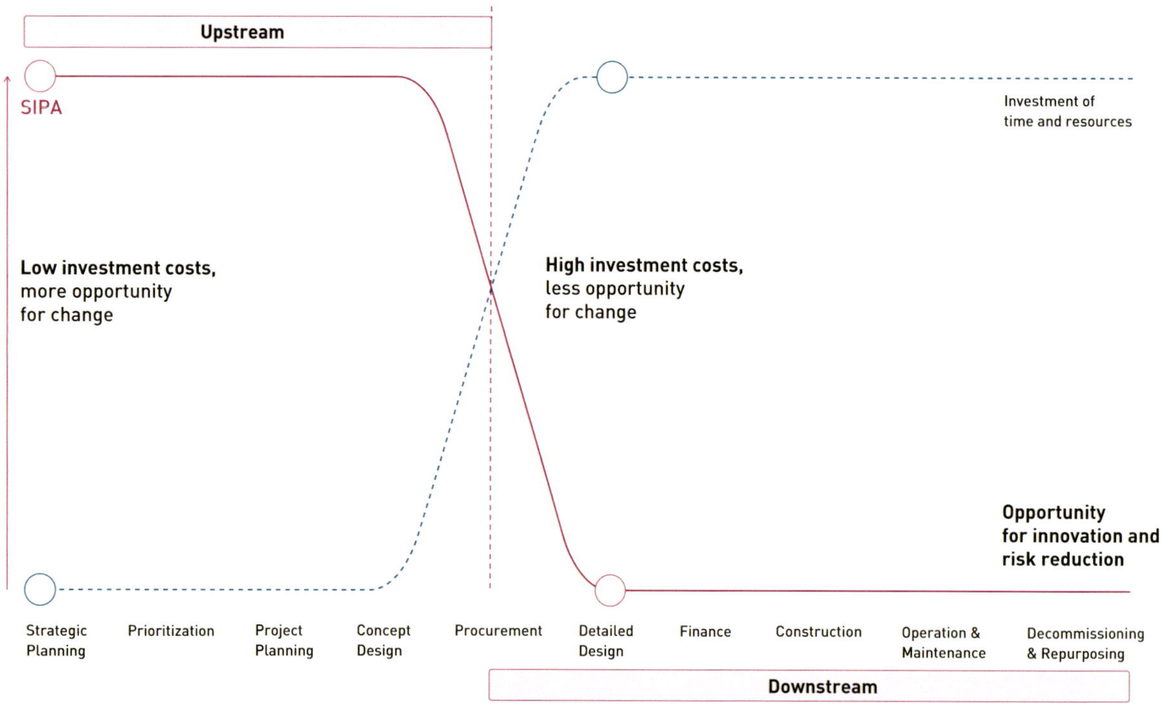

The infrastructure development lifecycle and the opportunity for influencing sustainability outcomes (Source: WWF, 2025)
As the cycle progresses, invested costs increase, making changes to planning or design to avoid major risks or impacts on nature less likely and more costly. SIPA refers to the Sustainable Infrastructure Program in Asia (SIPA) project, targeting the earliest stages of planning, described in more detail below.

development that is both net-zero and nature-positive.[7] With 60%–70% of the infrastructure anticipated to be built by 2050 not yet constructed—including approximately 3–5 million km of new paved roads, 90% of which will be in developing nations—infrastructure planning, design, and development practices are central to achieving national and global nature-positive and decarbonization goals.[8,9,10] NbS are critical to achieving these goals, but many current approaches focus too narrowly on restoring what is degraded rather than truly integrating intact nature and its values, especially at the earliest, upstream stages of development (see figure on previous page). Moving forward, to shift development onto a less destructive path, the engineering and design sectors must fully integrate nature into infrastructure development.

Barriers Preventing Greater Action: Policy and Regulatory Enabling Conditions

Current infrastructure standards, practices, and codes in many countries have remained largely unchanged for decades. Most were developed for a stationary climate and before global consensus emerged on the critical need to address the biodiversity crisis. At the policy level, strategic economic development planning still largely fails to adequately incorporate nature and biodiversity—and the significant multiple benefits they provide, including climate resilience.

A particularly well-known example of standards failing under non-stationarity is requirements for hydrological control infrastructure like reservoirs, bridges, and flood control systems to be designed for 100-year storms (that have a 1% chance of occurring in any given year). Precipitation frequency is typically based on historical data. Over the past 135 years, however, more than half of the globe has experienced significant changes in precipitation quantities, and by one estimate 78% of the Earth is now experiencing non-stationarity.[11] As a result, these projects now regularly fail under new climate extremes for which they were not designed.

The rapid loss of biodiversity intensifies the impacts of new climate extremes, driven by limited understanding of how infrastructure affects ecosystems—and how much it relies on them for essential services such as erosion control and flood risk reduction. For example, the deforestation and degradation impacts of first-cut roads in tropical forests have been shown to extend 17–68 km and 100–200 acres spatially for every kilometer constructed.[12] This type of degradation—and the resulting loss of ecosystem services—is rarely considered in the strategic economic development that often creates the access to these landscapes in the first place. Some countries have begun screening nationally strategic development projects for carbon intensity, but they are not similarly accounting for major impacts on, or dependencies on, nature and biodiversity and the valuable services they support.[13]

While this issue is critical in regions with existing infrastructure, it is even more consequential in areas where infrastructure is still being planned or built—particularly in many low- and middle-income countries. In these contexts, planners often adopt legacy approaches that favor grey infrastructure solutions ill-suited to local ecological and climate realities. This replication of outdated models not only risks locking in maladaptive infrastructure but also exacerbates vulnerability by sidelining nature-based and context-specific solutions that could support greater resilience in a changing climate.

Another impediment is current cost-benefit analysis practices and standards. Standard practice often reduces project evaluation to overly simplistic ratios that rarely encompass the multiple benefits provided by nature or account for long time horizons. Competitive procurement bids for infrastructure projects often do not allow designers and engineers to pursue low-impact solutions because design and construction costs appear to exceed those of traditional solutions. As part of a recent US Army Corps of Engineers (USACE) process for determining optimal coastal protection infrastructure for Miami-Dade County, for example, original designs proposed a six-mile-long, 13-foot seawall obstructing coastal views. Under significant backlash from residents who demanded inclusion of more nature-based approaches—such as mangroves and recreation access—the plan was revised to include $250 million for pilot studies to evaluate NbS, requiring a waiver from the regional USACE commander to bypass the typical cost-benefit ratio requirements.[14,15,16]

Solutions: Integrating Nature Upstream in Policy and Strategic Planning

In response to these challenges, World Wildlife Fund (WWF) has developed several projects with partners aimed at increasing the consideration of nature and biodiversity—including their role in supporting climate resilience—within the infrastructure sector at the earliest upstream stages of planning, when there is the greatest opportunity to integrate biodiversity and climate. We highlight two initiatives that tackle regulatory and policy barriers head-on.

Sustainable Infrastructure Program in Asia (SIPA)

To address national and local gaps in economic development planning that have historically overlooked nature's benefits, WWF has worked in collaboration with international civil society partners, governments, and academia in Indonesia and the Philippines to integrate NbS for climate resilience into strategic infrastructure planning. This work is part of the Sustainable Infrastructure

Program in Asia (SIPA), led by the Organisation for Economic Co-operation and Development (OECD).[17] WWF's work on the project defined national priorities for NbS that support climate resilience for both communities and road infrastructure through innovative geospatial analysis of priority benefits provided by intact or restored ecosystems: sediment retention and erosion control, coastal protection, flood risk reduction, and water supply.

The results are striking. Intact ecosystems on just 16% of Indonesia's lands most impacted by the road network (within 15 km of roads and 25 km of settlements) support climate resilience for 147 million people and 58% of roads (see figure below). Restoration offers a similarly significant opportunity for a relatively small area of the country. Restoring just 11% of the country's degraded lands to forests would improve resilience through NbS for 143 million people and roughly half the road network.[18] These assessments, built through years of technical collaboration with cross-ministerial working groups, now provide a foundation for the country to integrate NbS into strategic development plans so that future developments can avoid priority areas. This work serves as powerful initial guidance for targeted restoration opportunities—for example, to reduce road operations and maintenance costs.

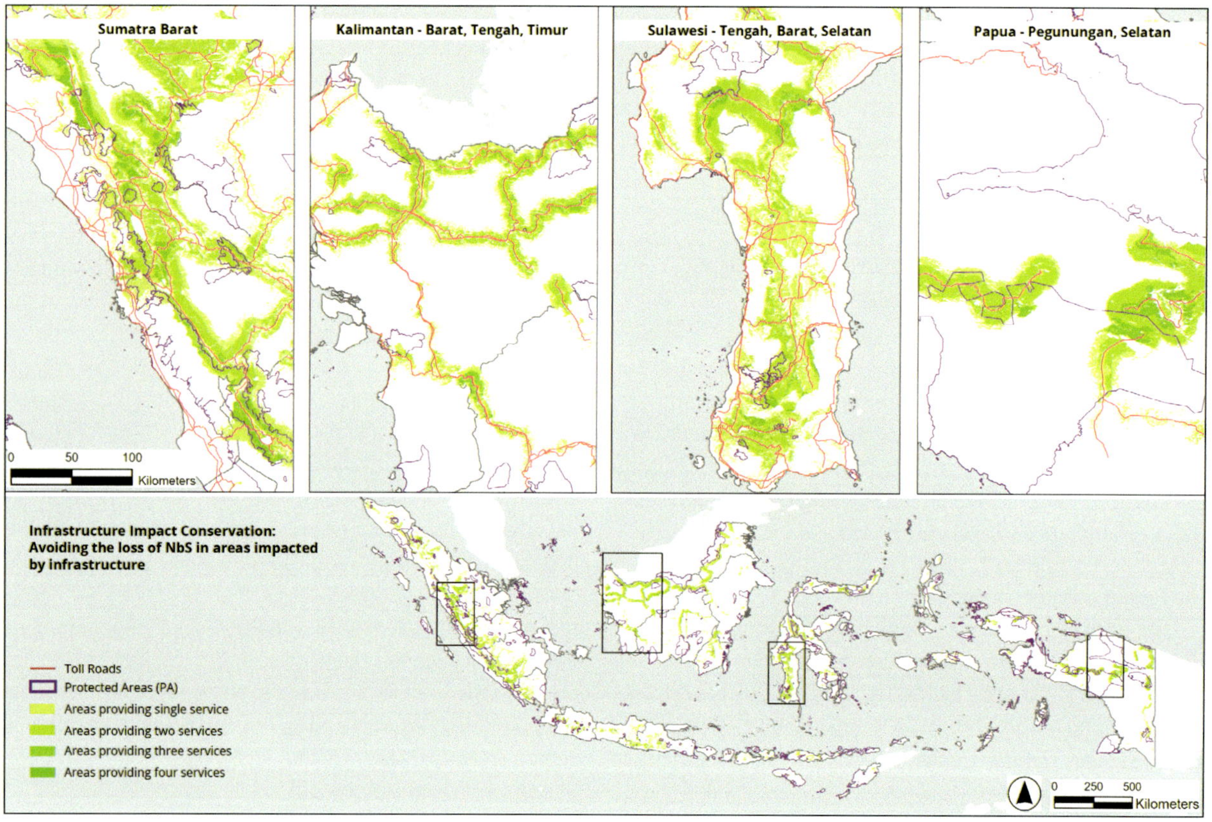

Example analysis of priority areas supporting the highest concentration of ecosystems services critical to climate resilience for the current road network in Indonesia. (Source: WWF, 2025)

Greening Transportation Infrastructure Development Integrated Program

Funded by the Global Environment Facility (GEF), WWF is also leading the Greening Transportation Infrastructure Development (GRID) Integrated Program in partnership with the Asian Development Bank, the UN Environment Programme (UNEP), and country governments in Malaysia, Nepal, the Philippines, Suriname, and Ukraine. GRID has two components—a global platform and government-led national projects—that will similarly mainstream nature considerations across multiple scales of transportation infrastructure planning and development. Through training programs, technical assistance, and policy and regulatory reforms (for example, integrating multicriteria project evaluation approaches into infrastructure procurement practices), GRID aims to shift the infrastructure sector toward greater integration of biodiversity and ecosystem services, including those that support resilience in the target countries and globally.[19]

Conclusion

This essay only begins to explore how humanity can better align infrastructure with nature's values, a challenge central to addressing worsening climate change and biodiversity loss. The examples presented here demonstrate how infrastructure development must evolve to meet the scale of the challenge. Analysis of NbS potential for mainstreaming in upstream strategic and spatial national and local development planning offers a sector- and region-agnostic model that can be replicated anywhere. Policy and regulatory reforms can improve procurement and design and incentivize the adoption of analytical approaches that yield more nature-positive projects. Strengthening the capacity of key technical staff within ministries and agencies at national and subnational levels is critical to guide planning and development of major projects toward these outcomes.

Ryan Bartlett
Director, Climate-Smart Planning and Design, WWF-US
Daniel Hornett
Senior Program Officer, Sustainable Infrastructure, WWF-US
Shail Joshi
Senior Program Officer, Climate-Smart Planning and Design, WWF-US

NOTES

1. Mark Poynting, Erwan Rivault and Becky Dale, "World's Hottest Year: 2024 First to Pass 1.5C Warming Limit," *BBC News*, January 10, 2025, https://www.bbc.com/news/articles/cd7575x8yq5o.
2. Asim Tanveer and Babar Dogar, "Pakistanis Displaced by Flooding in Punjab Return to Find Homes and Crops Destroyed," *AP News*, September 18, 2025, https://apnews.com/article/pakkistan-floods-punjab-damages-assessment-losses-6c28b31b8bbcbc47c0558fa7a2326f0c.
3. Jorge L. Ortiz, "'A New Reality': Price Tag for LA Fires Pegged at $65 Billion, Report Says," *USA Today*, July 22, 2025, https://www.usatoday.com/story/news/nation/2025/07/22/cost-la-fires-65-billion/85310181007/.
4. Ben Moore et al. *Summary of the Central Texas Floods in Early July 2025*. n.d. https://cw3e.ucsd.edu/wp-content/uploads/2025/07/14Jul2025_TexasFloods_EventSummary/TexasFloods_EventSummary.pdf.
5. World Resources Institute, "Fires Drove Record-Breaking Tropical Forest Loss in 2024 | World Resources Institute Research," accessed October 15, 2025, https://gfr.wri.org/latest-analysis-deforestation-trends.
6. World Wildlife Fund, "2024 Living Planet Report: A System in Peril," accessed August 27, 2024. https://livingplanet.panda.org/en-US/.
7. Martine Maron et al. "'Nature Positive' Must Incorporate, Not Undermine, the Mitigation Hierarchy," *Nature Ecology & Evolution* 8, no. 1 (2024): 14–17, https://doi.org/10.1038/s41559-023-02199-2.
8. Johan R Meijer et al. "Global Patterns of Current and Future Road Infrastructure," *Environmental Research Letters* 13, no. 6 (2018): 064006, https://doi.org/10.1088/1748-9326/aabd42.
9. Mohammed Alamgir et al. "Economic, Socio-Political and Environmental Risks of Road Development in the Tropics," *Current Biology* 27, no. 20 (2017): R1130–40, https://doi.org/10.1016/j.cub.2017.08.067.
10. International Energy Agency, *Net Zero by 2050: A Roadmap for the Global Energy Sector*, n.d. https://iea.blob.core.windows.net/assets/deebef5d-0c34-4539-9d0c-10b13d840027/NetZeroby2050-ARoadmapfortheGlobalEnergySector_CORR.pdf.
11. Mohammad Nazeri Tahroudi, "Comprehensive Global Assessment of Precipitation Trend and Pattern Variability Considering Their Distribution Dynamics," *Scientific Reports* 15, no. 1 (2025): 22458, https://doi.org/10.1038/s41598-025-06050-5.
12. Jayden E. Engert et al. "Explosive Growth of Secondary Roads Is Linked to Widespread Tropical Deforestation," *Current Biology* 35, no. 7 (2025): 1641-1648.e4. https://doi.org/10.1016/j.cub.2025.02.017.
13. World Wildlife Fund. "Biodiversity and Infrastructure Investing," August 22, 2024, https://www.worldwildlife.org/publications/biodiversity-and-infrastructure-investing-how-infrastructure-investors-are-factoring-biodiversity-impacts-into-decision-making/.
14. Alex Harris,"How Should Miami-Dade Be Protected from Future Storms? The County Wants Your Input." *Miami Herald*, November 14, 2022, https://www.miamiherald.com/news/local/environment/climate-change/article268730752.html.
15. U.S. Army Corps of Engineers, "Miami-Dade Back Bay Coastal Storm Risk Management Feasibility Study," accessed October 15, 2025, https://www.saj.usace.army.mil/MiamiDadeBackBayCSRMFeasibilityStudy/.
16. Norfolk District, "Miami-Dade County Moves Closer toward Coastal Resiliency with Key Project Endorsement," accessed October 15, 2025. https://www.nao.usace.army.mil/Media/News-Stories/Article/3884730/miami-dade-county-moves-closer-toward-coastal-resiliency-with-key-project-endor/.
17. World Wildlife Fund, "Sustainable Infrastructure Program in Asia: Nature Based Infrastructure Planning in Asia," accessed October 15, 2025, https://www.worldwildlife.org/our-work/sustainability/infrastructure/sustainable-infrastructure-program-in-asia/.
18. World Wildlife Fund, "Nature-Based Solutions for Climate Resilient Infrastructure Planning," May 30, 2025, https://www.worldwildlife.org/publications/nature-based-solutions-for-climate-resilient-infrastructure-planning-in-indonesia/.
19. World Wildlife Fund, "Greening Transportation Infrastructure Development Global Platform," accessed October 15, 2025, https://www.worldwildlife.org/our-work/funds/wwf-gef/projects/greening-transportation-infrastructure-development-global-platform/.

INFRASTRUCTURE AS A CATALYST FOR CLIMATE AND BIODIVERSITY ACTION

Introduction: The Infrastructure Paradox

Infrastructure sits at the heart of the world's response to the twin crises of climate change and biodiversity loss. It is both culprit and cure. Infrastructure is a primary source of greenhouse gas emissions and habitat disturbance, yet also the physical means through which societies can decarbonize, adapt, and restore ecological function. Every road, pipeline, or treatment plant represents a choice about how we balance human development with planetary limits. The urgency to be net-zero and nature-positive is increasingly shaping how those choices are made.

Through the Envision framework, the Institute for Sustainable Infrastructure (ISI) has observed this shift firsthand across hundreds of verified projects. These projects reveal a gradual but tangible movement from mitigating harm toward creating regenerative value, reducing carbon, building resilience, and enhancing ecosystems.

Still, the challenge remains one of scale and structure. Infrastructure choices are usually made locally, within the boundaries of budgets and jurisdictions, while climate and ecological systems function across entire landscapes. The path forward requires bridging those scales, linking project-level practice with ecological and regional planning—so that infrastructure becomes a unifying force in aligning climate ambition with the preservation of nature.

The Climate–Biodiversity Nexus in Infrastructure Practice

Climate change and biodiversity loss each accelerate the other. Rising temperatures and changing precipitation patterns degrade ecosystems, while the loss of forests, wetlands, and soils weakens the planet's capacity to regulate carbon and sustain life. Infrastructure exists at this intersection. Decisions about siting, materials, water management, and energy use determine both the emissions that drive climate change and the ecological pressures that erode biodiversity.

The Zofnass Program's framework of the three interrelated criteria of (1) climate mitigation, (2) climate adaptation, and (3) biodiversity action, captures this interdependence. In practice, however, infrastructure delivery remains largely compartmentalized. Design teams may pursue carbon reduction or resilience goals, but ecological outcomes are often treated as secondary or external to project scope. Compounding this challenge is a mismatch of scale. Infrastructure investments are planned and delivered locally, while biodiversity functions at regional and ecosystem levels. Thousands of small, well-intentioned projects can collectively fragment habitats and diminish species richness simply because no unifying ecological plan guides where infrastructure should, or should not, be built.

This disconnect underscores the need for integrative frameworks that bridge environmental science and engineering practice. The absence of coordinated ecological planning at broader scales remains a major blind spot. Until infrastructure and biodiversity are planned as parts of the same system, local improvements will struggle to offset the cumulative loss occurring across landscapes.

Evidence of Progress: Lessons from Envision-Verified Projects (2012–2025)

Over the past 13 years, hundreds of infrastructure projects have pursued Envision verification, offering a unique window into how the industry's priorities are evolving. Viewed collectively, these projects reveal a slow but unmistakable shift from compliance-driven changes toward genuine performance improvement. Carbon accounting, resilience planning, and ecological integration—once peripheral—are increasingly viewed as central indicators of project quality.

Across sectors, one of the clearest trends is the growing interest in lifecycle carbon accounting. Project teams are beginning to measure embodied emissions from materials and construction, not just operational energy use. Yet progress remains uneven. For many owners and agencies, lifecycle assessments are still treated as specialized exercises rather than standard practice. The data required are often difficult to obtain, and contracting mechanisms rarely incentivize full transparency. The trajectory is positive, but the pace is cautious—a reflection of the institutional inertia that still governs how infrastructure is planned and procured.

The same gradual pattern is visible in the use of nature-based solutions. From living shorelines and constructed wetlands to vegetated corridors that reconnect fragmented habitats, projects are increasingly turning to ecological processes to deliver engineering functions. However, their deployment is still limited by permitting constraints, cost perceptions, and uncertainty about long-term maintenance. These barriers are procedural rather than conceptual. Most practitioners recognize the value of these approaches, but funding and governance systems have not yet evolved to reward them.

Even with these constraints, the data shows clear movement. Projects are quantifying their greenhouse gas footprints, integrating resilience thinking into design, and documenting measurable community and ecosystem benefits. Perhaps most importantly, they are learning from one another. Progress may appear incremental, but it is cumulative. Each project demonstrates that sustainable infrastructure is achievable, credible, and replicable—building a foundation from which higher ambition can emerge.

Raising Ambition: From Compliance to Contribution

Sustainability in infrastructure has long been defined by the absence of harm—avoiding, reducing, or offsetting impacts. Yet the urgency of the climate and biodiversity crises demands a more ambitious paradigm. Projects must not only limit their footprint but actively contribute to environmental regeneration. The question is how to move from incremental improvement to transformative change within institutions that remain structured around compliance.

Through the work of the Institute for Sustainable Infrastructure, several recurring barriers have become evident. The first is fragmented accountability. Ownership, design, construction, and operation often fall under different entities, each optimizing for cost and schedule rather than long-term performance. Second is the absence of ecological and carbon data at the planning stage. Without clear baselines or accessible metrics, decision-making defaults to precedent instead of performance. Third, procurement and funding mechanisms rarely reward the added time and effort required to innovate.

Overcoming these barriers requires reframing sustainability as a value proposition, not a constraint. The Envision framework offers one pathway, translating global ambitions into quantifiable project-level actions. Its structure—plan, do, check, act—aligns with the Zofnass Program conceptual framework and provides a practical means to integrate climate, biodiversity, and social equity within one system.

To raise ambition, infrastructure owners and agencies must do three things.

- **Integrate**: align project objectives with science-based carbon and biodiversity targets
- **Transform**: use pilot projects to demonstrate what "net-zero" and "nature-positive" performance look like in practice, creating new norms for the industry
- **Include**: engage communities early so that local priorities and ecological benefits reinforce one another

Ambition grows through evidence, not aspiration alone. Each project that exceeds conventional practice expands the boundary of what is considered possible. The goal is not perfection but momentum.

Bridging Scales: From Local Projects to Regional Systems

The challenge of aligning infrastructure development with biodiversity protection is as much about scale as it is about intent. Most infrastructure decisions are made at the local or project level—where a municipality replaces a roadway, an agency upgrades a wastewater plant, or a utility adds transmission lines. Each decision seems minor in isolation, justified by immediate public need. Yet biodiversity operates on regional and ecosystem scales. Species migration, watershed integrity, and habitat connectivity depend on patterns far larger than any single project. When thousands of small interventions proceed without coordination, they collectively fragment landscapes, diminish ecological function, and erode the natural capital upon which communities depend.

This scale mismatch is not a failure of motivation but of governance. Infrastructure planning, permitting, and finance systems are rarely structured to account for cumulative effects or to integrate ecological priorities into spatial development strategies. The result is an incremental loss of ecosystem value—death by a thousand cuts—where no individual project is responsible, yet the collective outcome is significant.

Bridging this divide requires coupling project-level frameworks, like Envision, with regional ecological planning that guides where and how infrastructure should evolve. By integrating ecological data, carbon mapping, and resilience targets into long-range capital planning, governments can direct investment toward areas where infrastructure reinforces both human and environmental resilience. Such an approach would transform infrastructure from a reactive service provider into a proactive land-use tool—shaping growth patterns, steering density, and preserving intact habitats.

Ultimately, reconciling local action with regional systems is the missing link in sustainable infrastructure. Without that alignment, progress achieved one project at a time risks being undone at the landscape scale.

Infrastructure as a Systemic Lever

Infrastructure is more than a collection of assets; it is the physical manifestation of a society's priorities. The roads we build, the systems we upgrade, and the corridors we preserve determine how resources move, how communities grow, and how ecosystems endure. Because of this centrality, infrastructure holds unparalleled leverage in addressing both climate and biodiversity challenges. Every project represents not only an engineering decision but a policy decision—one that can reinforce or reimagine the relationship between human and natural systems.

When viewed collectively, infrastructure becomes a system of systems that can either accelerate or correct unsustainable trajectories. Transit networks influence land use and emissions; water systems determine watershed health; energy grids dictate the pace of decarbonization.

To harness this leverage, institutional structures must evolve in parallel with technical innovation. Performance-based standards need to be embedded in procurement, finance, and regulation so that climate and biodiversity outcomes are rewarded, not incidental. The promise of sustainable infrastructure lies precisely in this systemic influence: through consistent, verifiable practice, individual projects can collectively redefine what progress means in a changing world.

Conclusion: Toward Regenerative Infrastructure

The transition to a sustainable future depends not only on new technologies or materials, but on how we conceive of infrastructure itself. Every project—whether a highway interchange, a water treatment plant, or a renewable energy facility—embodies a decision about the relationship between human development and the living systems that support it. The experience of the past decade shows that progress is real but gradual. Tools like Envision have influence in reshaping expectations for performance and accountability. Yet institutional structures and fragmented decision-making still constrain how quickly these practices can scale.

To meet the urgency of the climate and biodiversity crises, infrastructure must evolve from minimizing damage to restoring value—becoming regenerative rather than merely sustainable. This requires aligning project delivery, finance, and policy around measurable outcomes for both climate and ecosystems. The frameworks now in place provide the foundation. The next step is collective commitment. Thousands of coordinated decisions that, together, rebuild the resilience of the planet we depend on.

Anthony Kane
President and CEO
Institute for Sustainable Infrastructure

Urgency

for Climate and Biodiversity Action

Chapter 1

THE TWIN CLIMATE AND BIODIVERSITY CRISES AS THREATS

The world is in urgent need of addressing the climate and biodiversity crises. This state of urgency is driven by global scientific assessments that indicate record changes in key climate and biodiversity indicators; the underlying drivers of climate change and biodiversity loss reaching new observed highs; projections of future outcomes under specific development and policy scenarios; and identified gaps in the actions required to address these crises. The rapidly closing window to 2030—a decisive milestone for climate action, biodiversity conservation, and the Sustainable Development Goals—makes this urgency call even more alarming.

RECOGNITION AS GLOBAL THREATS

Climate change and biodiversity loss are not new threats. The discussion of climate change and biodiversity decline as global challenges is long-lasting and continues to this day, driven by emerging scientific evidence and observable impacts. Global conversation on the two issues formally began during the United Nations Conference on Environment and Development (the Rio Earth Summit) in 1992, where they were recognized as intrinsically linked to the long-term survival of the planet and the quality of life of its inhabitants. The Summit highlighted how different social, economic and environmental factors are interdependent, and how progress in one area cannot be sustained over time without action in others. In other words, it recognized the need to integrate and balance the economic, social, and environmental dimensions of development. The Summit provided the political impetus for an action agenda at global, national and local levels aimed at achieving "development that meets the needs of the present without compromising the ability of future generations to meet their own needs."[1] The Summit produced two landmark treaties: The United Nations Framework Convention on Climate Change (UNFCCC) and the Convention on Biological Diversity (CBD). These conventions remain the two most important international legal instruments for advancing climate change and biodiversity conservation goals.

The UNFCCC raised concerns and awareness about the rising atmospheric concentrations of greenhouse gases (GHGs) caused by human activity, their effects on natural ecosystems and humankind, and the critical role of terrestrial and marine ecosystems as carbon sinks and reservoirs. Its ultimate objective is to stabilize greenhouse gas concentrations "at a level that would prevent dangerous human interference with the climate system."[2] It further states that "such a level should be achieved within a time-frame sufficient to allow ecosystems to adapt naturally to climate change, to ensure that food production is not threatened, and to enable economic development to proceed in a sustainable manner."[3]

The CBD, meanwhile, represents "a dramatic step forward in the conservation of biological diversity, the sustainable use of its components, and the fair and equitable sharing of benefits arising from the use of genetic resources."[4]

The establishment of the two Conventions was a remarkable step for a time when scientific evidence on climate and biodiversity threats was far more limited than it is today.

SCIENTIFIC EVIDENCE ON CLIMATE CHANGE AND BIODIVERSITY LOSS

The discussion on the two issues is supported by relevant research and the generation and dissemination of credible, legitimate knowledge intended to inform policy formation and decision-making. Over recent decades, global awareness and understanding of risk exposure has been largely confined to climate change, creating a "climate-first" approach; however, biodiversity loss is equally urgent.

A milestone in scientific evidence on climate change was the creation in 1988 of the Intergovernmental Panel on Climate Change (IPCC) by the World Meteorological Organization (WMO) and the United Nations Environmental Programme (UNEP). The IPCC's objective is to provide policymakers with regular scientific information on the current state of knowledge on climate change, its implications and potential future risks, and to put forward adaptation and mitigation options.[5] Since 1990 the IPCC has published six Assessment Reports—the most comprehensive scientific reports developed on climate change—each of which has paved the way for significant steps toward reducing global warming. IPCC reports are prime examples of the process of building consensus within the climate science community and establishing scientific facts. The IPCC is widely regarded as "the most authoritative and important common knowledge base for international and local climate policy making."[6]

The first IPCC Assessment Report (FAR, 1990) emphasized the need for international cooperation and helped lead to the creation of the UNFCCC.[7] In the aftermath of UNFCCC's establishment, cooperation with IPCC was strengthened, with the latter providing scientific and technical information requested by the Convention—particularly by its supreme decision-making body, the Conference of Parties (COP). IPCC assessments have repeatedly stimulated governments to intensify negotiations, clarify the importance of limiting greenhouse gas emissions, set binding targets for developed countries, and lay the groundwork for defining temperature limits that eventually culminated in the Paris Agreement (2015)—which sets the goal of keeping global temperature rise this century well below 2°C above pre-industrial levels while pursuing efforts to limit the increase to 1.5°C.[8] The 2018 IPCC Special Report[9] highlighted the dramatic reduction in impacts by limiting warming to 1.5°C rather than 2°C, and established 1.5°C as the more appropriate scenario to pursue in alignment with sustainable development.

The biodiversity counterpart—the Intergovernmental Science-Policy Platform on Biodiversity and Ecosystem Services (IPBES)—was established in 2012, nearly 25 years after the IPCC. Although earlier syntheses, such as the Millennium Ecosystem Assessment (2000) and the CBD's Global Biodiversity Outlooks, documented unprecedented rates of biodiversity loss, strong momentum for action did not emerge until IPBES was created as an independent intergovernmental scientific body for biodiversity. Modeled on its counterpart, the IPBES was designed to bring the same attention to nature that the IPCC brought to climate.[10] The release of IPBES's first Global Assessment Report on Biodiversity and Ecosystem Services in 2019 gained substantial public visibility and media attention.

KEY FACTS INDICATING THREAT LEVELS

The world is warming faster than at any point in recorded history.[11] There is tangible evidence of global warming, coupled with definitive scientific consensus on the issue. The global average temperature for the most recent ten-year period (2014 to 2023) is estimated to be the warmest on record—at around 1.2°C above the pre-industrial average.[12] This is radically altering the Earth's climate and releasing a series of extreme weather events, including intensifying extreme weather events, floods, droughts, and alarming glacial retreat, causing widespread harms to people, economies, and nature.

The World Meteorological Organization (WMO) State of Global Climate 2024 presented trends in key climate indicators based on latest available records:[13]

- "In 2023, the atmospheric concentration of carbon dioxide, as well as those of methane, and nitrous oxide, reached the highest levels in the last 800,000 years. Real-time data from specific locations show that levels of three main greenhouse gases continued to increase in 2024"
- "The annually averaged global mean near-surface temperature in 2024 was 1.55°C ± 0.13°C above the 1850–1900 average, which is used to represent pre-industrial conditions." (This does not mean that the Paris target of 1.5°C has been exceeded, because the Paris targets refer to sustained warming over extended periods, not individual years)
- "2024 was the warmest year iWn the 175-year observational record, clearly surpassing the previous warmest year, 2023 at 1.45°C ± 0.12°C above the 1850–1900 average.[...] Each of the past ten years, 2015 to 2024, were individually the ten warmest years on record"
- "In 2024, ocean heat content reached the highest level in the 65 year observational record, exceeding the previous record high set in 2023. Over the past eight years, each year has set a new record for ocean heat content." The rate of ocean warming from 2005 to 2024 is more than twice that of 1960 to 2005
- "In 2024, global mean sea level reached a record high in the satellite record (from 1993 to present)," with the rate "increasing from 2.1 mm per year between 1993 and 2002 to 4.7 mm per year between 2015 and 2024"
- "Acidification of the ocean surface has continued over the past 39 years"
- "Glacier mass loss from 2021/2022 to 2023/2024 represents the most negative three-year glacier mass balance on record, and seven of the ten most negative annual glacier mass balances since 1950 have occurred since 2016"

If no urgent action is taken, "climate change will cause average global temperatures to increase by around 3.2°C and will adversely affect every ecosystem."[14] Simply scanning climate change-related headlines—referring to climate action as a "race" or "the defining crisis of our time, happening even more quickly than we feared,"[15]—makes it clear that we are in a state of climate emergency. The risks of climate-related disasters are not a distant future scenario but a reality affecting millions of people.

The scientific community has repeatedly warned that warming beyond 1.5°C risks unleashing far more severe climate change impacts and extreme weather events—and that "every fraction of a degree of warming matters."[16]

Exceeding 1.5°C could trigger multiple climate tipping points—"conditions beyond which changes may become abrupt, irreversible, and dangerous—with serious implications for humanity,"[17] such as the breakdown of major ocean circulation systems, abrupt permafrost thaw, and the collapse of tropical coral reefs.[18]

Past and present greenhouse gas emissions have already committed the world to long-lasting climate change, with multiple and growing environmental and society-wide risks.[19] Many climate impacts, such as sea level rise, are essentially permanent for many generations to come and will take centuries to return even to today's conditions. Climate change and its cascading effects threaten to erode decades of development gains.[20]

At the same time, there is mounting evidence of unprecedented and accelerating biodiversity loss on a global scale. The 2019 Global Assessment Report on Biodiversity and Ecosystem Services of the Intergovernmental Science-Policy Platform on Biodiversity and Ecosystem Services (IPBES) stated that "biodiversity is declining faster than at any time in human history."[21] The report represented a critical assessment of the status and trends of the natural world—the first since the release of the Millennium Ecosystem Assessment in 2005.[22] Some of the key messages of the Assessment Report, which highlighted the significant alteration and rapid decline of ecosystems and biodiversity, include:

- "The global rate of species extinction is already at least tens to hundreds times higher than the average rate over the past ten million years and is accelerating," with the total biomass of wild mammals declining by 82% compared to historical records[23]
- "An average of 25% of species in assessed animal and plant groups are threatened, suggesting that around 1 million species already face extinction"[24]
- "75% of the land surface is significantly altered, 66% of the ocean area is experiencing increasing cumulative impacts, and over 85% of wetlands (by area) have been lost"[25]
- "Natural ecosystems have declined (in extent and condition) by 47% compared to their earliest estimated states"[26]
- "Biotic integrity—the abundance of naturally-present species—has declined by 23% on average in terrestrial communities" compared to historical records, potentially affecting ecosystem processes and, hence, nature's contributions to people[27]
- Today, only 15% of land and 8% of the ocean are protected[28]

The IPBES report highlighted the urgent need for action to reduce the intensity of drivers of biodiversity loss. Without such action, the global rate of species extinction will continue to accelerate.

Biological diversity is recognized as a global asset of tremendous value to present and future generations. Its decline represents not only an irreversible loss for the planet but also a direct threat to humanity's life support systems. The services that nature provides encompass everything from the food we eat to the water we drink and the air we breathe.[29] According to the World Economic Forum's Global Risks Report 2025, biodiversity loss is ranked as the second most impactful risk for the next decade.[30]

The twin crises of climate change and biodiversity loss are recognized as pressing global issues and "a common concern of humankind."[31,32] At their core, they are existential threats.

THREATS TO SUSTAINABLE DEVELOPMENT GOALS

Climate change and biodiversity loss are not just standalone issues; they are intrinsically linked to multiple aspects of sustainable development. The far-reaching consequences of these crises are best understood when viewed in the context of the UN 2030 Agenda for Sustainable Development and its Sustainable Development Goals (SDGs). The SDGs provide an integrated, universal framework for addressing the world's most pressing social, economic, and environmental challenges by 2030. Agreed upon by the United Nations in 2015, the 17 SDGs and their associated targets aim to eradicate poverty and achieve sustainable development by 2030 while protecting the global environment and ensuring that no one is left behind. Global environmental challenges are directly addressed in Goal 13 (Climate Change), Goal 14 (Life below Water), and Goal 15 (Life on Land). These environmentally focused SDGs, however, underpin the achievement of many other goals.[33]

Climate change undermines progress made toward nearly all sustainable goals. It is increasingly regarded as a "threat multiplier."[34] Climate change affects agricultural productivity, threatening food security and access (SDG 2: Zero Hunger), and exacerbating poverty (SDG 1: No Poverty). It contributes to health issues by worsening air quality, exacerbating heat waves, spreading vector-borne diseases, and increasing wildlife risks (SDG 3: Good Health and Well-being). Changing precipitation patterns threaten the availability or quality of water resources (SDG 6: Clean Water and Sanitation) and increase the potential for water-related conflicts. Moreover, climate-induced disasters have already led to internal displacements and migration, a trend expected to escalate in the future.

Biodiversity is foundational to the Agenda 2030. While it is explicitly featured in SDG 14 (Life below Water) and SDG 15 (Life on Land), it also underpins several other goals, including SDG 2 for food security and improved nutrition, SDG 6 for provision of clean water and sanitation, or SDG 3 for good health and well-being, through the direct provisioning of goods, or indirectly through multiple regulating services. It is worth highlighting that according to estimates, "$44 trillion of economic value generation – more than half of the world's total GDP – is moderately or highly dependent on nature and its services and is therefore exposed to nature loss."[35]

At the same time, achieving the Sustainable Development Goals contributes to the conservation and sustainable use of biodiversity. "Some goals address the drivers of biodiversity loss, such as climate change (SDG 13), pollution (SDGs 6, 12, and 14), and overexploitation (SDGs 6, 12, 14, and 15). Others address sustainable production and consumption, efficient use of natural resources, and reduced food waste (SDG 12: Responsible Consumption and Production)."[36]

URGENT NEED FOR ADDRESSING THE ACTION GAP

Climate action refers to "efforts taken to combat climate change and its impacts. These efforts involve reducing greenhouse gas emissions (climate mitigation) and/or taking action to prepare for and adapt to both the current effects of climate change and its predicted future impacts (climate adaptation)."[37]

Biodiversity action refers to efforts to halt and reverse biodiversity loss and to mainstream biodiversity-inclusive activities. These efforts include reducing threats to ecosystems and species, conserving and enhancing biodiversity, ensuring its sustainable use and benefit-sharing, and fully integrating its values into planning and decision-making at all levels and across all sectors.[38]

The previous section already referred to some of the key actors that have shaped the understanding of climate change and biodiversity loss and have organized international action. The two domains are structured in a similar way: Both climate change and biodiversity have their own conventions—international legal instruments and binding agreements—the convention's governing bodies (the Conferences of the Parties, or COPs), the governments that have ratified the treaties, and their independent intergovernmental bodies that assess the state of available knowledge.

COMPARISON OF THE STRUCTURE OF ACTION ACROSS THE CLIMATE CHANGE AND BIODIVERSITY DOMAINS

Structural Components	Climate Change	Biodiversity
International Conventions: the international legal instruments, binding agreements	UN Framework Convention on Climate Change (UNFCCC)	Convention on Biological Diversity (CBD)
Supporting Secretariats (for the implementation of the Conventions)	UNFCCC Secretariat	CBD Secretariat
Parties of Conventions	198 countires	196 countries
Conference of the Parties (COP): the meetings of the convention's governing bodies	Most recent COP held: COP29	Most recent COP held: COP15
Intergovernmental Panels which assess available knowledge	Intergovernmental Panel on Climate Change (IPCC) established in 1988	Intergovernmental Science-Policy Platform on Biodiversity and Ecosystem Services (IPBES) established in 2012
Assessment Reports on the state of knowledge of climate change and biodiversity	IPCC Assessment Reports published every 5 to 7 years	IPBES Global, Regional and Thematic Assessment Reports
Landmark Binding Agreement	Paris Agreement, 2015	Kunming-Montreal Global Biodiversity Framework, 2022
Time-bound Goals: 2030 and 2050 critical years	Limit global warming to 1.5°C by 2030 Net zero emissions by 2050 for climate action	Nature positive by 2030 to halt and reverse biodiversity loss "Living in harmony with nature' by 2050
Development of National Plans under commitment to Convention agreements	Nationally Determined Contributions (NDCs) and long-term strategies (LTS)	National Biodiversity Strategy and Action Plans (NBSAPs)
Taskforces established to develop guidance for businesses	Task Force on Climate-related Financial Disclosures (TCFD)	Task Force on Nature-related Financial Disclosures (TNFD)

THE CURRENT FRAMEWORK FOR ACTION

2015 was a landmark year for climate. For the first time, a legally binding agreement brought all nations together in a common cause to undertake ambitious efforts toward a sustainable, low-carbon future to combat climate change and adapt to its effects: The Paris Agreement.[39,40] The Paris Agreement has become synonymous with climate action. Its central aim is to cut GHG emissions to keep the global temperature rise this century well below 2°C above pre-industrial levels, while pursuing efforts to limit the increase to 1.5°C.[41] Global temperature will stabilize when carbon dioxide emissions reach net zero. To achieve this, net-zero CO_2 emissions must be reached globally in the early 2050s for the 1.5°C pathway and in the early 2070s for the 2°C pathway.[42]

For biodiversity action, the decisive year was 2022, when the Kunming-Montreal Global Biodiversity Framework (GBF), also known as the Post-2020 Global Biodiversity Framework, was adopted under the Convention on Biological Diversity (CBD). The Framework outlines the CBD's 2050 vision of a world "living in harmony with nature," where "biodiversity is valued, conserved, restored and wisely used, maintaining ecosystem services, sustaining a healthy planet, and delivering benefits essential for all people."[43] To operationalize this vision, four goals for 2050 and 23 supporting action-oriented targets for 2030 were established. The Framework aims to accelerate progress following the failure to meet the 20 Aichi Biodiversity Targets of the Strategic Plan for Biodiversity 2011-2020, and in light of the IPBES 2019 assessment findings.

The GBF builds on the Aichi targets and establishes new, more ambitious targets for 2030, along with a monitoring framework comprising a set of indicators to complement mechanisms under other biodiversity-related conventions.[44]

The GBF sets benchmarks for improvements across indirect and direct drivers of biodiversity loss, the state of biodiversity, the benefits derived from it, policies, and other enabling conditions. Key quantitative targets for 2030 include:

- Close to zero loss of areas of high biodiversity importance
- Restoration of 30% of degraded ecosystems
- Conservation of 30% of terrestrial and inland water areas, and of marine and coastal area
- Halting human-induced extinction of known threatened species
- Reduction of the rates of introduction and establishment of known or potential invasive alien species by at least half

It is worth highlighting that the GBF is also referred to as "the Paris Agreement for nature."[45]

THE NARROWING WINDOW FOR ACTION

The year 2030 marks a decisive milestone for global action on both climate change and biodiversity, coinciding with the deadline for achieving the UN SDGs. It is also the timeline that scientists have identified as the last chance to prevent catastrophic changes.

While collective climate action is underway, it still lags behind what science is necessary, making the next few years critical. With five years remaining until 2030, the time to act is undoubtedly now.

Global progress against climate goals is being tracked based on countries' Nationally Determined Contributions (NDCs), the mechanism for implementing the Paris Agreement. UNEP's annual Emissions Gap Reports (EGRs)—issued for over a decade—assess the gap between countries' pledges on greenhouse gas emissions reductions and the reductions required to limit global temperature increase to below 2°C by the end of this century.[46] Even if all signatories fully implement their commitments, projected warming would still reach 2.6–3.1°C by 2100 unless additional, substantial commitments are made and fulfilled. Without such action, the resulting temperature rise would have catastrophic consequences for both biodiversity and humanity.[47]

The alarming titles of UNEP's recent Emission Gap Reports—The Heat is On (2021),[48] The Closing Window (2022),[49] Broken Record (2023),[50] No More Hot Air... Please! (2024)[51]—underscore the urgency of addressing the action lag. The recently released report for 2025, Off Target, states that current NDCs have limited effect on narrowing the emissions gap by 2030 and 2035. Similar urgency is reflected in UNEP's Adaptation Gap Reports (AGRs): Underfinanced. Underprepared – Inadequate investment and planning on climate adaptation leaves world exposed (2023)[52] and Come Hell and High Water (2024).[53]

According to the 2022 IPCC Assessment Report, keeping the 1.5°C temperature target within reach requires "global GHG emissions to peak before 2025 at the latest and to be reduced by 43% by 2030."[54] Emission cuts of 42% by 2030 and 57% by 2035 are needed to stay on track for 1.5°C.[55] These mitigation efforts must occur alongside climate adaptation measures.[56] The narrowing window for action makes the fight against climate change a true "race against time." Without immediate and deep emissions reductions across all sectors, limiting global warming to 1.5°C will be beyond reach. As the IPCC notes, "Delivering on the mitigation potential would require immediate global mobilization on a scale and pace only ever seen following a global conflict."[57]

COP28, held in Dubai in 2023, marked the halfway point toward the 2030 goals and served as the first Global Stocktake of progress under the Paris Agreement. It revealed that current emission reductions still fall short

TIMELINE OF SETTING CLIMATE AND BIODIVERSITY TARGETS

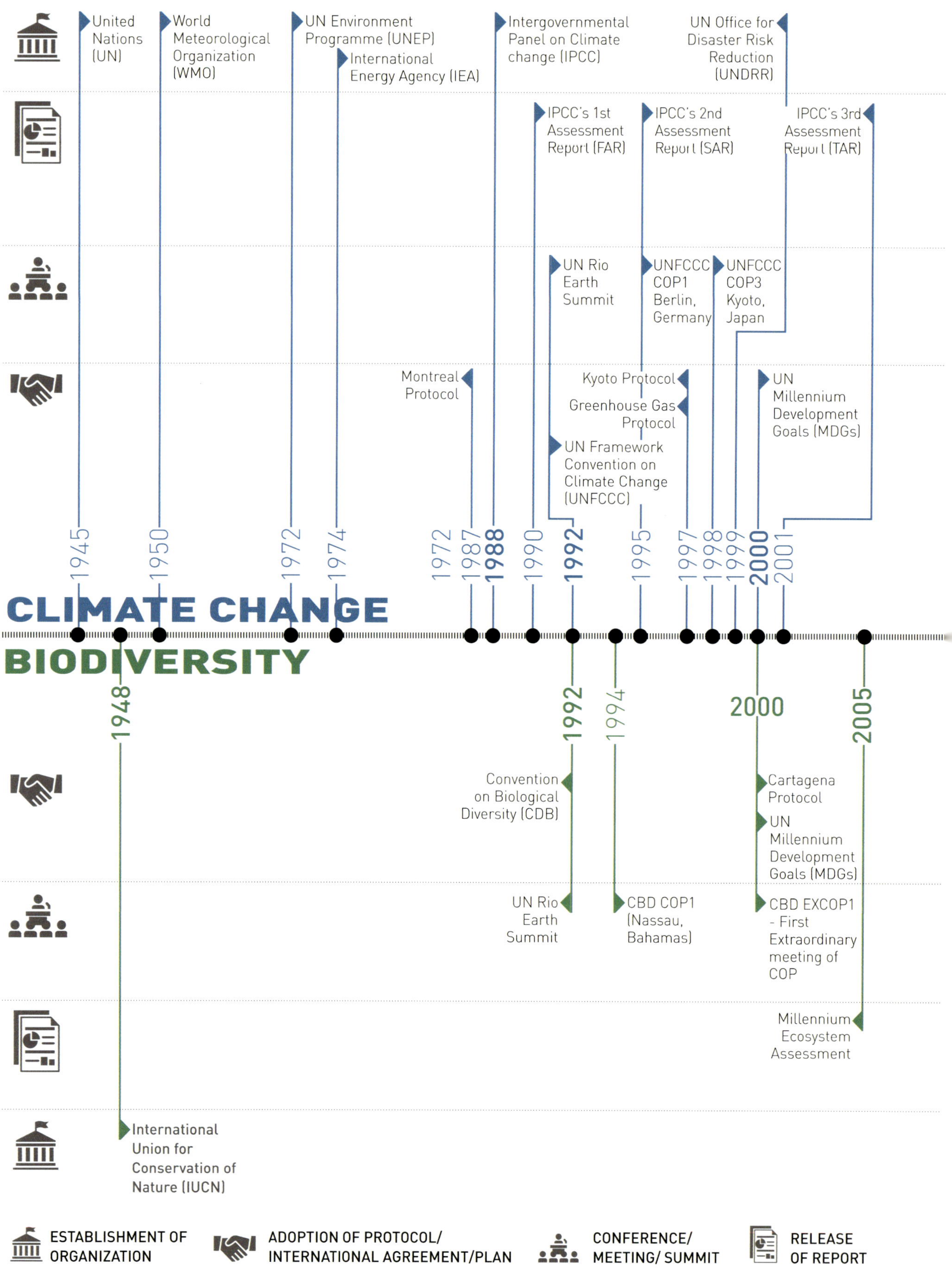

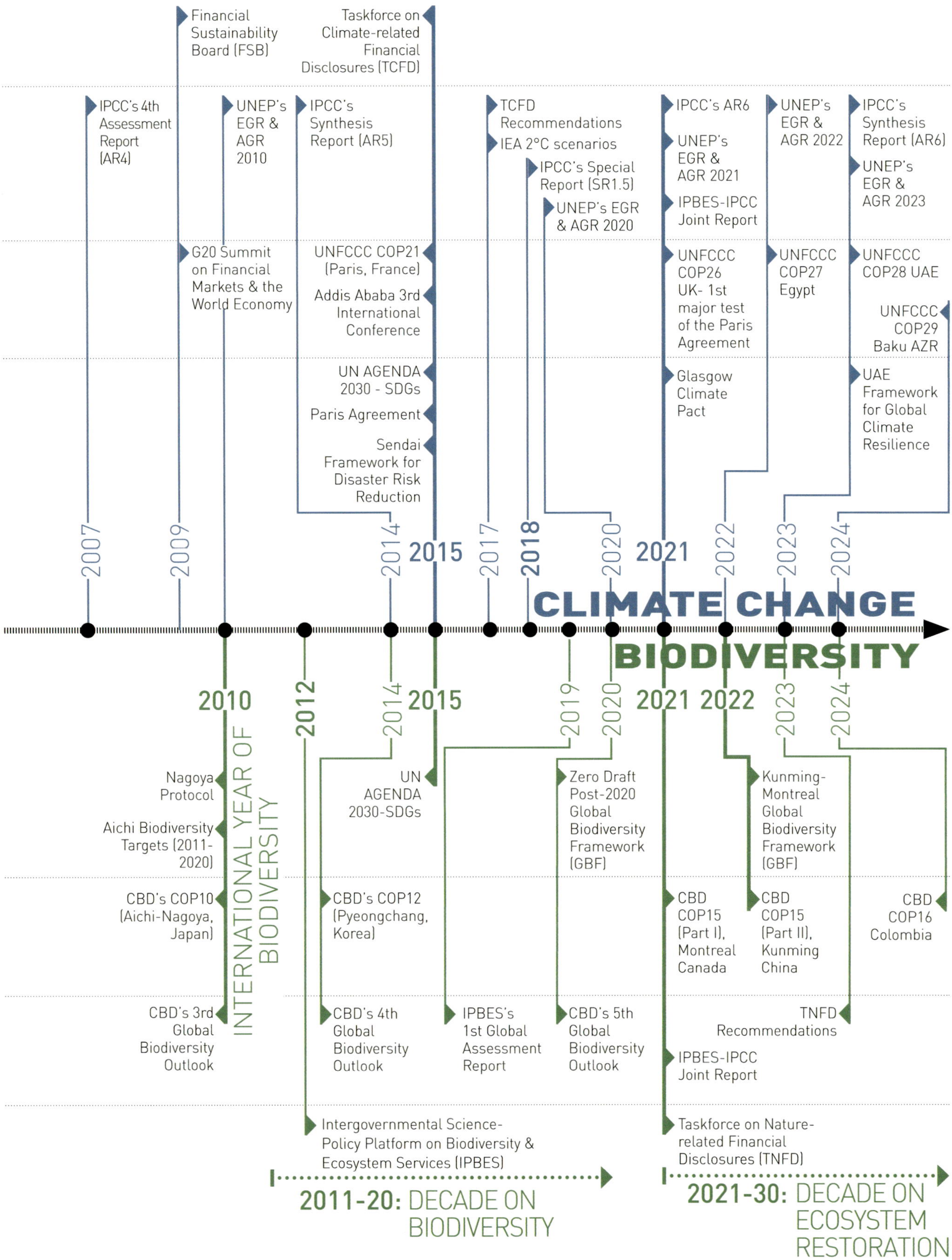
CLIMATE CHANGE
2007
IPCC's 4th Assessment Report (AR4)
2009
Financial Sustainability Board (FSB)
G20 Summit on Financial Markets & the World Economy
UNEP's EGR & AGR 2010
2014
IPCC's Synthesis Report (AR5)
2015
Taskforce on Climate-related Financial Disclosures (TCFD)
UNFCCC COP21 (Paris, France)
Addis Ababa 3rd International Conference
UN AGENDA 2030 - SDGs
Paris Agreement
Sendai Framework for Disaster Risk Reduction
2017
TCFD Recommendations
IEA 2°C scenarios
2018
IPCC's Special Report (SR1.5)
2020
UNEP's EGR & AGR 2020
2021
IPCC's AR6
UNEP's EGR & AGR 2021
IPBES-IPCC Joint Report
UNFCCC COP26 UK- 1st major test of the Paris Agreement
Glasgow Climate Pact
2022
UNEP's EGR & AGR 2022
UNFCCC COP27 Egypt
2023
IPCC's Synthesis Report (AR6)
UNEP's EGR & AGR 2023
UNFCCC COP28 UAE
UAE Framework for Global Climate Resilience
2024
UNFCCC COP29 Baku AZR
BIODIVERSITY
2010
INTERNATIONAL YEAR OF BIODIVERSITY
Nagoya Protocol
Aichi Biodiversity Targets (2011-2020)
CBD's COP10 (Aichi-Nagoya, Japan)
CBD's 3rd Global Biodiversity Outlook
2012
Intergovernmental Science-Policy Platform on Biodiversity & Ecosystem Services (IPBES)
2014
CBD's COP12 (Pyeongchang, Korea)
CBD's 4th Global Biodiversity Outlook
2015
UN AGENDA 2030-SDGs
2019
IPBES's 1st Global Assessment Report
2020
Zero Draft Post-2020 Global Biodiversity Framework (GBF)
CBD's 5th Global Biodiversity Outlook
2021
CBD COP15 (Part I), Montreal Canada
IPBES-IPCC Joint Report
Taskforce on Nature-related Financial Disclosures (TNFD)
2022
Kunming-Montreal Global Biodiversity Framework (GBF)
CBD COP15 (Part II), Kunming China
2023
TNFD Recommendations
2024
CBD COP16 Colombia
2011-20: DECADE ON BIODIVERSITY
2021-30: DECADE ON ECOSYSTEM RESTORATION

of national commitments. In response, governments renewed their pledges with a series of new 2030 targets, including the tripling renewable energy capacity and doubling of energy-efficiency improvements.[58] Nevertheless, despite warnings and pledges, emissions have continued to rise in 2024, increasing by 2.3% from 2023 levels.[59] The next COP, to be scheduled for November 2025, must deliver a decisive breakthrough in ambition and implementation.

For biodiversity, where the framework for action is more recent, comprehensive stocktakes of progress are not yet available. A partial view of the steps made since the GBF was adopted, is provided by the UNEP's and IUCN's Protected Planet Report 2024,[60] the first official evaluation of global progress on one of the GBF's flagship targets: Protecting 30% of land and seas for nature by 2030, known as "30 by 30." This target encompasses not only coverage but also management, governance, social safeguard mechanisms, as well as strategic conservation and connectivity of priority areas for biodiversity (key biodiversity areas).

According to official data by governments and other stakeholders, despite progress made in expanding national networks of protected and conserved areas, significant gaps remain:[61]

- Global coverage of protected and conserved areas "has only just passed 17% on land and 8% in the marine realm"
- Protected and conserved areas are not always established where they are most needed. "A third of the areas most important for biodiversity are not yet protected"

The challenge of meeting biodiversity targets for 2030 becomes clearer when seeing what happened with the Aichi targets. The 20 Aichi Biodiversity Targets, structured around five Strategic Goals, set benchmarks for improvements across indirect and direct drivers of biodiversity loss, the state of biodiversity, the benefits derived from it, policies, and other enabling conditions. The CBD's Fifth Global Biodiversity Outlook[62] reported that despite progress by nations, none of the 20 targets was fully achieved at a global level, and only six were partially met. It further highlighted the risks of inaction: "The projected decline in biodiversity will affect all people, but it will have a particularly detrimental effect on indigenous peoples and local communities, and the world's poor and vulnerable, given their reliance on biodiversity for their wellbeing."[63]

The failure to meet the Aichi Targets is attributed not only to a lack of necessary ambition from governments but also the absence of clear, quantified targets and metrics to assess and document progress, and limitations in up-to-date local biodiversity data. Unlike climate action—where GHG emissions provide a globally-accepted common metric—in biodiversity there is no single metric that can comprehensively define impact and document progress. Its higher complexity makes it difficult to set limited, streamlined metrics.

Moreover, climate change tends to be perceived as having a more global configuration in terms of impacts, than biodiversity. A key reason is that biodiversity loss is experienced locally and "reflects place-based socio-ecological systems." In reality, climate change effects are also experienced locally, even though the issue needs to be tackled with a global approach.[64]

The UN Biodiversity Conference COP16 provides a crucial opportunity for governments to strengthen their actions and bring about critically important changes to meet the commitments of the Global Biodiversity Framework.

THE FINANCING GAP AND THE ROLE OF CORPORATE ACTION

A vital part of the solution is securing finance. In 2018, it was estimated that the world should invest an average of US$2.4 trillion in clean energy annually through 2035 to prevent global temperature increases from exceeding 1.5°C.[65] In 2022 IPCC announced that yearly climate finance flows must increase three- to six-fold, globally, to meet average annual needs between 2020 and 2030.[66] Current mitigation and adaptation financing flows are insufficient to cover average needs across all regions and sectors, creating a significant climate finance gap. "Yet investments in climate action far outweigh the costs of inaction."[67] A more recent World Economic Forum (WEF) report, states that "annual investment in the low-carbon transition needs to multiply sevenfold by 2030, if the world is to stand a chance of steering the climate towards a Paris-aligned pathway."[68] Moreover, the Climate Policy Initiative (CPI) reports that annual climate finance requirements are projected to rise to US$9 trillion by 2030 and to exceed US$10 trillion from 2031 to 2050. Yet only $1.2 trillion was invested in 2021/2022.[69]

Not all nations have the capacity to contribute equally to action. This was acknowledged in the Paris Agreement through provisions for support to developing countries to implement climate actions and adaptation measures. Importantly, these nations, which contribute the least to climate change, are the most vulnerable to and disproportionately affected by its impacts. The New Collective Quantified Goal for climate finance, agreed at COP29, called for developed nations to provide at least US$300 billion per year for climate action in developing countries by 2035.[70] However, it is still not considered sufficient to meet developing countries' needs. Updated estimates place the adaptation costs within a range of US$310–365 billion per year for developing countries by 2035.[71]

Under the Global Biodiversity Framework, countries committed to increase investment in biodiversity from all sources to at least US$200 billion per year by 2030. At the same time, the biodiversity finance gap is estimated at US$700 billion per year.[72]

Private investment is critical to closing the investment gap generated by the insufficient government financial commitments—whether due to budget constraints, conflicting priorities, or other limitations.

Over the past few years, investor and corporate awareness of environmental issues has grown rapidly but has largely focused on climate change. This trend is reflected in initiatives, such as the Global Investor Statements, as well as the ongoing revision of Environmental, Social, and Governance (ESG) systems. The development of ESG systems is driven by investors' demand and thus is representative of the shifts in their priorities for investment. Recent developments mark a transition toward global-level goals alignment—whether with the UN SDGs, the Paris Agreement, or more recently, biodiversity. This, by extension, is transforming corporate action.

Though it is a still new ESG consideration for investors, the biodiversity crisis is climbing up the investors' agenda as the next priority. After all, biodiversity loss is a material risk for investors. Biodiversity is a fundamental component of long-term business sustainability since businesses rely on natural resources as inputs and depend on healthy ecosystems.[73] A 2020 research by Swiss-Re found that 55% of our global GDP is "moderately or highly dependent on nature and its service."[74]

Therefore, the incentive for companies to contribute towards global solutions is greater than ever.[75]

The Paris Agreement referred to the voluntary cooperation of public and private entities to allow for higher ambition, calling them to "apply robust accounting."[76] The Global Biodiversity Framework refers directly to the role to be played by businesses, requesting that they "assess and transparently disclose their risks, dependencies and impacts on biodiversity" across their operations, supply chains, and portfolios "to progressively reduce negative impacts, increase positive impacts, reduce biodiversity-related risks to business and financial institutions, and promote actions to ensure sustainable patterns of production."[77] It is worth mentioning that the Aichi Targets did not require biodiversity-related reporting. This highlights the current high focus on mainstreaming biodiversity in business practices.

Investors' "climate-first" approach was supported by the availability of meaningful metrics to demonstrate a company's exposure to risks. A major milestone in raising investors' awareness of climate change as a financial risk was the establishment of the Task Force for Climate-related Financial Disclosures (TCFD) in 2015. In June 2017, the TCFD published its Recommendations on Climate-related Financial Disclosures[78] and an Implementation Guide for companies, including example metrics. The TCFD recommendations were welcomed by the investor community, with investors committing to assist their implementation. The TCFD's work acted as a catalyst for bringing climate risk to the mainstream for the financial sector and "resulted in increasing investor scrutiny of corporate disclosures on the topic and growing momentum towards mandatory climate risk disclosure in many jurisdictions around the world."[79]

A similar step toward mainstreaming biodiversity in corporate accounting is the recent formation of the Taskforce on Nature-related Financial Disclosures (TNFD) in 2020. TNFD has a mission similar to that of the TCFD, to develop a framework for organizations to "incorporate nature-related risks and opportunities into their strategic planning, risk management and asset allocation decisions," which "can help shift the flow of global capital to more positive outcomes for nature and society."[80] According to TNFD, nature-related financial risks and opportunities are "all financial risks and opportunities to an organization as a result of impacts and/or dependencies on nature."[81]

Integrating biodiversity-related risks and opportunities in the investment process does not necessarily mean starting from scratch. Existing climate-related practices can be extended or adapted to integrate biodiversity.[82]

NOTES

1. World Commission on Environment and Development, "Chapter 2: Towards Sustainable Development," in *Our Common Future* (United Nations, 1987), http://www.un-documents.net/ocf-02.htm#I.
2. United Nations, *United Nations Framework Convention on Climate Change* (New York: United Nations, 1992), 4.
3. Ibid.
4. Secretariat of the Convention on Biological Diversity, United Nations Environment Programme, *Convention on Biological Diversity (CBD): Text and Annexes* (Secretariat of the Convention on Biological Diversity, 2011), https://www.cbd.int/doc/legal/cbd-en.pdf.
5. "About the IPCC," Intergovernmental Panel on Climate Change (IPCC), accessed November 3, 2025, https://www.ipcc.ch/about/.
6. Pamela McElwee, "A Tale of Two Panels: Learning and Coordinating across IPCC, IPBES, and Other Science-Policy Interfaces," *Climatic Change* 178, no. 3 (March 1, 2025): 45, https://doi.org/10.1007/s10584-025-03869-9.
7. "IPCC - Intergovernmental Panel on Climate Change: History," IPCC, accessed November 1, 2025, https://archive.ipcc.ch/organization/organization_history.shtml.
8. United Nations, *Paris Agreement*, December 12, 2015, in *United Nations Treaty Collection*, 3, https://unfccc.int/sites/default/files/english_paris_agreement.pdf.
9. IPCC, *Global Warming of 1.5°C: An IPCC Special Report on the Impacts of Global Warming of 1.5°C above Pre-industrial Levels and Related Global Greenhouse Gas Emission Pathways, in the Context of Strengthening the Global Response to the Threat of Climate Change, Sustainable Development, and Efforts to Eradicate Poverty*, ed. V. Masson-Delmotte et al. (Cambridge University Press, 2022), https://doi.org/10.1017/9781009157940.
10. McElwee, "A Tale of Two Panels", 45.
11. "The Sectoral Solution to Climate," United Nations Environment Programme (UNEP), accessed November 5, 2025, https://www.unep.org/interactive/sectoral-solution-climate-change/.
12. "1.5°C: what it means and why it matters," Climate Action, United Nations, accessed November 24, https://www.un.org/en/climatechange/science/climate-issues/degrees-matter.
13. World Meteorological Organization, *State of the Global Climate 2024*, No. 1368. (WMO, 2025), 1-11, https://wmo.int/sites/default/files/2025-03/WMO-1368-2024_en.pdf.
14. IPCC, *Summary for Policymakers*, in *Climate Change 2023: Synthesis Report. Contribution of Working Groups I, II and III to the Sixth Assessment Report of the Intergovernmental Panel on Climate Change*, ed. Core Writing Team, H. Lee, and J. Romero (Geneva, Switzerland: IPCC, 2023), 1–34, https://doi.org/10.59327/IPCC/AR6-9789291691647.001.
15. "The Climate Crisis – A Race We Can Win," United Nations, accessed November 20, 2025, https://www.un.org/en/un75/climate-crisis-race-we-can-win.
16. "Jim Skea: Every Fraction of a Degree of Warming Matters," United Nations, November 15, 2022, https://www.un.org/en/climatechange/voices-from-cop27/jim-skea.
17. David I. Armstrong McKay et al., "Exceeding 1.5°C global warming could trigger multiple climate tipping points," *Science* 377 (2022): eabn7950, https://doi.org/10.1126/science.abn7950.
18. "Data Show Global Cities Impact of 3°C Rise Trajectory," Climate Adaptation Platform, October 10, 2024, https://climateadaptationplatform.com/data-show-global-cities-impact-of-3c-rise-trajectory/.
19. UNEP, *GEO-6 Key Messages* (2019), accessed November 20, 2025, https://wedocs.unep.org/rest/api/core/bitstreams/87ef3956-dfc6-4acf-b102-2c6f81951d85/content.
20. "The Road to Rio+20 and Beyond: 20 Years on From the Earth Summit," World Bank Group, November 16, 2011, https://www.worldbank.org/en/news/feature/2011/11/16/road-to-rio-twenty-years-after-earth-summit.
21. Intergovernmental Science-Policy Platform on Biodiversity and Ecosystem Services (IPBES), *Summary for Policymakers of the Global Assessment Report on Biodiversity and Ecosystem Services of the Intergovernmental Science-Policy Platform on Biodiversity and Ecosystem Services*, ed. S. Díaz et al. (IPBES secretariat, 2019), 10, https://zenodo.org/records/3553579.
22. Ibid., 2.
23. Ibid., 24-25.
24. Ibid., 11-12.
25. Ibid., 11.
26. Ibid., 25.
27. IPBES, *Global Assessment Report on Biodiversity and Ecosystem Services of the Intergovernmental Science-Policy Platform on Biodiversity and Ecosystem Services*, ed. E. S. Brondizio, et al. (Bonn, Germany: IPBES secretariat, 2019), 25.
28. "Why 30%," Campaign for Nature, accessed November 20, 2025, https://www.campaignfornature.org/why-30-1.
29. Science for Environment Policy, *Ecosystem Services and the Environment, In-depth Report 1*, produced for the European Commission, DG Environment by the Science Communication Unit, UWE Bristol, 2015, 3, https://uwe-repository.worktribe.com/output/834982.
30. World Economic Forum (WEF), *The Global Risks Report 2025: Insight Report* (WEF, 2025) https://reports.weforum.org/docs/WEF_Global_Risks_Report_2025.pdf.
31. Secretariat of the Convention on Biological Diversity, United Nations Environment Programme, *Convention on Biological Diversity (CBD). Text and Annexes*, 2.
32. United Nations, *United Nations Framework Convention on Climate Change*, (1992).
33. Secretariat of the Convention on Biological Diversity, *Global Biodiversity Outlook 5 – Summary for Policy Makers* (Montréal: Secretariat of the Convention on Biological Diversity, 2020), 3, https://www.cbd.int/gbo/gbo5/publication/gbo-5-spm-en.pdf.
34. "Climate change recognized as 'threat multiplier', UN Security Council debates its impact on peace," United Nations, UN News, January 25, 2019, https://news.un.org/en/story/2019/01/1031322.
35. WEF, *Nature Risk Rising: Why the Crisis Engulfing Nature Matters for Business and the Economy* (WEF and PWC, 2020), 8, https://www3.weforum.org/docs/WEF_New_Nature_Economy_Report_2020.pdf.
36. Secretariat of the Convention on Biological Diversity, *Global Biodiversity Outlook 5-Summary for Policy Makers*, 3.

37. "Climate Action," *EUR-Lex Glossary of European Union Policies*, European Union, accessed November 25, 2025, https://eur-lex.europa.eu/EN/legal-content/glossary/climate-action.html.
38. Convention on Biological Diversity (CBD), *Decision Adopted by the Conference of the Parties to the Convention on Biological Diversity: CBD/COP/DEC/15/4* (December 19, 2022), https://www.cbd.int/doc/decisions/cop-15/cop-15-dec-04-en.pdf.
39. "Background - Cooperation with the IPCC," United Nations Climate Change, accessed November 21, 2025, https://unfccc.int/topics/science/workstreams/cooperation-with-the-ipcc/background-cooperation-with-the-ipcc.
40. United Nations, *Paris Agreement*.
41. Ibid., 3.
42. IPCC, *Global Warming of 1.5°C*, 2018.
43. CBD, *Decision Adopted by the Conference of the Parties to the Convention on Biological Diversity: CBD/COP/DEC/15/4*, 8.
44. Neil D. Burgess et al., "Global Metrics for Terrestrial Biodiversity," *Annual Review of Environment and Resources* 49, no. 1 (October 18, 2024): 673–709, https://doi.org/10.1146/annurev-environ-121522-045106.
45. Clifford Chance, "Nature's Paris? Key Outcomes and Measures of the Kunming-Montreal Global Biodiversity Framework (COP15)," January 20, 2023, https://www.cliffordchance.com/briefings/2023/01/nature-s-paris--key-outcomes-and-measures-of-the-kunming-montrea0.html.
46. UNEP, *Emissions Gap Report 2019* (Nairobi: UNEP, 2019), xiv, https://wedocs.unep.org/rest/api/core/bitstreams/30f3e088-bd80-4c7e-a338-3840156783db/content.
47. Corey J. A. Bradshaw et al., "Underestimating the Challenges of Avoiding a Ghastly Future," *Frontiers in Conservation Science* 1 (January 13, 2021), https://doi.org/10.3389/fcosc.2020.615419.
48. UNEP, *Emissions Gap Report 2021: The Heat Is On – A World of Climate Promises Not Yet Delivered* (Nairobi: UNEP, 2021), https://www.unep.org/emissions-gap-report-2021.
49. UNEP, *Emissions Gap Report 2022: The Closing Window – Climate crisis calls for rapid transformation of societies* (Nairobi: UNEP, 2022), https://www.unep.org/emissions-gap-report-2022
50. UNEP, *Emissions Gap Report 2023: Broken Record – Temperatures hit new highs, yet world fails to cut emissions (again)* (Nairobi: UNEP, 2023), https://doi.org/10.59117/20.500.11822/43922.
51. UNEP, *Emissions Gap Report 2024: No More Hot Air ... Please! With a massive gap between rhetoric and reality, countries draft new climate commitments* (Nairobi: UNEP, 2024), https://doi.org/10.59117/20.500.11822/46404.
52. UNEP, *Adaptation Gap Report 2023: Underfinanced. Underprepared. Inadequate investment and planning on climate adaptation leaves world exposed* (Nairobi: UNEP, 2023), https://doi. org/10.59117/20.500.11822/43796.
53. UNEP, *Adaptation Gap Report 2024: Come Hell and High Water – As fires and floods hit the poor hardest, it is time for the world to step up adaptation actions* (Nairobi: UNEP, 2024), https://doi.org/10.59117/20.500.11822/46497.
54. IPCC, "The Evidence Is Clear: The Time for Action Is Now. We Can Halve Emissions by 2030," press release, April 4, 2022, https://www.ipcc.ch/2022/04/04/ipcc-ar6-wgiii-pressrelease/.
55. UNEP, *Emissions Gap Report 2024 Key Messages* (UNEP, 2024), https://wedocs.unep.org/rest/api/core/bitstreams/e7d6bc1f-f83b-492c-8fb4-01f574a1adfd/content.
56. UNEP, "The Sectoral Solution to Climate Change."
57. UNEP, *Emissions Gap Report 2024*.
58. "About the Climate Action Coalition," Climate Action Coalition, accessed November 25, 2025, https://coalition.climateaction.org/about/.
59. UNEP, *Emissions Gap Report 2025 Key Messages* (UNEP, 2025), https://wedocs.unep.org/rest/api/core/bitstreams/162130ba-17df-4f8d-80be-fed7384c8442/content.
60. "Protected Planet Report 2024", Protected Planet, accessed November 25, 2025, https://digitalreport.protectedplanet.net/
61. Ibid.
62. Secretariat of the Convention on Biological Diversity, *Global Biodiversity Outlook 5 – Summary for Policy Makers*, 4.
63. Ibid., 12.
64. McElwee, "A Tale of Two Panels."
65. IPCC, *Global Warming of 1.5°C: An IPCC Special Report*.
66. IPCC, ed., "Investment and Finance," in *Climate Change 2022: Mitigation of Climate Change*, 1547-1640 (Cambridge University Press, 2023), https://doi.org/10.1017/9781009157926.017.
67. UNEP, *Adaptation Gap Report 2025: Running on Empty. The World Is Gearing Up for Climate Resilience — Without the Money to Get There*, ed. H. Neufeldt et al. (Nairobi: UNEP, 2025), https://wedocs.unep.org/20.500.11822/48798.
68. WEF, *Bridging the Gap: How to Finance the Net-Zero Transition WHITE PAPER* (January 2025, https://reports.weforum.org/docs/WEF_Bridging_the_Gap_How_to_Finance_the_Net_Zero_Transition_2025.pdf.
69. Allen & Overy and Climate Policy Initiative (CPI), *How big is the Net Zero financing gap?* (September 2023), https://www.climatepolicyinitiative.org/wp-content/uploads/2023/09/How-big-is-the-Net-Zero-financing-gap-2023.pdf.
70. UNEP, "Slow climate adaptation threatening lives and economies", press release, October 29,2025, https://www.unep.org/news-and-stories/press-release/slow-climate-adaptation-threatening-lives-and-economies#:~:text=A%20worrying%20gap&text=International%20public%20adaptation%20finance%20flows%20to%20developing%20countries%20were%20US,as%20much%20as%20current%20flows.
71. UNEP, *Adaptation Gap Report 2025: Running on Empty. The World Is Gearing Up for Climate Resilience — Without the Money to Get There*, ed. H. Neufeldt et al. (Nairobi: UNEP, 2025), viii.
72. United Nations Environment Programme Finance Initiative (UNEP FI), "Governments Adopt First Global Strategy to Finance Biodiversity: Implications for Financial Institutions," press release, March 3, 2025, https://www.unepfi.org/themes/ecosystems/governments-adopt-first-global-strategy-to-finance-biodiversity-implications-for-financial-institutions/.
73. "Can ESG awakening end the biodiversity crisis?," PA Future, September 29, 2020, https://future.portfolio-adviser.com/can-esg-awakening-end-the-biodiversity-crisis/.
74. United Nations, *Paris Agreement*, 7.

75. "Biodiversity crisis emphasizes need for corporate transparency," GRI, June 4, 2021, https://www.globalreporting.org/news/news-center/biodiversity-crisis-emphasizes-need-for-corporate-transparency/.
76. United Nations, *Paris Agreement*, 7.
77. CBD, *Decision Adopted by the Conference of the Parties to the Convention on Biological Diversity: CBD/COP/DEC/15/4*.
78. Task Force on Climate-Related Financial Disclosures, *Recommendations of the Task Force on Climate-related Financial Disclosures* (TCFD, 2017), https://assets.bbhub.io/company/sites/60/2021/10/FINAL-2017-TCFD-Report.pdf.
79. KPMG IMPACT, *The time has come: The KPMG Survey of Sustainability Reporting 2020* (KPMG, 2020), 36, https://assets.kpmg.com/content/dam/kpmg/my/pdf/time-has-comes.pdf.
80. "About us: Mission," Taskforce on Nature-related Financial Disclosures, accessed November 27, 2025, https://tnfd.global/about/#mission.
81. Taskforce on Nature-related Financial Disclosures, *Proposed Technical Scope Recommendations for the TNFD* (June 2021), 9, https://tnfd.global/wp-content/uploads/2021/07/TNFD-%E2%80%93-Technical-Scope-3.pdf.
82. "Integrating Biodiversity Considerations into Infrastructure," Principles for Responsible Investment, August 1, 2023, https://public.unpri.org/infrastructure/integrating-biodiversity-considerations-into-infrastructure/11611.article.

Need for Integration

Chapter 2

NEED FOR AN INTEGRATED APPROACH TO BIODIVERSITY AND CLIMATE CRISES

Calls for action on climate and biodiversity refer to the detrimental effects of "business-as-usual" pathways. Meaningful action requires "substantial changes and innovations, implemented on a short timescale and involving a wide range of actors at all scales"[1]—a "whole-of-society" approach. However, even the most aggressive goals and efforts in each of these areas will not succeed unless they are tackled together. At the same time, combining actions holds the potential to make each separate action or goal easier to achieve by leveraging synergies.

Although biodiversity loss and climate change are recognized as two of the most pressing issues the world is currently facing and "as interconnected in both scientific and policy-making circles," they are still largely addressed within their separate domains.[2] "This functional separation creates a risk of incompletely identifying, understanding, and dealing with the connections between the two. In the worst case, it may lead to taking actions that inadvertently prevent the solution of one or the other, or both issues."[3]

This separation is also reflected in the separate national plans designed to each set of commitments, the National Determined Contributions (NDCs) under the UNFCCC Paris Agreement, and the National Biodiversity Strategies and Action Plans (NBSAPs) under the CBD Global Biodiversity Framework.

INTERCONNECTED CRISES, SHARED DRIVERS, AND COMMON SOLUTIONS

"Human-caused climate change is increasingly threatening nature and its contributions to people, including its ability to mitigate climate change. Changes in biodiversity, in turn, affect climate, especially through impacts on nitrogen, carbon and water cycles."[4] In short, biodiversity and climate change mutually reinforce each other.

Despite the fact that the two overarching frameworks for climate and biodiversity action—the Paris Agreement and Kunming-Montreal Global Biodiversity Framework—both refer to the role of biodiversity or climate in achieving their respective goals, their implementation has not been integrated.

The Paris Agreement recognizes the importance of "the conservation and enhancement, as appropriate, of sinks and reservoirs of the greenhouse gases"[5] and "of ensuring the integrity of all ecosystems, including oceans, and the protection of biodiversity."[6] It refers to the need for:

- "Policy approaches and positive incentives for activities relating to reducing emissions from deforestation and forest degradation" (Article 5)[7]
- "Alternative policy approaches, such as joint mitigation and adaptation approaches for the integral and sustainable management of forests, while reaffirming the importance of incentivizing, as appropriate, non-carbon benefits associated with such approaches" (Article 5)[8]
- The Kunming-Montreal Global Biodiversity Framework, in turn, refers to the need to:
- "Minimize the impacts of climate change on biodiversity and increase its resilience through mitigation, adaptation, and disaster risk reduction actions, including through nature-based solution and/or ecosystem-based approaches" (Target 8)[9]
- "Restore, maintain, and enhance nature's contributions to people, including ecosystem services, such as regulation of climate, and protection from natural hazards and disasters, through nature-based solutions and/or ecosystem-based approaches" (Target 11)[10]
- "Optimize co-benefits and synergies of finance targeting the biodiversity and climate crises" (Target 19)[11]

The call for integrated action is also supported by scientific studies demonstrating the potential benefits of jointly addressing climate change and biodiversity. For example, a UNEP World Conservation Monitoring Centre (WCMC) study found that conserving 30% of priority biodiversity areas "could safeguard more than 500 gigatons of carbon" stored in vegetation and soil, "while reducing the extinction risk of almost 88% of the species considered."[12] The study also states that "when prioritizing areas for conservation, accounting for biodiversity and carbon together can secure 95% of the biodiversity benefits and nearly 80% of the carbon stocks that could be obtained by prioritizing either value alone."[13]

There are significant opportunities to mitigate and

adapt to climate change while enhancing the biodiversity conservation. However, realizing these synergies requires "a conscious attention to biodiversity concerns."[14]

Discussions about which types of solutions should be prioritized for joint climate-biodiversity action—in addition to policy, research, and enabling financial mechanisms—generally revolve around nature-based solutions (NbS). NbS are widely recognized for their potential contribution to climate mitigation, although "the extent of that contribution is debated, and they can only be effective alongside ambitious reductions in all human-caused greenhouse gas emissions."[15] NbS are characterized by broad scope and multi-functionality and can provide more than one benefit at a time, including carbon sequestration, disaster risk reduction, and a range of parallel contributions to human well-being.

From a longer-term perspective, "climate change and biodiversity decline share the same indirect drivers."[16] These are the underlying forces that shape the extent, severity, and combination of direct anthropogenic drivers that operate in each place. They are rooted in societal values and reinforced by institutional and governance structures, as well as social, economic, and cultural contexts that shape behavioral patterns, including consumption and energy use.[17] In short, the transformative action identified as essential for a livable planet is, by definition, integrated climate-biodiversity action.

THE JOINT IPCC-IPBES WORKSHOP

The importance of integrated solutions for climate and biodiversity re-emerged as a prominent topic of discourse with the publication of a IPBES-IPCC Joint Report on biodiversity and climate change in June 2021.[18] The Intergovernmental Science-policy Platform on Biodiversity & Ecosystems Services (IPBES) and the Intergovernmental Panel on Climate Change (IPCC)—the two independent intergovernmental science-policy bodies—collaborated for the first time to organize a joint workshop on biodiversity and climate change, exploring the multiple and complex connections between the two.[19] The workshop took place in 2020, following the alarming findings of the first IPBES Global Assessment Report. The resulting scientific outcome report highlighted that "in the coming decades, biodiversity and climate targets will be mutually dependent on one another to be achievable. They must also be dynamic and adaptable to accommodate synergies and uncertainties."[20]

The scientific outcome report of the IPBES-IPCC workshop reaffirmed the urgency of taking action on both climate and biodiversity, through "a rapid but careful ramp-up of solutions that deliver co-benefits for both."[21] It emphasized that "actions to enhance the adaptive capacity of ecosystems are placed at risk by unabated climate change exceeding adaptation limits—highlighting the importance of keeping climate warming well below 2°C—and by high levels of other pressures such as land-use change, overexploitation, and pollution."[22]

The report explores how plausible future changes in climate could affect biodiversity and potentially lead to irreversible changes in ecosystems and species—exceeding their capacity to adapt—as well as how plausible changes in biodiversity could affect climate and mitigation actions. Through a series of examples, it discusses the risks associated with narrow-focused measures, technical and technology-based measures for climate change mitigation or adaptation, as well as short-term land- and ocean-based solutions that have long-term negative impacts on biodiversity. It argues that risks can be transformed into opportunities through a combined approach to climate and biodiversity. Such an approach can secure win-win solutions and prevent future lock-ins. An integrated approach is considered essential to provide a complete overview of climate change risks and the potential unintended trade-offs of mitigation and adaptation measures on ecosystems and biodiversity, as illustrated by examples such as biofuel crop production, afforestation of biodiversity-rich habitats, and monocultures.

An integrated approach must also be established in scenario-analysis and policy pathways for climate and biodiversity—which drive goal setting—in order to avoid conflicts. As explained in the IPBES-IPCC Report, existing scenarios tend to have a specific focus on climate change mitigation while paying less attention to biodiversity. For example, the Deep Decarbonization Pathways Project primarily focuses on energy sources. Moreover, current scenarios used by IPCC, when referring to the contribution of natural carbon sequestration enhancement actions, "do not differentiate between natural forest regrowth, reforestation with plantations, or afforestation of land not previously tree-covered, making it difficult to assess biodiversity impacts."[23]

The report acknowledges that developing scenarios for both biodiversity and climate presents challenges related to increasing complexity, nonlinearity, and uncertainty. Ecosystems are inherently complex, with interdependent components and processes, and there will always be a degree of uncertainty in how they will respond to specific interventions or external changes. The complexity of these relationships is illustrated by various examples in the report, showing that interactions differ depending on ecosystem type, geographic location, condition, extent, and connectivity to other ecosystems. The report therefore highlights the need for a new science-practice relationship to enable purposeful interventions that can initiate and accelerate the transition to a new paradigm.[24]

COMMON MILESTONES OF INTEGRATION

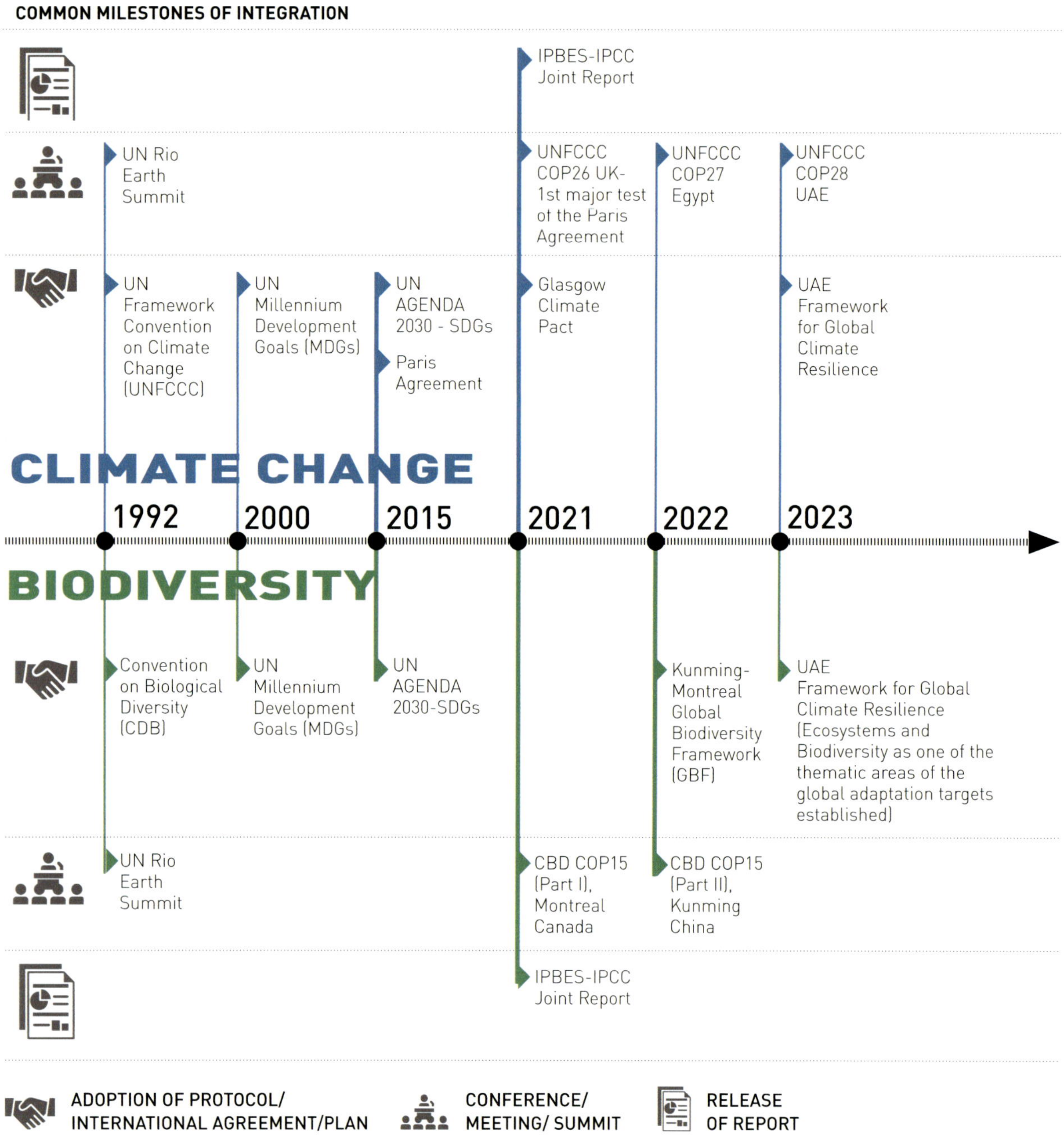

PAIRING OBJECTIVES

The UNFCCC Climate Change Conference COP25 in Chile-Madrid was the first climate negotiations where biodiversity was also part of the agenda. Taking placing after the publication of the 2019 IPBES Global Assessment—outlining the extent of the biodiversity crisis—the COP discussions also focused on ecosystems and biodiversity. The connection among ecosystems, climate change, and biodiversity and the need to consider them jointly, was recognized in the COP25 decision of December 2019, which underlines "the essential contribution of nature to addressing climate change and its impacts and the need to address biodiversity loss and climate change in an integrated manner."[25]

The year 2021 can be considered a landmark for an integrated approach to the objectives of the Conferences of Parties for climate and biodiversity. Following the publication of the IPBES-IPCC Joint Report, efforts were made to integrate biodiversity into discussions under the UNFCCC Climate Change Conference (COP26) in Glasgow, UK and to integrate climate change into the UN Biodiversity Conference (COP15) in Kunming, China. The two COPs were held only weeks apart: COP15 (Part I) in October and COP26 in November 2021.

COP15 was expected to produce a negotiated new

Post-2020 Biodiversity Framework as a successor to the 2010 CBD Aichi Targets for addressing biodiversity loss. It represented a critical opportunity to embed nature-based solutions within the international framework for global environmental action through 2030 and beyond.

COP26 aimed to review progress towards meeting the Paris Agreement climate targets. The urgency for improved biodiversity reporting gained additional significance at COP26, where one of four goals was to ensure adaptation measures that protect natural habitats and restore ecosystems. The UK, as COP 26 president, emphasized that one of its objectives was to maximize the potential of nature-based solutions to enhance prosperity, reduce emissions, and strengthen resilience, building on the related decision of COP 25.

COP26 negotiations resulted in the adoption of the Glasgow Climate Pact. Although "the pact and other commitments made during the summit fall short of limiting global warming to the Paris Agreement's 1.5°C stretch target, full implementation of the commitments made throughout COP26 could limit heating to 1.8°C. Importantly, several nature-related elements feature in the final agreed text of the Glasgow Climate Pact."[26] The explicit link between the climate and biodiversity agendas is more pronounced than in the Paris Agreement. Moreover, the share of countries whose climate plans include nature-based solutions increased from 82% to 92%.[27]

"The prominence of nature and biodiversity in the Glasgow Climate Pact and throughout the climate talks marks the first time nature was meaningfully incorporated into global climate negotiations."[28]

COP27—held in Sharm el-Sheikh in 2022, right before COP15 (Part II), when the Kunming-Montreal Global Biodiversity Framework was adopted—also underlined "the urgent need to address, in a comprehensive and synergetic manner, the interlinked global crises of climate change and biodiversity loss in the broader context of achieving the Sustainable Development Goals, as well as the vital importance of protecting, conserving, restoring and sustainably using nature and ecosystems for effective and sustainable climate action."[29] Moreover, the decision calls for consideration of nature-based solutions or ecosystem-based approaches for mitigation and adaptation action.

COP28 in the United Arabic Emirates marked the moment of agreement on 11 new global adaptation targets for 2030 to serve as complements to the UN SDGs, the Sendai Framework, and the Convention on Biological Diversity. The targets include seven thematic areas: "water, food and agriculture, health, ecosystems and biodiversity, infrastructure, poverty and livelihoods, and cultural heritage."[30] Urges were made to "reduce climate impacts on ecosystems and biodiversity, and accelerating the use of ecosystem-based adaptation and nature-based solutions, including through their management, enhancement, restoration and conservation and the protection of terrestrial, inland water, mountain, marine and coastal ecosystems."[31] Thus, ecosystems and biodiversity are highlighted as integral components of adaptation action.

Stewarding biodiversity is a core part of the agenda for the upcoming COP30 in Belém, Brazil. The very choice of hosting the climate COP "at the gateway to the Amazon rainforest"[32] is itself highly symbolic. "COP30 must be the moment when halting and reversing deforestation and ecosystem degradation becomes fully incorporated into climate negotiations."[33]

BIODIVERSITY AS AN INTEGRAL COMPONENT OF CLIMATE ACTION

"There is no climate solution without the full contribution from nature."[34] Nature-based solutions—those actions that capitalize on nature's contributions—can provide up to 37% of net reductions in GHG emissions needed to meet the Paris climate goal[35]—about one third of the climate solution.[36] In particular, those related to the protection 30% of high priority biodiversity areas of global importance—the 30x30 target of the Global Biodiversity Framework—could absorb 12 gigatons of carbon dioxide each year[37] while preventing additional land-based emissions from deforestation.

Carbon sequestration—the process of storing carbon in a carbon pool—is a key ecosystem service, or contribution of nature, to climate change mitigation. However, there is a distinction between short-term carbon flows (e.g., diurnal CO_2 exchange between vegetation and the atmosphere) and long-term sequestration.

The provision of services by ecosystems is "underpinned by various aspects of biodiversity, although the relationship is a complex one. 'Biodiversity' is a composite, multidimensional term and there is no simple relationship between biodiversity and ecosystem services."[38] Nevertheless, high-diverse ecosystems tend to be better able to adapt to climate change and climate variability than functionally impoverished ones,[39] and are thus more capable of delivering services. As biodiversity is degraded or lost, ecological communities and human societies become more vulnerable because options for adaptation diminish and ecosystem services decline.

Long-term carbon sequestration varies by ecosystem type and condition. Different types of ecosystems have different qualities as carbon stocks. Biodiversity-rich ecosystems such as forests, rangelands, croplands, peatlands, and wetlands represent globally significant carbon stores. "Their conservation, restoration, and sustainable use are included in many Nationally Determined Contributions and are therefore critical components to fulfilling the Paris Agreement."[40] Their loss, fragmentation, or decline, therefore, can undermine climate change mitigation efforts.

These points indicate the complexity of the interrelation of biodiversity and the services ecosystems provide, particularly under a changing climate. They also highlight how critical ecosystem condition is for enabling ecosystems to function effectively as carbon sinks.

Biodiversity and healthy ecosystems are also vital for increasing resilience and reducing the risks and damages associated with the negative impacts of climate change. "They can serve as natural buffers against extreme climate and weather events, such as changing rainfall patterns, droughts, storms, and other disasters."[41] However, "the ability of ecosystems to either resist or return to their former state following a disturbance may depend on certain levels of functional diversity."[42] Maintaining this ability is an important consideration when designing mitigation and adaptation actions. "The persistence of biodiversity under changing climates is essential for the success of NbS for climate mitigation and adaptation,"[43] and strategies that support biodiversity's adaptability must be incorporated into NbS.

Therefore, the conservation of biodiversity—along with the reduction of habitat loss, fragmentation, and degradation—supports the long-term persistence of ecosystems and strengthens their potential to adapt to climate change. At the same time, "while some natural pest-control, pollination, soil-stabilization, flood-control, water purification and seed-dispersal services can be replaced when damaged or destroyed by climate change, technical alternatives may be costly and therefore not feasible to apply in many situations."[44] Thus, conserving biodiversity ensures that options remain available to help human societies adapt better to climate change, anticipate increasing pressures or demand, and compensate for likely future losses.[45]

NOTES

1. Secretariat of the Convention on Biological Diversity, *Global Biodiversity Outlook 5 – Summary for Policy Makers* (Secretariat of the Convention on Biological Diversity, 2020), 15, https://www.cbd.int/gbo/gbo5/publication/gbo-5-spm-en.pdf.
2. H.O. Pörtner et al., *Scientific outcome of the IPBES-IPCC co-sponsored workshop on biodiversity and climate change;* IPBES secretariat (Zenodo, 2021), 4, DOI:10.5281/zenodo.4659158.
3. Ibid.
4. "Tackling Biodiversity & Climate Crises Together and Their Combined Social Impacts," press release issued by the Secretariat

of the Intergovernmental Science-Policy Platform on Biodiversity and Ecosystem Services (IPBES), June 10, 2021, https://www.un.org/sustainabledevelopment/blog/2021/06/tackling-biodiversity-climate-crises-together-and-their-combined-social-impacts/.
5. United Nations, *Paris Agreement*, December 12, 2015, in *United Nations Treaty Collection*, 2, https://unfccc.int/sites/default/files/english_paris_agreement.pdf.
6. Ibid.
7. Ibid., 6.
8. Ibid.
9. Convention on Biological Diversity, *Decision Adopted by the Conference of the Parties to the Convention on Biological Diversity: CBD/COP/DEC/15/4* (December 19, 2022), 10, https://www.cbd.int/doc/decisions/cop-15/cop-15-dec-04-en.pdf.
10. Ibid.
11. Ibid., 12.
12. X. De Lamo et al., *Strengthening synergies: how action to achieve post-2020 global biodiversity conservation targets can contribute to mitigating climate change* (UNEP-WCMC, 2020), 8, https://www.unep.org/resources/report/strengthening-synergies-achieve-biodiversity-goals.
13. Ibid., 3.
14. Secretariat of the Convention on Biological Diversity, *Interlinkages between Biological Diversity and Climate Change: Advice on the Integration of Biodiversity Considerations into the Implementation of the United Nations Framework Convention on Climate Change and Its Kyoto Protocol* (Montreal, SCBD,2003), CBD Technical Series no. 10, iii, https://www.cbd.int/doc/publications/cbd-ts-10.pdf.
15. Pörtner et al., *Scientific outcome of the IPBES-IPCC co-sponsored workshop on biodiversity and climate change*, 16.
16. Ibid., 28.
17. Ibid.
18. Ibid.
19. "IPBES-IPCC Co-Sponsored Workshop on Biodiversity and Climate Change," IPBES, June 10, 2021, https://www.ipbes.net/events/ipbes-ipcc-co-sponsored-workshop-biodiversity-and-climate-change.
20. Pörtner et al., *Scientific outcome of the IPBES-IPCC co-sponsored workshop on biodiversity and climate change,* 48-49.
21. Ibid., 146.
22. Ibid., 15.
23. Ibid., 60.
24. Ibid., 174.
25. United Nations, *Report of the Conference of the Parties on its Twenty-Fifth Session, held in Madrid from 2 to 15 December 2019, Addendum, Part Two: Action Taken by the Conference of the Parties at its Twenty-Fifth Session*, FCCC/CP/2019/13/Add.1 (United Nations, March 16, 2020), Decisions adopted by the Conference of the Parties, 2, https://unfccc.int/sites/default/files/resource/cp2019_13a01E.pdf.
26. "After COP26: Nature Positive Set to Become Key Component of Net Zero," Taskforce for Nature-related Financial Disclosures, November 15, 2021, https://tnfd.global/after-cop26-nature-positive-set-to-become-key-component-of-net-zero/.
27. Ibid.
28. "COP26: Nature Features Prominently at COP26," Campaign for Nature, November 17, 2021, https://www.campaignfornature.org/nature-features-prominently-at-cop26.
29. United Nations, *Report of the Conference of the Parties on its Twenty-Seventh Session, held in Sharm el-Sheikh from 6 to 20 November 2022, Addendum, Part Two: Action Taken by the Conference of the Parties at its Twenty-Fifth Session,* FCCC/CP/2022/10/Add.1 (United Nations, March 17, 2023), Decisions adopted by the Conference of the Parties, 2, https://unfccc.int/sites/default/files/resource/cp2019_13a01E.pdf.
30. United Nations Environment Programme (2024). *Adaptation Gap Report 2024: Come hell and high water — As fires and floods hit the poor hardest, it is time for the world to step up adaptation actions.* Nairobi. https://doi.org/10.59117 /20.500.11822/46497.
31. United Nations Framework Convention on Climate Change (UNFCCC), *Decision -/CMA.5: Glasgow–Sharm el-Sheikh Work Programme on the Global Goal on Adaptation Referred to in Decision 7/CMA.3* (Bonn: UNFCCC, 2023).
32. Marcene Mitchell, "What to expect at COP30: Closing the gaps in Belém," *World Wildlife Fund*, accessed November 4, 2025, https://www.worldwildlife.org/news/stories/what-to-expect-at-cop30-closing-the-gaps-in-bel%C3%A9m/.
33. Ibid.
34. "Research Reveals Benefits of Joint Action on Climate and Nature," UNEP-WCMC, November 2020, https://www.unep-wcmc.org/en/news/research-reveals-major-benefits-of-joint-action-on-climate-and-nature.
35. "COP26 Hub," Campaign for Nature, accessed November 1, 2025, https://www.campaignfornature.org/cop26-hub.
36. "Nature 'One of Most Effective Ways' of Combatting Climate Change," UN News, September 19, 2019, https://news.un.org/en/story/2019/09/1046752#:~:text=The%20bottom%20line%20is%20that,it%20the%20chance%20it%20deserves.
37. Ibid.
38. Secretariat of the Convention on Biological Diversity, *Interlinkages Between Biological Diversity and Climate Change*, 20.
39. Ibid., 2.
40. Secretariat of the Convention on Biological Diversity (CBD), Food and Agriculture Organization of the United Nations, World Bank, United Nations Environment Programme, and United Nations Development Programme, *Biodiversity and the 2030 Agenda for Sustainable Development: Technical Note* (CBD, December 2016), 16, https://www.cbd.int/development/doc/biodiversity-2030-agenda-technical-note-en.pdf.
41. Ibid.
42. Secretariat of the Convention on Biological Diversity, *Interlinkages Between Biological Diversity and Climate Change*, 2.
43. Pörtner et al., *Scientific outcome of the IPBES-IPCC co-sponsored workshop on biodiversity and climate change*, 49.
44. Secretariat of the Convention on Biological Diversity, *Interlinkages Between Biological Diversity and Climate Change*, 77.
45. Ibid.

B

FRAMING CLIMATE AND BIODIVERSITY ACTION IN INFRASTRUCTURE PROJECTS

Part B represents the core of the book. Chapter 3 outlines the scope and methodology of the research behind this publication, providing a guide for the reader throughout the book. Individual chapters are then dedicated to the criteria for framing climate, biodiversity, and integrated action in infrastructure projects.

CREDIBLE CLIMATE LEADERSHIP IN INFRASTRUCTURE

All organizations are responsible for taking measures for climate action as part of their corporate responsibility. Engineering companies have an even greater responsibility as they lead in planning, designing, and supervising the construction of low-emission and resilient infrastructures.

At IDOM, we address climate change through mitigation and adaptation as a part of sustainability in four axes: through our projects, in our workplace, in the way/means we use to develop our projects, and with our contributions at a professional and personal level.

This includes supporting the transition to renewable energy, decarbonizing the transportation sector through mobility plans or railway electrification, promoting green hydrogen projects for electricity generation and mobility, reducing CO_2 emissions, and contributing to national strategies on disaster risk management, climate-resilient city planning, and ecosystem restoration and conservation, among other aspects.

Infrastructure as Both Solution and Problem in Climate Change Challenge: What Kind of Infrastructure Are We Going to Leave for Future Generations?

Infrastructure development has historically contributed to climate change but also offers a pathway to mitigate its global effects. By adopting sustainable practices and innovative technologies, infrastructure can transition from being part of the problem to becoming vital to the solution, contributing to SDG 13.

Infrastructure plays a dual role in terms of climate change. First, it significantly contributes to greenhouse gas emissions, generating "around 60% of global greenhouse gas emissions."[1] Still, it is the sector that suffers—and will continue to suffer—from climate impacts. Second, infrastructure is a crucial part of the solution which can only be achieved through sustainable development planning and innovation. According to a report published by the United Nations Office for Project Services (UNOPS), infrastructure is linked to achieving up to 92% of the UN SDGs.[2]

Traditional infrastructure projects, such as highways, bridges, and buildings, often rely on materials like concrete and steel, which have high carbon footprints due to their energy-intensive production processes. Expanding infrastructure can lead to urban sprawl, increasing reliance on automobiles and contributing to higher emissions. Large-scale infrastructure projects can also lead to the depletion of natural resources, deforestation, and biodiversity loss.

However, building infrastructure for renewable energy sources—such as bioenergy, geothermal, hydropower, ocean, solar, wind, nuclear, green hydrogen, and others—can significantly reduce carbon footprints. Investing in efficient and extensive public transportation systems can also reduce reliance on personal vehicles, thereby lowering emissions.

At the same time, infrastructure projects play a crucial role in adapting to climate change by enhancing resilience and reducing vulnerability. For example, the construction of levees and sea walls can protect coastal and riverine areas and help protect communities and their population from flooding and rapid storm surges. Projects such as Sustainable Urban Drainage Systems manage rainwater runoff, reducing the risk of flooding and improving water quality. Increasing green spaces and protecting forests minimizes the probability of an urban heat island effect, lowers temperatures, and improves surrounding air quality. IDOM strives to design infrastructure to withstand extreme weather events and ensure long-term functionality and safety.

The Journey Towards Sustainable Infrastructures: A Lifecycle Approach

Sustainable infrastructure refers to "infrastructure projects that are planned, designed, constructed, operated, and decommissioned in a manner to ensure economic and financial, social, environmental (including climate resilience), and institutional sustainability over the entire lifecycle of the project."[3]

The infrastructure lifecycle vision is applied by incorporating climate change integration criteria in the early stages and adapting to the needs of each phase along the way. Since this approach provides a better understanding of trade-offs and expected long-term impacts, it has proven to be one of the most cost-effective mechanisms for reducing future impacts. Our projects are tailored to meet the needs of the environment and the community they serve. Inspired by the Sustainable Development Goals (SDGs), each project is designed to be climate- and risk-resilient while respecting the environment and its biodiversity, prioritizing nature-based solutions.

Participatory processes are employed during the planning and design phases, fostering a culture of generating and evaluating multiple alternatives. This approach enables us to identify innovative solutions that minimize resource usage, reduce carbon footprints, and mitigate vulnerability to climate risks. Key examples include:

- Infrastructure planning as the basis of our preventive sustainability model, utilizing advanced analytics and decision-making tools to ensure robust, informed, and consensual outcomes
- Climate scenario analysis, allowing the assessment of various climate situations and their possible impacts on different infrastructure assets to enhance their resilience
- Circular economy principles, including rethinking, reducing, reusing, recycling/recovering resources
- Material optioneering, based on the technology available in the market, exploring materials that significantly lower greenhouse gas emissions and resource depletion, such as locally sourced or with eco-labeled materials, considering both initial costs and long-term savings

The development is supervised during the construction phase to meet all necessary objectives and ensure compliance with planned sustainability standards. If required, modifications are made to support sustainable development, such as incorporating additional locally sourced materials or energy-efficient technologies. During the operation and maintenance process, defining how to operate and manage new and existing assets to extend the life expectancy of these infrastructures is of utmost importance. In addition, it is necessary to consider possible deconstruction and how it could contribute to the circular economy, incorporating the reuse, recycling, and revaluation of materials.

Our perspective is integrated into sustainability strategies based on our tailor-made methodology, which aligns with the vision of sustainable infrastructure assessed by internationally recognized certifications, such as Envision or Building Research Establishment Environmental Assessment Method for Infrastructure (BREEAM Infrastructure).

In conclusion, the sooner climate change integration criteria are applied during the project lifecycle, the greater the value delivered. As the project timeline progresses, the ability to implement effective changes diminishes. In contrast, the cost of making such changes escalates. For this reason, we assess and design mitigation and adaptation measures in the early stages of the project, where modifications are easier and more cost-effective.

MITIGATION—Infrastructures for a Low-carbon Future

While taking mitigation measures, we apply the hierarchy of emission reductions: don't build, build less, build smart, build efficiently, and optimize the operation phase to better align with global net-zero targets. Once the need for building is justified, we utilize the carbon reduction hierarchy (avoid, switch, improve) to reduce emissions throughout all stages of the project lifecycle. We adopt methodologies, standards, and databases that increase the consistency of the carbon footprint calculation. However, we must highlight that carbon footprint is a decision-making tool which we use to compare different scenarios and select materials with the principle of "as low as reasonably possible" (ALARP).

However, it should be noted that each infrastructure has a different capability in terms of technology development and social/client acceptance. For instance, incorporating prefabricated systems will reduce onsite emissions, avoid misapplications, and generate efficient use of materials. Unfortunately, this is not always available or cost-effective when building bridges and structures. Recycled materials could be rejected because they can be considered not to be as durable as new ones. For example, this happens with recycled plastic railway sleepers in some regions.

ADAPTATION—Building Resilience Against Disaster and Climate Change

Risk assessment is considered good practice in identifying the most significant hazards to which the project could be vulnerable. Careful consideration is made when analyzing the climate scenarios and projections during the planning and design phases. This enables the designer to understand what impacts can be found in earlier stages of the project to adapt the most critical assets towards an easy and cost-effective strategy. Some cases can include:

- Damage and erosion of slopes, causing possible landslides due to heavy rainfall. This can also affect the stability of bridges
- Increase the frequency of breakwater overtopping due to sea level rise, therefore causing a possible decrease in the operability of ports
- Damage to railways due to higher temperature fluctuations

All possible natural hazards must be addressed during the design phase by re-evaluating the design for different climate scenarios. Otherwise, potential risks would be impossible to avoid or minimize. Therefore, the planning phase is a crucial preventive (non-structural) measure.

Regarding other measures, we support and prioritize nature-based solutions, such as Sustainable Urban Drainage Systems. Yet, we also strive to improve slope stabilization with vegetation to increase the height of the levees or propose preventive maintenance tasks.

In some cases, due to the uncertainty and lack of information, we advocate for "no regret," "low regret," and "win-win" measures to achieve guaranteed results that can provide benefits even if the unforeseen impacts of climate change do not materialize, such as reducing flood risks by avoiding building developments in high-risk areas. On the other hand, improving water efficiency helps address drought risks.

Ultimately, the goal is to harness nature's ability to deliver essential and cost-effective infrastructure services while providing multiple benefits for humanity and the planet. Thus, great importance is placed on working to prevent biodiversity loss and avoid an environmental impact on infrastructure systems while investing in natural infrastructure.

Climate Action at the Project Level: Regional Differences

The demand for climate action at the project development level is highly dependent on geography:

1. Participation in international agreements: For example, the Paris Agreement can drive demand for net-zero carbon infrastructure projects and sustainability services as countries commit to meeting specific climate targets.
2. Regulatory environment: Different countries and regions have a variety of regulations and policies related to climate action. For example, the European Green Deal pressures governments and companies more than other areas.
3. Climate and environmental conditions: Regions more vulnerable to the effects of climate change, such as coastal areas or regions prone to extreme weather, have greater demand and urgency for resilient infrastructure. Additional measures are necessary to mitigate these risks to existing assets.
4. Economic development: More developed economies often have more resources to invest in climate action. Despite having a greater need for climate adaptation and urgent infrastructure development than underdeveloped countries, they face additional challenges. Challenges are economic and include limitations to advanced technologies and governance structures. While global awareness is increasing, some progress is being made. Such barriers impact the ability of developing nations to implement practical climate-resilient infrastructure projects.
5. Cultural attitudes: Public awareness and cultural attitudes towards climate change can differ widely. In some regions, a strong public demand for climate action influences public sector policies and initiatives.

Green Finances—Closing the Infrastructure Investment Gap

Possible higher costs are a challenge for implementing sustainable infrastructure projects; therefore, it is sometimes necessary to incorporate other dimensions or externalities into the analysis and decision-making: urban, environmental, and social. In this way, infrastructures generate co-benefits: environmental (reduction of GHGs or lower consumption of natural resources), of a social nature, such as reduction of social inequalities or generation of green jobs, and economic, such as improved competitiveness or energy security.

In this way, the growing awareness of the effects of climate change has led society to demand more responsible and sustainable investments. This translates into increased demand for investment funds in sustainable infrastructure. This shift helps mitigate climate risks and offers opportunities to close the investment gap necessary to achieve a sustainable future, as outlined within the SDGs.

Final Message

To summarize, the present climate change emergency and society's increasing need to address mitigation and adaptation of infrastructures. This is not merely an option but an obligation and a commitment to humanity to improve the future.

As a pledge to "the future will be sustainable, or it won't be," the knowledge and experience of our team focuses on developing infrastructure projects around the world, driven by the four dimensions of sustainability: environmental (Including climate change mitigation and adaptation), social, economic, and governance. We are confident that this approach is the most effective way to improve our clients' quality of service. We prioritize the professional development of our people to maximize the benefits to the environment, the planet, and society.

María Álvarez Mingorance
Chief Sustainability Officer (CSO) at IDOM
Elena Calcerrada Romero
Infrastructure Sustainability Manager at IDOM
Daniel Rubio Blanco
Climate Change Director at IDOM

NOTES

1. World Economic Forum, "Could Infrastructure Investment Help Tackle Climate Change?" February 25, 2016, https://www.weforum.org/agenda/2016/02/could-infrastructure-investment-help-tackle-climate-change/.
2. S. Thacker et al. *Infrastructure: Underpinning Sustainable Development* (Copenhagen: UNOPS, 2018), https://content.unops.org/publications/Infrastructure_underpining_sustainable_development_EN.pdf.
3. IDB Invest and Inter-American Development Bank, "What Is Sustainable Infrastructure? A Framework to Guide Sustainability Across the Project Cycle" (March 26, 2018), 3, https://doi.org/10.18235/0001043.

INFRASTRUCTURE, CLIMATE LEADERSHIP, AND THE PATH BEYOND NET-ZERO

Where We Stand—and Why This Matters

We live in a world where the weather now feels like news we already knew. Summers arrive a little earlier and stay a little longer. Flood maps must be redrawn. Smoke turns sunsets strange. We measure these changes with models and metrics, but most people feel them first at the bus stop, in the electricity bill, or in the park that once stayed green and cool but now bakes by noon. Climate change is no longer a headline—it's the background of daily life.

At the same time, the living fabric that steadies our world—biodiversity—is thinning. When wetlands disappear, cities lose their buffers. When trees vanish, heat and energy demand rise together. When soils degrade, floods hit harder and infrastructure fails faster. Climate and biodiversity are not separate crises, but one system expressing its imbalance.

For those of us who build and manage infrastructure, these truths are not abstractions—they are job descriptions. Every design choice sits at the intersection of physical risk, human need, and ecological limit. Our decisions can either compound exposure and cost or create resilience and value across generations.

I came to this work not chasing a trend but following a shift in purpose. I began as a geologist, drawn to the Earth's deep processes. When my family moved to the United States, the world's energy priorities were changing. The path I imagined—finding oil, powering growth—was no longer the path the planet needed. I didn't abandon systems thinking; I redirected it. Instead of extracting, I chose to conserve. Instead of balancing reserves, I learned to rebalance relationships—between people and place, between built and natural systems.

That has been my vocation at the Los Angeles County Metropolitan Transportation Authority (LA Metro) and my passion through the Philippine Development Foundation (PhilDev): turning large, complicated systems toward stewardship.

This essay shares what that looks like in practice—how ambition becomes delivery, how leadership sustains progress through uncertainty, and why "net-zero" cannot be the finish line. I write from a public-agency vantage point—where budgets, timelines, and trust define reality—and from a Filipino lens that has shaped how I lead: bayanihan (communal unity), pakikipagkapwa (shared humanity), and resilience. These are not slogans; they are working tools—quiet, durable, and effective.

Because this book centers nature-based solutions (NbS), I place them at the core. Done right, NbS are not decoration; they are cost control, risk reduction, and dignity—shade where people wait, water that has a place to go, habitat that stitches the city back together. The most enduring infrastructure is the kind that works with nature and with people, not against either.

Translating Big Goals Into Things That Work

Ambition is easy to announce and hard to deliver. The world is full of pledges—net-zero by 2050, biodiversity protection, circular economies. The challenge is translation: turning lofty intent into grounded, operational change.

At Metro—an agency of 11,000 employees and one of the largest transit capital programs in the nation—translation is everything. It means moving from vision to specification, from policy to project, from speeches to functioning systems. Over time, we learned a simple framework: align, measure, embed.

Align means linking climate goals to core operational priorities: safety, reliability, lifecycle cost, and service quality. A sustainability measure that weakens those anchors will not survive. That's why our Moving Beyond Sustainability plan integrates energy, water, heat, flood, and habitat goals directly into performance metrics.

Measure means using credible standards so that results are visible and trusted. We use the Envision framework because it captures what project accounting often misses: how a bioswale reduces flood risk, how a green corridor cools neighborhoods, or how material choices affect long-term maintenance. Measurement is not ideology; it is stewardship made visible.

Embed means building sustainability into process, not personality. If it lives only in a champion, it dies with a reorganization. When every single job description has sustainability as one of its core pillars, and the whole framework is written into procurement, design criteria, and staff training, it becomes culture. That's how we protected progress to withstand any leadership changes—by making the better way the normal way.
Let me visualize this through some examples.

Zero-Emission Buses. Often described as a vehicle story, it is really a depot story. Power availability, charging layout, and workforce skills make or break the program. We coordinated with utilities, reconfigured yards, trained workers, and used transitional fuels like renewable natural gas to clean the air immediately while building capacity for full electrification.

Energy Master Planning. Our energy loads are large and sensitive—traction power, maintenance, and stations. Without planning, cost volatility rules. We built a framework to balance utility supply, distributed generation, storage, and demand management so reliability and cost could be managed over the asset's lifetime, not just the budget year.

Green Infrastructure and NbS. A bioswale may look ornamental, but in a storm it protects assets and neighborhoods. Trees cool surfaces, lower heating, ventilation, and air conditioning (HVAC) demand, and extend pavement life. Native landscapes bring pollinators and civic pride. What seems like landscaping often becomes the cheapest insurance we own.

Across these examples, one lesson endures: sustainability sells itself when it reduces risk per dollar. NbS outperform when we count what was once ignored—heat stress, flood damage, social cohesion—and when communities are treated as partners, not afterthoughts.

A Filipino Lens for Systems Thinking

Systems thinking did not come to me through textbooks. It arrived through the values I grew up with.

Bayanihan—the instinct to show up and share the load—teaches multi-functionality. A tree is shade, habitat, carbon capture, and mental health. A park is flood control, cooling, and community. A right-of-way is mobility, biodiversity corridor, and neighborhood identity. That is systems design through culture.

Pakikipagkapwa—shared humanity—reminds me that every line on a drawing touches real lives. Engagement is not a checkbox; it is discovery. Listening reveals the lived criteria that never appear in bid sets—a shaded bench, a school dismissal schedule, a vendor's livelihood. When people feel seen, they become co-stewards. NbS thrive because living systems need living relationships.

Resilience is our practiced habit. In the Philippines, it means rebuilding together after every storm. In infrastructure, it means absorbing shocks and maintaining function. Marrying the two creates adaptive systems—designed for recovery, trained for response, and trusted in crisis.

Together, these values turn sustainability from a technical objective into a way of working. They anchor decisions when trade-offs tighten and leadership must endure beyond headlines.

Leadership That Lasts When Conditions Don't

Leadership in infrastructure is not about grand announcements; it is about staying the course. Who keeps the better specification intact when budgets strain? Who mentors the next engineer? Who sustains the culture after the spotlight moves on?

I never set out to be a "sustainability leader." I was a geologist who loved systems. At Metro, I began in compliance—important but insufficient. Meeting regulations is the floor; endurance is the ceiling. True progress came from reframing sustainability as risk control, lifecycle savings, and uptime. Once we spoke that language, the right allies leaned in.

Recognition helps—the Engineering News Record (ENR) Award of Excellence, global board roles, and the Pamana ng Pilipino Award—but the most meaningful validation is a cooler corridor or a community that says, "This serves us." Policies, manuals, and procurement standards outlast people. If I leave any legacy, I hope it is scaffolding strong enough for others to climb higher: sustainability in our standards, resilience in every scope, and equity as an operational principle, not a press release.

No agency can do this alone. That's why I build coalitions—with universities for modeling, contractors for innovation, utilities for co-investment, and communities for lived knowledge. It's slow work, but it's how scale is achieved.

Leadership under uncertainty demands honesty about trade-offs. Electrification strains budgets; embodied carbon complicates procurement. Progress comes when we name constraints early and design around them together. Pilots, data, and mid-course corrections are not setbacks; they are discipline.

Net-Zero Is a Milestone. Regeneration Is a Mission.

"Net-zero" gave the world direction, but neutrality is not healing. A city can meet its carbon budget and still overheat; a depot can run on renewables and still flood. Neutrality alone leaves systems cleaner but fragile.

The next horizon is climate-positive, nature-positive infrastructure—systems that sequester more carbon than they emit, reduce risk more than they displace, and create more value than they consume. NbS are essential to that future.

Urban Cooling. Tree canopy in underserved neighborhoods is not just beautification—it's emissions policy, health policy, and equity policy at once.

Stormwater. Pairing grey infrastructure with green systems—bioswales, permeable surfaces, native plantings—reduces floods, recharges groundwater, and restores habitat.

Corridors as Habitat. Transit and roadway rights-of-way can become ecological threads that reconnect landscapes and support species migration.

Transformative approaches bring real challenges—financing long-life benefits, valuing ecosystem services, maintaining living assets—but these are solvable through the same tools we already use: standardization, training, and transparent metrics.

Technology amplifies our options. Digital twins can reveal where a single tree or curb extension prevents costly damage. Advanced materials reduce embodied carbon without compromising safety. Distributed energy and storage make systems both cleaner and more resilient. Yet even with these tools, success depends on one timeless factor: people.

People Are the Operating System

We can automate almost everything except trust. The best design will fail if the people who live beside it don't feel ownership. The best model is useless if operators lack training. Living infrastructure fails when maintenance is everyone's job and nobody's priority.

So we design for relationships as much as for loads. That means listening before drawing, specifying for maintainability, budgeting for stewardship, and telling the truth about cost—the cheapest option today often becomes the most expensive to sustain.

Equity is not an add-on; it's how systems remain legitimate. Shade at bus stops in heat-prone neighborhoods, flood relief where basements fill, local hiring for green jobs—these are not "extras." They are what make public systems durable.

Mentorship fits here, too. The projects we start will be completed by younger professionals. Through Metro's Sustainability Youth Program, I try to make space for them to lead early—to present, negotiate, and experiment. They bring new tools; we give them context and compass. That exchange is how institutions learn.

A Global Network, a Local Promise

The problems ahead are larger than any city, but their solutions are local. Networks bridge that gap. The Filipino diaspora I belong to—scientists, entrepreneurs, civil servants, students—shares ideas that move faster than projects can. Through PhilDev, I've seen scholarships become labs, labs become start-ups, and start-ups become social infrastructure. Bayanihan travels under many names; pakikipagkapwa quietly guides design. These values are universal across any diaspora; the collective contribution of each diaspora show how they work in practice.

What Comes After the Finish Line

People sometimes ask when we'll be "done." When the buses are electric? When the dashboards are green? I hope never. Systems that serve life must keep learning. The climate will change; designs must mature. Communities evolve; our engagement must, too.

Net-zero is a milestone. The mission is regeneration—depots that produce more energy than they use, corridors that cool and connect, contracts that reward circularity, plans that fund stewardship because it saves money and lives.

If we keep that focus, our arguments will be within a shared story: infrastructure exists to protect people, steward nature, and build prosperity that lasts.

Legacy, in Practice

When I think about legacy, I think of my son waiting at a shaded stop in a neighborhood that once sweltered, a bus that still runs during a heat wave, a small business that survives construction because we planned for it. And I think of trees—always the trees.

If our generation does one unarguably good thing, let it be planting and protecting them where they hold everything together. Years from now, no one will recall the meetings or memos. They will simply feel the shade.

Infrastructure is not just concrete and steel. It is a promise—of safety, reliability, and community. Nature provides the blueprint. People hold the tools.

If we keep both in view—aiming beyond neutrality toward systems that repair and regenerate—we will leave behind places that work, and keep working, for people we may never meet. That is the path beyond neutrality. That is the legacy worth leaving.

Infrastructure is not just concrete and steel.

It is hope made tangible.

Dr. Cris B. Liban
Chief Sustainability Officer at LA Metro
Board Chair of the Philippine Development Foundation (PhilDev)

General
Framework

Chapter 3

Framework

provides an overview of the scope of the criteria for framing climate and biodiversity action, and outlines their common characteristics, structural components, overarching concepts, and cross-cutting principles.

SCOPE

Climate change and biodiversity loss are global crises, and the need for action against their impacts is recognized as a shared global responsibility. Addressing these global challenges requires commitment and action at multiple levels—national, sectoral, regional, city, corporate, system-level, portfolio-of-projects, and project-by-project level—highlighting both the complexity of these issues and the interconnection of actions across levels. Climate and biodiversity action require contributions at multiple scales that can later be aggregated to collectively address climate change and biodiversity loss. Moreover, the two crises are deeply interlinked, calling for an integrated approach.

In the current context of urgent need to accelerate climate action to avoid the catastrophic effects of climate change, climate-compatible and climate-resilient infrastructure projects have a key enabling role to play. The transition to a net-zero future cannot be achieved without infrastructure's contribution in various sectors. However, if poorly planned, designed, and managed, these climate-action-focused projects may result in additional loss of biodiversity and decline of associated ecosystem services upon which humans and businesses depend. Infrastructure project development "often involves the destruction or fragmentation of wildlife habitat, and the extraction of the raw materials needed for renewable energy technologies carry substantive biodiversity risks."[1] There are opportunities for action across the entire infrastructure lifecycle, involving different practitioners groups—including planners, designers, financiers, suppliers, manufacturers, contractors, operators, and decommissioning and repurposing specialists, among others.

The present publication aims to contribute to the various types of guidance that emerge for different levels and scales of climate and biodiversity action. It proposes a high-level conceptual framework to holistically capture how climate and biodiversity action are defined at the infrastructure project level. More importantly, the publication introduces a set of criteria for integrated climate-biodiversity action for infrastructure projects, enabling a move beyond narrowly focused solutions aimed solely at rapid outcomes.

This work was triggered by various efforts to define and streamline corporate climate action and by more recent approaches for corporate biodiversity action,

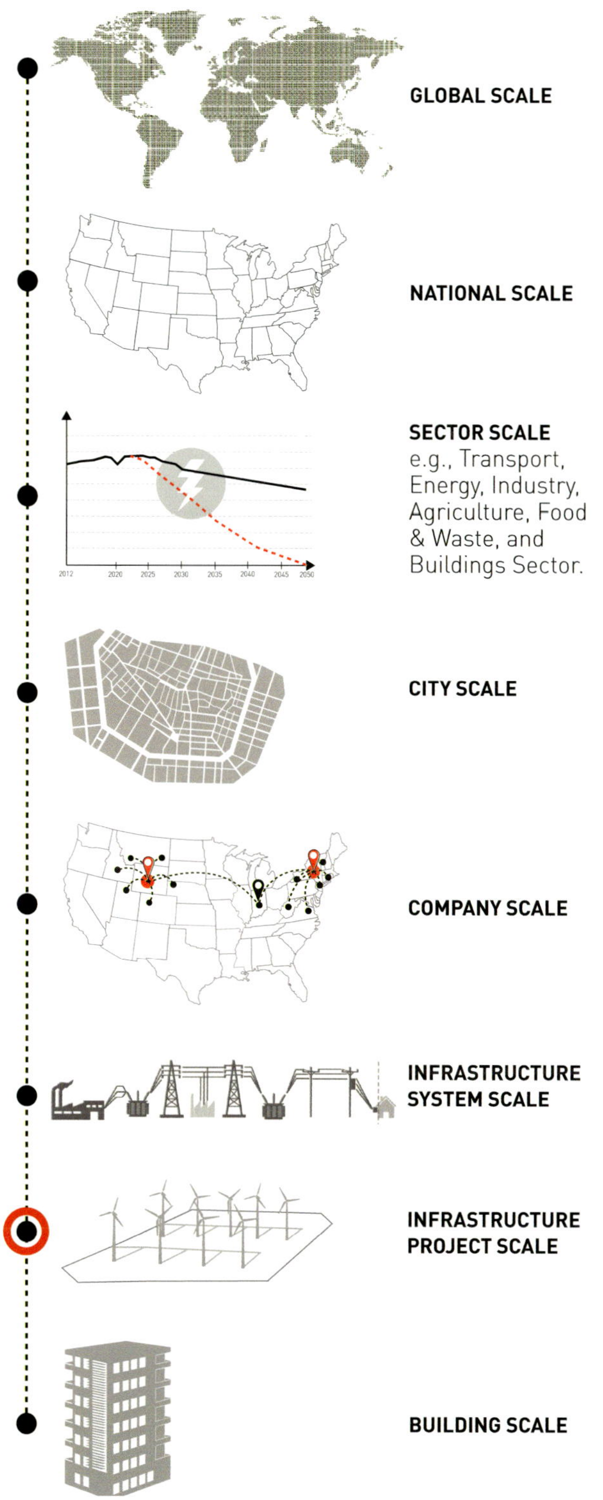

developed or under development through the work of Environmental, Social, and Governance (ESG) standards and frameworks, as well as other corporate action initiatives. ESG standards and frameworks have been continuously updated in recent years to keep pace with the fast-growing demand for alignment with global goals—from the Sustainable Development Goals to the Paris Agreement climate goals, and, most recently, the expanding global agenda for biodiversity goals. The publication explores the relevance of selected guidance and climate and biodiversity frameworks to infrastructure projects and proposes a set of criteria for framing infrastructure project-level action against climate change and biodiversity loss.

According to the United Nations Office for Project Services (UNOPS), "to address the challenges faced by practitioners, and to deliver on the potential of infrastructure for climate compatible development, it is essential that the right infrastructure is done well."[2] Identifying the right infrastructure requires an integrated, systemic approach that can extend to a regional or national scale. The criteria come into play after the right infrastructure projects have been identified, ensuring that projects are done well by managing negative trade-offs regarding their climate impacts and enhancing co-benefits. The criteria focus on "how an infrastructure project is done well" to achieve both climate- and biodiversity- positive outcomes.

The criteria represent a framing and scoping approach for infrastructure projects that aims to contribute to integrated climate and biodiversity action, by:

- Identifying and prioritizing the key issues and concepts associated with climate change and biodiversity loss
- Outlining the level of ambition needed
- Identifying options and opportunities
- Guiding decision-making and justifying the final project choices

They bring forth the crucial issues that projects must address in the current context of urgency for climate and biodiversity action. The criteria break down key concepts related to climate and biodiversity action as they appear in key publications and high-profile initiatives—national, sector-level, city-level, industry-level, and corporate—and translate them into infrastructure- and project-specific language to facilitate a better understanding of how these concepts can be applied. A key question that emerges is how global and national targets can be translated into project-level targets. Achieving ambitious climate and biodiversity goals will require disaggregating targets into actionable components that can be addressed through clear sets of actions.

The criteria outline key expectations and considerations underpinning climate and biodiversity action. A key difference between the criteria and the corporate action frameworks and standards is that while the latter refer to risks and opportunities in financial terms, the criteria translate risks and opportunities into project performance terms. Therefore, they provide a performance-based framework for making decisions on required strategies and actions. They can act as a checklist to identify gaps and opportunities for improvement toward win-win projects.

It is important to highlight that reference to win-win projects does not imply the absence of trade-offs between climate and biodiversity outcomes. Rather, it contrasts with lose-lose situations, when actions fail to address the interconnection between climate change and biodiversity loss and, in the worst case, inadvertently hinder progress on one or both. While discussion of trade-offs between climate and biodiversity is inevitable, the focus here is on the feasibility of achieving multiple sustainable outcomes. After all, both climate and biodiversity goals use the prefix "net," which recognizes and assumes the presence of negative impact, provided that equal or even greater positive impacts occur in parallel.

To provide a replicable framework for various types of infrastructure projects, the criteria are generic and infrastructure type-agnostic. They represent a conceptual framework that provides a high-level, holistic view of what climate and biodiversity action entails for infrastructure projects. On a project-by-project basis, the criteria must be informed and supplemented by location-specific and activity-specific considerations to sufficiently frame a project's climate and biodiversity action. This is particularly important in the case of climate change adaptation and biodiversity action, as the effects of climate change and biodiversity loss are experienced locally. Yet, even in climate change mitigation, a project's location determines the availability of options for driving ambition and accelerating action (e.g., market conditions, proximity to energy sources, climate conditions, etc.).

The proposed criteria represent a flexible, modular framework that can be adapted to specific locations or project types and applied based on project relevance. They intentionally avoid rigid, prescriptive thresholds for what projects must achieve or how to achieve it. Moreover, as many concepts are still under development—even at a definition level—and research on new possibilities for transformative action is ongoing, the criteria are an open-ended resource designed to support projects in outlining a solid plan, learning, and continuously improving the effectiveness of their actions.

Particular emphasis is given to nature-based solutions (NbS) as key responses towards integrated performance. The benefits of nature-based solutions are typically presented in the literature through case studies of projects that implemented specific types of solutions, which tend to be location- and ecosystem-type-specific. While these examples are valuable as practical applications of the NbS concept that can be replicated under similar conditions, they usually do not provide a high-level, comprehensive view of the inherent potential of NbS. In this publication, NbS are presented through their connection with ecosystem services. A comprehensive list of ecosystem services, as defined by the United Nations System of Environmental Economic Accounting—Ecosystem Accounting (SEEA-EA) classification, is used to capture all potential contributions that NbS support and to clarify linkages with specific components of climate action. This aims to further explain why NbS are a critical component of integrated climate and biodiversity action.

Finally, it is important to highlight that the criteria encompass concepts that are familiar to infrastructure projects, particularly through sustainability approaches. Climate and biodiversity action are integral components of sustainability. Yet, in the current context of urgency, higher ambition is critical. The criteria show how familiar sustainability-related concepts are expanded and how new concepts emerge. This alignment with existing project sustainability management approaches facilitates the integration of the proposed criteria into current sustainable practice. Project teams can use the criteria to map the impact drivers they are already considering and the strategies they adopt in their projects, identifying which impacts they address and to what extent. Building on this, they can raise their standards to strengthen performance on specific impacts, broaden the scope of their planned interventions, or identify interconnections between impacts and strategies to achieve more integrated outcomes.

The criteria do not aspire to serve as a one-stop guide on how to plan, design, and build infrastructure projects that address two of the greatest challenges of our times—climate change and biodiversity loss—nor are they intended to replace existing sustainability frameworks already used in the industry. Rather, they aim to encourage the mainstreaming of biodiversity into climate action, as opposed to viewing biodiversity as a marginal trade-off to be mitigated or a supplementary contribution. This approach enables the full potential of integrated climate-biodiversity action to be realized.

METHODOLOGY

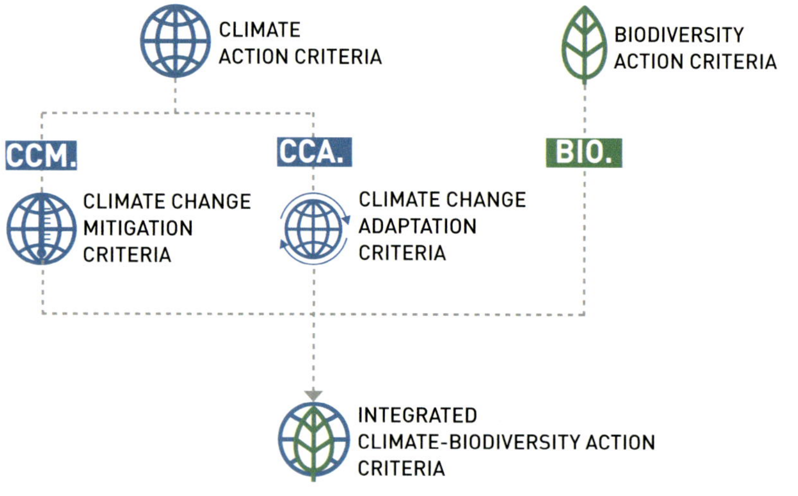

In a process to understand and frame the key components and concepts that determine integrated project action toward both climate and biodiversity goals, three sets of recommended criteria were identified:

- The climate change mitigation criteria
- The climate change adaptation criteria
- The biodiversity action criteria

The three sets of criteria were initially built as standalone sets to provide a high-level overview of the key components of climate change mitigation, adaptation, and biodiversity action. However, the different sets of criteria have linkages and significant overlaps. These overlaps constitute the core argument for claiming the potential of integrated climate and biodiversity action. Therefore, the standalone sets were then combined to provide guidance for and demonstrate the benefits of integrated climate-biodiversity action. The three sets of criteria must be applied together as one approach that collectively addresses the issues of climate change and biodiversity loss.

The work builds upon existing resources that were explored for relevance to infrastructure projects, synthesized, and adapted as needed to reflect the specific needs and requirements of this level of action.

Moreover, an effort was made to better illustrate the linkages that exist between climate change mitigation, adaptation, and biodiversity action so that they become more evident.

A similar structured methodology was followed for both climate and biodiversity, comprising the following steps:

- Extensive literature review of key components, concepts, and definitions of action as included in high-profile organizations' publications, flagship assessment reports, and recommendations, as well as in key treaties that set the goals and timeframes for action
- Review and analysis of climate- and biodiversity-related indicators as introduced by well-established and widely used Environmental, Social, and Governance (ESG) reporting systems
- Review of an infrastructure project sustainability framework, the Institute for Sustainable Infrastructure's (ISI) Envision Rating System (Envision).[3] Envision was used to ensure the criteria are relevant for infrastructure projects
- Application of the criteria to selected types of infrastructure projects. Projects were analyzed to test the real-world relevance of the criteria and provide additional insights based on different infrastructure project types

It is important to highlight that ESG and corporate reporting frameworks were used as part of the research methodology due to their continuous efforts (recent revisions, major updates, new initiatives) to keep track of the latest and emerging developments and trends in climate change-related data—and, more recently, biodiversity-related data—that are aligned with global climate and nature goals and targets, the Sustainable Development Goals (SDGs), and are relevant to investors.

A targeted analysis of the systems' principles, approaches, and their sets of indicators and metrics highlighted the climate change and biodiversity-related topics and commonly used indicators. This analysis resulted in a comprehensive list of criteria. The frameworks, standards, and initiatives reviewed included:

- The Taskforce for Climate-related Financial Disclosures (TCFD) and its Climate-related Financial Disclosure Recommendations.[4] TCFD's work provides input on climate risks and opportunities, financial impacts, and climate-related risk management
- The approach and work in progress of the recently established Taskforce for Nature-related Financial Reporting (TNFD)[5]
- The Science Based Targets Network's (SBTN) draft guidance on science-based targets for nature[6]
- The Climate Disclosure Standards Board's (CDSB) Framework,[7] which reflects the intrinsic interconnection between climate and other environmental matters, and its Application Guidance for Biodiversity-related Disclosures[8]
- The Global Reporting Initiative (GRI) Standards,[9] which are among the most widely used ESG standards. It is considered one of the most comprehensive tools for reporting the impacts of a company on the environment, society, and economy
- The Global Real Estate Sustainability Benchmark (GRESB) Infrastructure Asset Assessment ESG benchmark and reporting framework[10]
- The recently developed World Economic Forum (WEF) and the International Business Council (IBC) ESG Reporting Metrics and Disclosure Standards,[11] which after reviewing hundreds of available ESG metrics, highlighted 21 core metrics that are well-established, universal, industry-agnostic, and considered material to sustainable value creation

For biodiversity, in parallel to the ESG systems, ecosystem assessment and accounting systems were analyzed to answer the question: "How is biodiversity-related performance being assessed?" The analysis focused on the theoretical frameworks that underlie these approaches and their ecosystem services classification systems. Six approaches to the classification of ecosystem services were analyzed to finally select one system to be used for a detailed analysis of ecosystem services and their relevance to climate change mitigation and adaptation:

- The Millennium Ecosystem Assessment (MA) framework (2005),[12] the first major global assessment to adopt and establish ecosystem services as a central methodological framework
- The US Environmental Protection Agency (EPA's) National Ecosystem Services Classification System (NESCS) (2020)[13]
- The European Environmental Agency's Common International Classification of Ecosystem Services (CICES) (2018)[14]
- The United Nations System of Environmental Economic Accounting—Ecosystem Accounting (SEEA-EA) framework (2021)[15]
- The United Nations Environment Programme (UNEP)'s "The Economics of Ecosystems & Biodiversity" (TEEB) (2013)[16]
- The IPBES Nature's Contributions to People (NCPs) framework (2017)[17]

The cross-examination of the six classification systems led to the selection of the UN SEEA classification as the reference list for ecosystem services. It is among the most recently updated frameworks and is also used by the TNFD.

The following graph provides a brief overview of key references and resources for the development of the criteria.

METHODOLOGY OVERVIEW

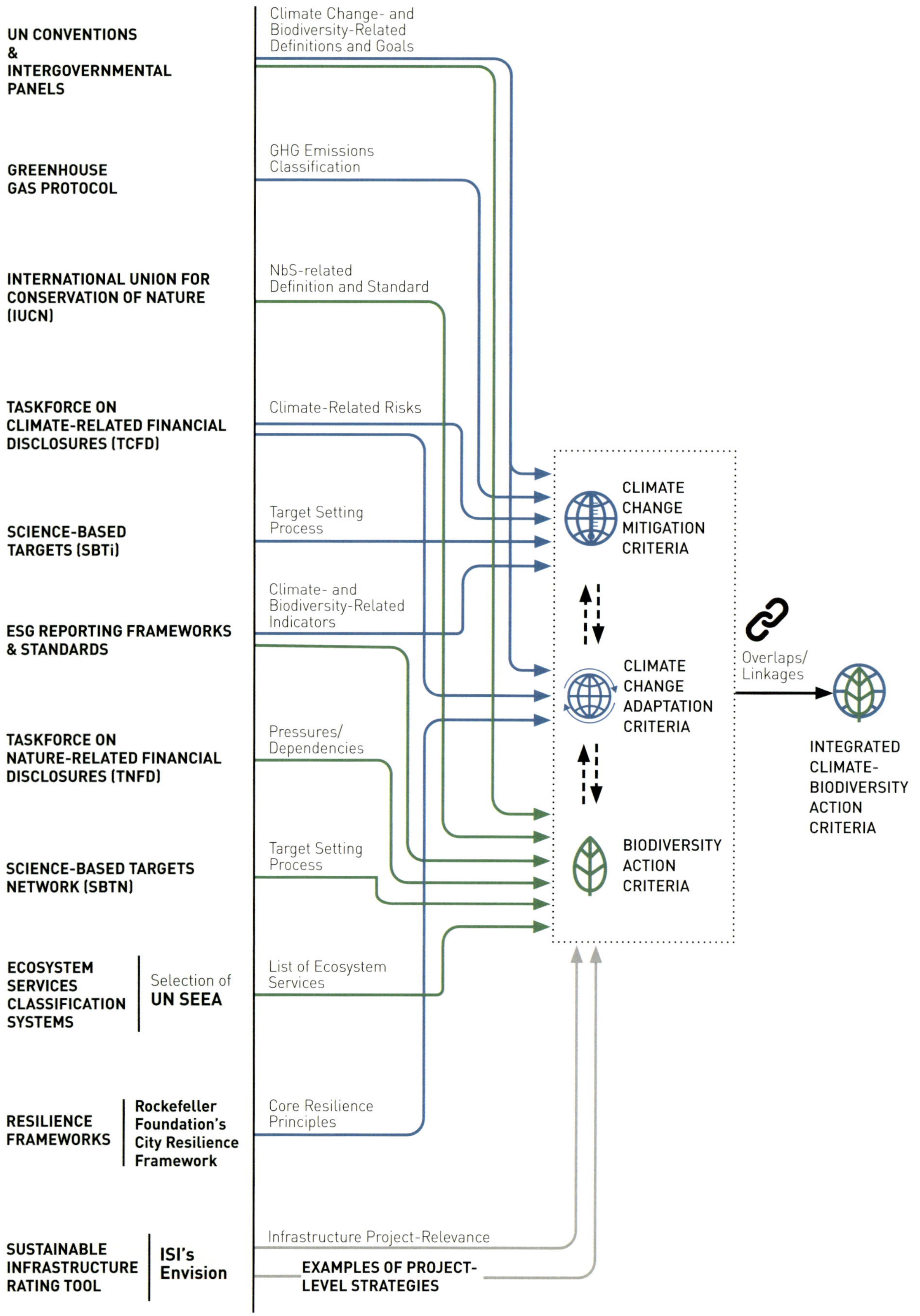

STRUCTURE OF THE CRITERIA

The key components of all sets of criteria are:

- The action steps, which are common to all sets of criteria
- The subsets of criteria
- The criteria

Each set of criteria is organized into four broader action steps:

1. Assess
2. Commit
3. Respond
4. Track

The action steps, therefore, organize the criteria into four key groups:

- The assessment criteria
- The commitment criteria
- The response criteria
- The tracking/monitoring criteria

Each action step includes one or more subsets of criteria. Each subset groups criteria of the same issue or topic of action.

GENERIC STRUCTURE OF THE CRITERIA

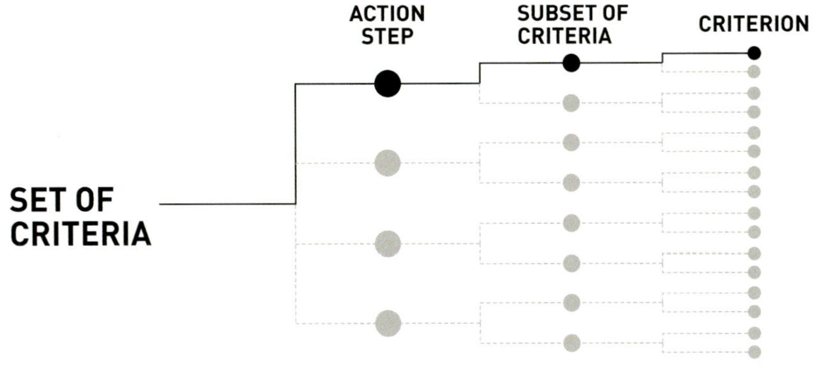

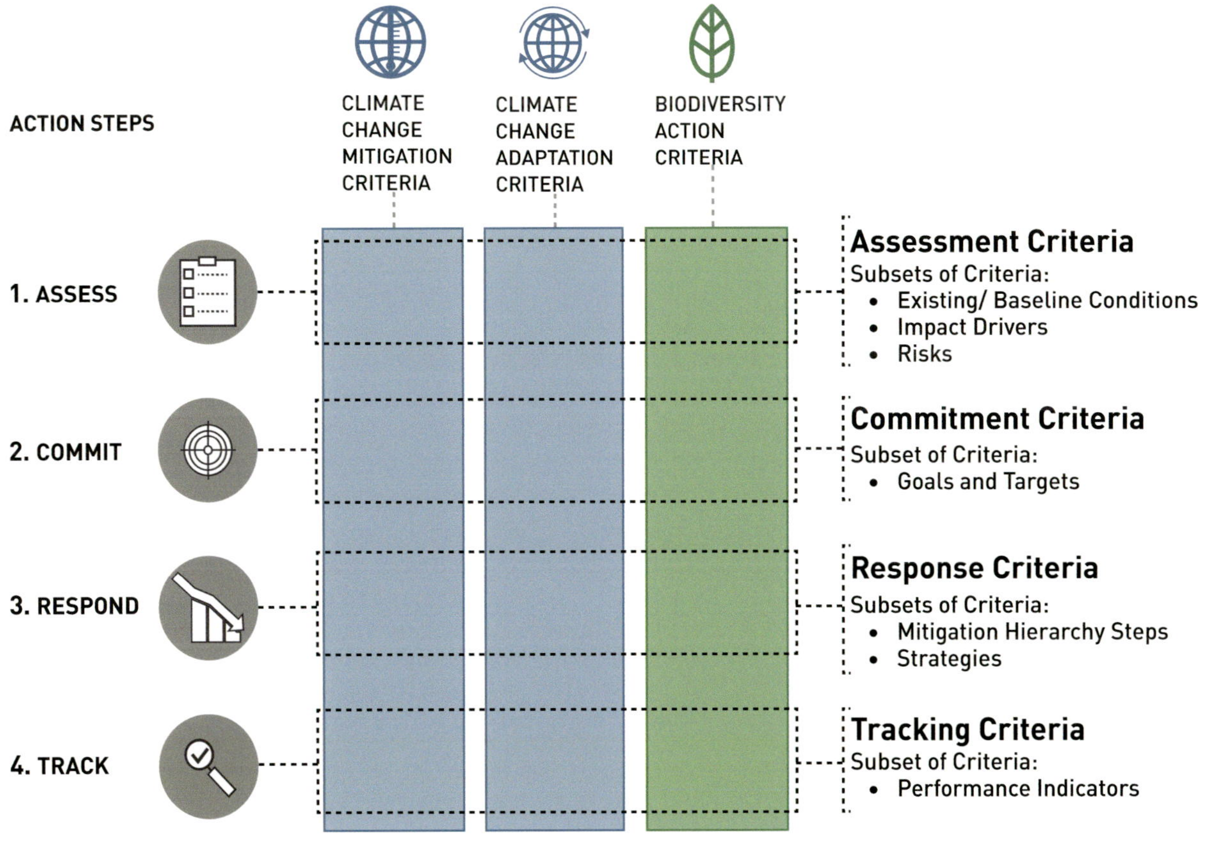

ACTION STEPS

The first action step, "Assess," is where the baseline against which performance will be assessed is set—either in relation to the identification and mapping of existing conditions or to a reference scenario that serves as the project baseline. The step identifies how a project may potentially impact these conditions (negatively or positively). It identifies key impact drivers of climate change and biodiversity loss and climate change-driven threats and risks to which the project will need to adapt.

Quantification of potential impacts is an essential first action in line with the "measure to manage" principle. However, a project cannot measure and manage its climate or biodiversity performance unless it is clear what it is setting out to achieve. Therefore, the second step of action, the "Commit" step, involves the development of a vision and sets the level of ambition through measurable goals and targets to achieve it. It includes setting performance targets for each of the identified project impact drivers and risks. These goals and targets will guide decisions on the selection of strategies to deliver them. The third action step, "Respond," compiles project strategies that collectively address all identified impact drivers of climate change and biodiversity loss and safeguard against the identified hazards and risks.

Finally, the "Track" step monitors and verifies progress against the targets set. It supports the use of "a robust plan-do-check-act methodology to identify priorities, evaluate progress, and make adjustments to continually improve project performance."[18]

The action steps are typical process steps followed to inform action decision-making. More specifically, they build on the Pressure-State-Response model, a framework commonly used for environmental management. According to this model, "a change in one component is expected to lead to a change in another."[19] For example, a change in response is expected to lead to a change in pressure on the environment, which leads to a change in state (environmental condition), which in turn results in more benefits for people. Therefore, a key element of the Pressure-State-Response framework is the relationship between its components. This can enable projects to gain a more holistic overview of their performance at multiple levels and understand how and where to act in response to the results obtained. "It allows an understanding of how project strategies, actions, and responses are performing, how they relate to any change in pressures (impact drivers), and how they lead to improvements."[20] Thus, these steps can help projects structure and deliver their contributions to climate and biodiversity.

Although these process steps entail a sequential order, they represent an iterative process for assessing required efforts and available options toward more ambitious climate and biodiversity goals. The objective is to explore potential combinations of options that are manageable and, where possible, set higher ambition. Revising and adapting initial plans is critical in the context of climate change and biodiversity loss, given the long timeframes—spanning decades—and inherent uncertainty about the timing and severity of impacts. It is important that projects avoid lock-in to carbon-intensive and other impactful technologies and leverage the potential of active research on new innovative solutions. There are technologies currently at their pilot to near-commercial stage that, as they become available, can step up project ambitions and outcomes. Similarly, pilot projects, once tested, can share lessons learned and lead to the development of new best practices.

At the same time, this structure of the criteria helps break down the high-level global goals for climate and biodiversity into actionable project steps. Following the approach of the Science-based Targets Network (SBTN) target-setting process for companies,[21] the criteria propose that infrastructure project climate and biodiversity targets be impact driver-specific and strategy-specific. This approach enables the disaggregation of the overall global goal into manageable, measurable targets. This will eventually allow the tracking of progress toward each target set, as impacts and related strategies will provide the performance indicators for which monitoring data will be collected. It is important to note that the criteria do not include performance benchmarks, nor do they describe methods for validating that the selected targets are aligned with global goals.

These impact driver- and strategy-specific targets are similar to thematic targets for key themes that are typical of sustainable project performance, such as GHG emissions, energy, materials, waste, water use, and water quality, among others. These themes should be seen as an integral part of the framework and consistent with overarching goals for climate and biodiversity, not as standalone elements, to achieve meaningful outcomes.

The framework suggests setting specific targets for each relevant impact driver and using the mitigation hierarchy as a mechanism for setting and implementing the targets while also recognizing that no net loss or net gain may not be feasible for every impact.

This approach aims to drive ambition for "doing better," taking into consideration all impacts and ensuring that relevant, expected impacts are not overlooked in the process of setting overall goals. Especially when using the different sets of criteria together, this approach serves as a safeguard for action that minimizes adverse side effects and maximizes mutual benefits.

AN ITERATIVE PROCESS

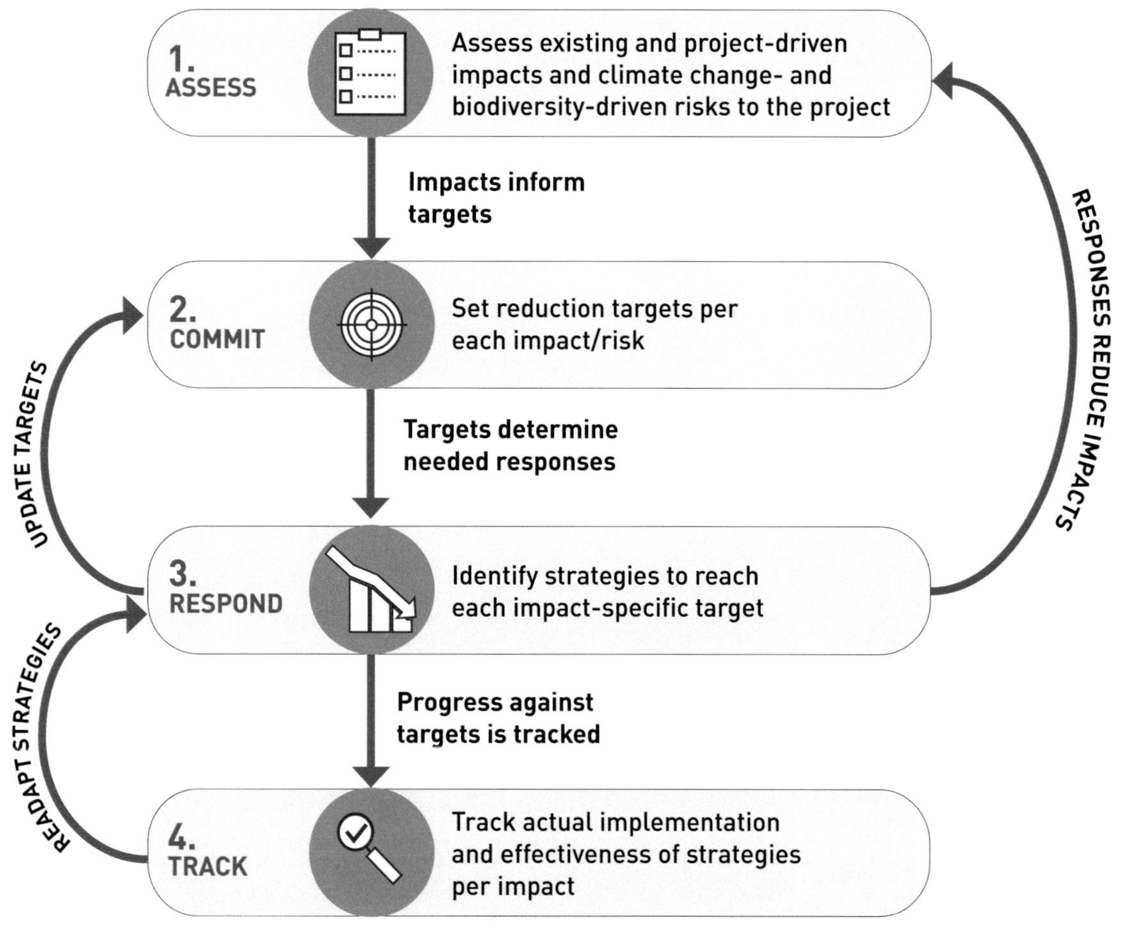

LINKAGES BETWEEN THE CRITERIA

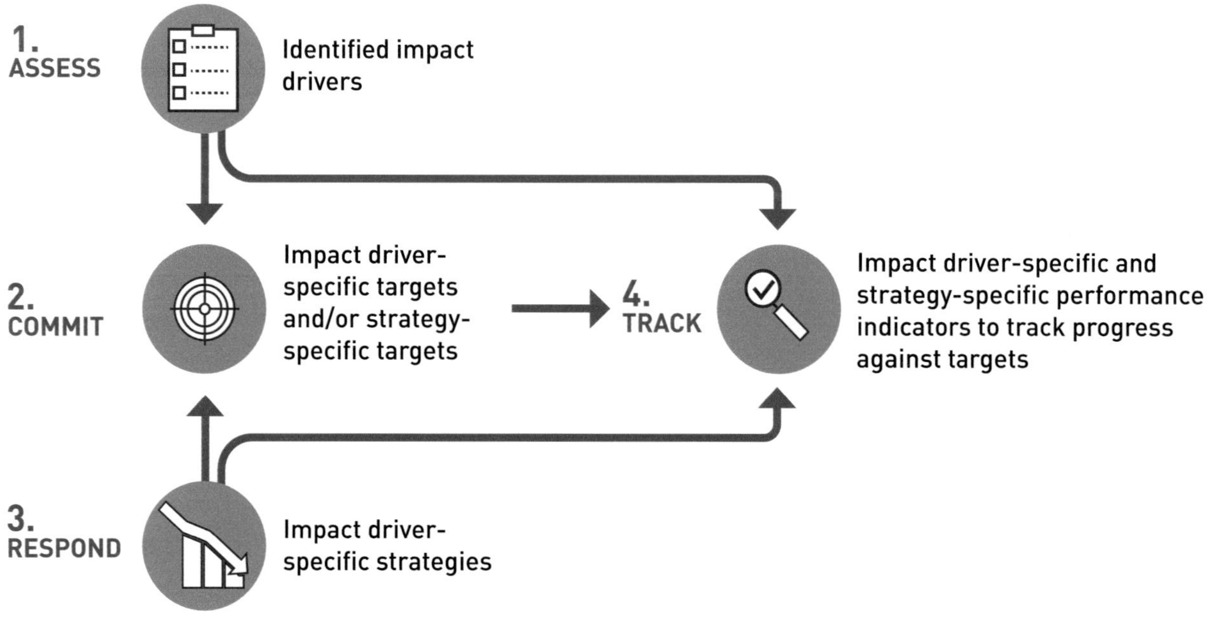

VARIATIONS OF SUBSETS OF CRITERIA PER TYPE OF ACTION

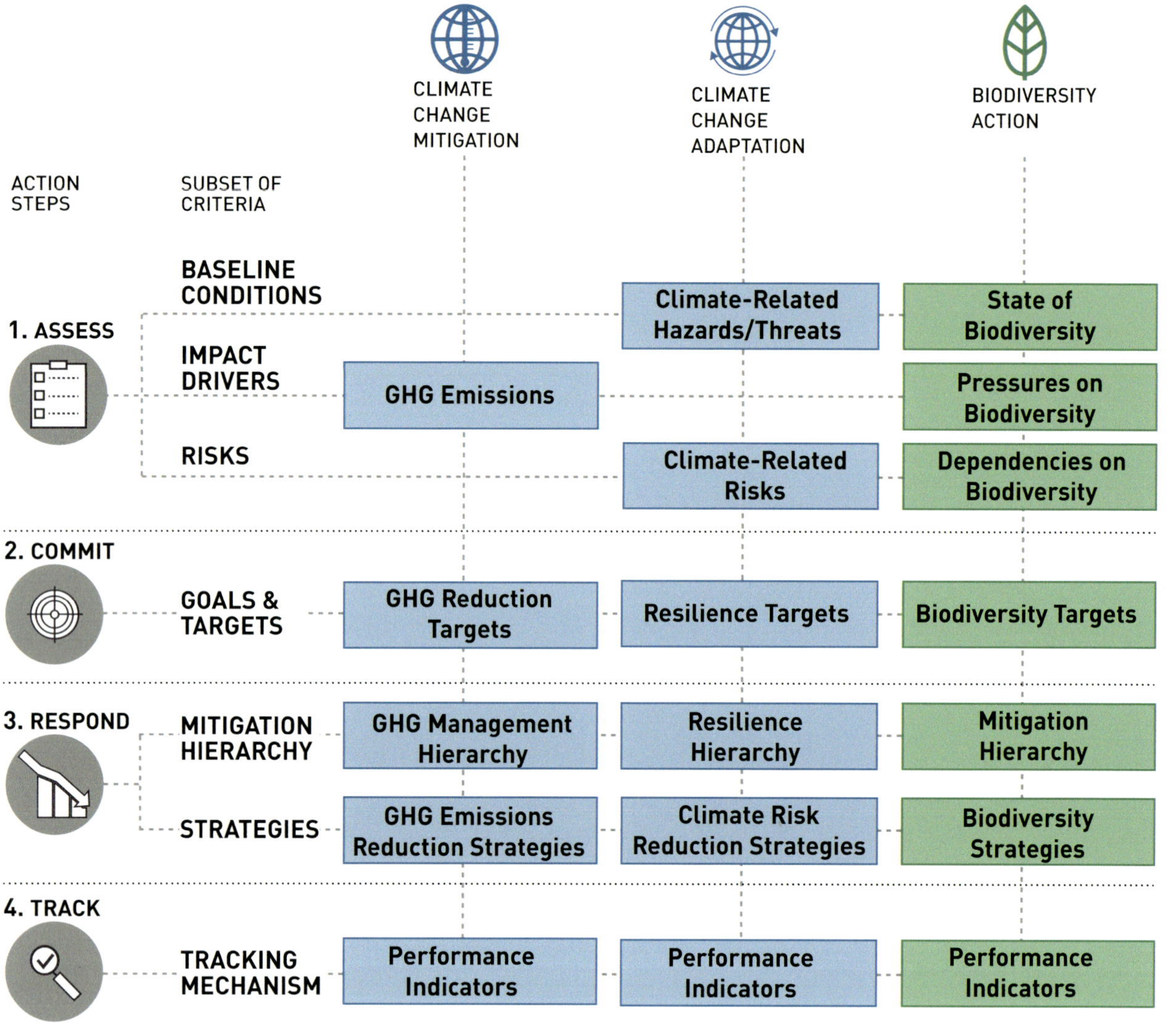

SUBSETS OF CRITERIA

In addition to the action steps that organize all criteria, further efforts have been made to ensure consistency across the different sets through the use of subsets of criteria. The graph above provides an overview of the generic topics these subsets address and their variations by action type. The subsets are not identical across climate change mitigation, adaptation, and biodiversity action; rather, they are tailored to the specific characteristics of each. These variations reflect the unique features of each action type. Different terminology is used for each subset, depending on the corresponding set of criteria, to facilitate easier navigation within the framework. More detailed information on the rationale behind the selection of subsets is provided in the respective criteria chapters.

As noted in the scope of the criteria, risks—and opportunities—are captured only in terms of project performance, not financial terms. This is why TCFD's "transition risks" are not included in the assessment subsets for climate change mitigation. Moreover, the criteria focus on new infrastructure projects, which explains why mitigation does not include an assessment of baseline conditions. In the case of a replacement project, however, this assessment should also be conducted.

The following graph provides a brief preview of the overall structure of the criteria, including the color coding and naming conventions used (e.g., CCM for climate change mitigation, CCA for adaptation, and BIO for biodiversity). This will help maintain clarity when the criteria are later merged in an integrated set—the integrated climate-biodiversity criteria—allowing users to track the origin of each criterion.

COLOR CODING AND NAMING CONVENTIONS OF THE CRITERIA

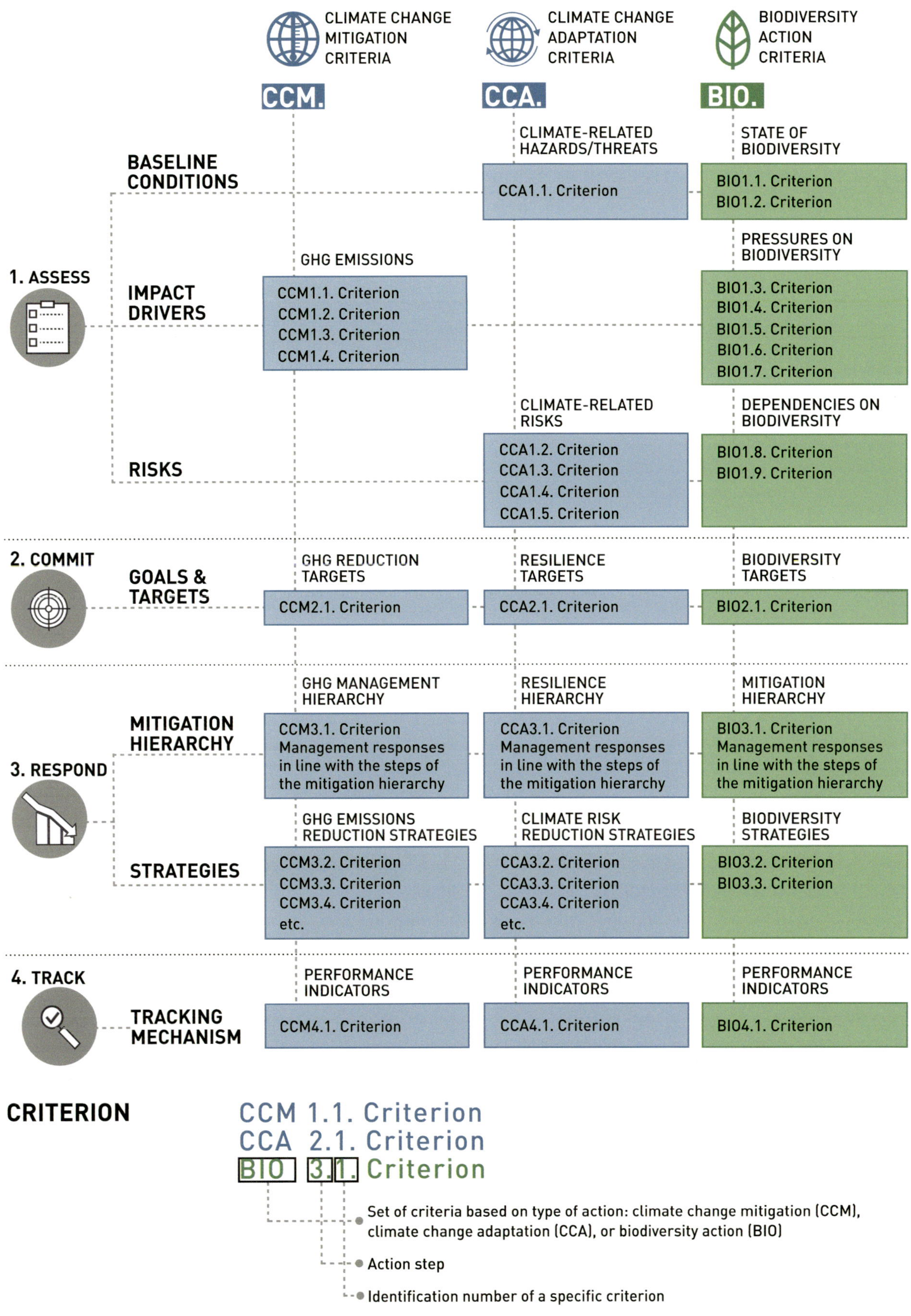

CROSS-CUTTING PRINCIPLES THAT GOVERN THE CRITERIA

There are some cross-cutting principles that govern all sets of criteria:

- Project lifecycle boundary of the assessment and responses (including supply chain)
- Adherence to the mitigation hierarchy
- Expansion beyond the project site to account for infrastructure system-wide, ecosystem-wide, and community-wide impacts and risks
- Stakeholder engagement and meaningful participation in the identification of potential issues and decision-making
- Compliance with minimum social safeguards

These principles aim to introduce wider sustainability considerations into the criteria and serve as a reminder that climate and biodiversity action cannot be effective if it ignores social factors, focuses only on one project stage, or overlooks the wider infrastructure system to which the project is part of or is connected.

They apply to all types of action. In some cases—such as the mitigation hierarchy—they are integral to the criteria themselves. In others, like the minimum social safeguards, they complement the criteria as key prerequisites that support effective and meaningful climate and biodiversity action.

CONSIDERING THE ENTIRE PROJECT LIFECYCLE BOUNDARY

Project-driven impacts on climate change and biodiversity loss can occur at all stages of an infrastructure project's lifecycle and throughout its associated supply chain. These impacts are not limited to the construction or operation phase but can emerge during material sourcing, maintenance (including future minor or major rehabilitation works), and eventually during decommissioning or deconstruction. Therefore, it is essential that the criteria are applied across the entire project lifecycle.

To support this, the criteria include a structured list of potential impact drivers of climate change and biodiversity loss that have to be assessed, if relevant, at each project stage. Lifecycle considerations must be embedded into the assessment of project impacts, the target-setting process, and the selection of strategies that will be implemented as part of the project. Targets and strategies must collectively address whole-life impacts of a project to support claims of net-zero and net positive performance.

This whole-lifecycle approach ensures coherent action that is not concentrated in one phase of the project while neglected in others, and ensures credible net-zero GHG emissions and net positive biodiversity outcomes.

THE CRITERIA APPLY TO THE ENTIRE PROJECT LIFECYCLE

Product stage
- **Raw material extraction, harvesting, refinement**
- **Transport to manufacturer**
- **Manufacturing into products**

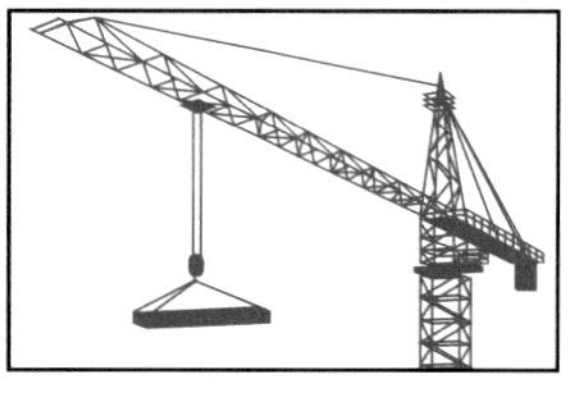

Construction stage
- **Permanent and temporary land occupation and alteration**
- **Transport of materials to the project site**
- **Construction-installation process**
- **Workforce commuting**

Operation stage
- **Operation processes**
- **Operation resources**
- **Fleet**
- **Routine maintenance works and resources**
- **Minor or major rehabilitation works and resources**

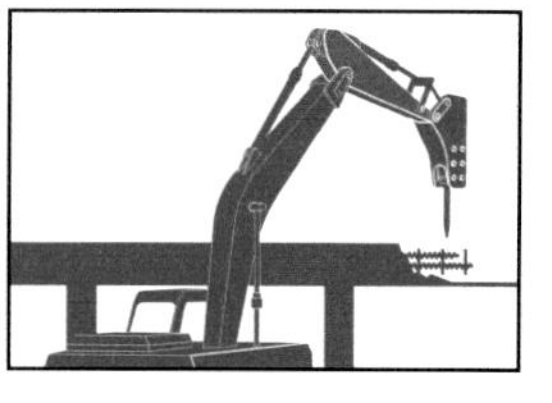

End-of-Life stage
- **Project replacement**
- **Deconstruction**
- **Transport to landfill, or waste processing facility**
- **Waste processing**
- **Disposal**

climate change-related and biodiversity-related impacts can occur during all stages of the project lifecycle

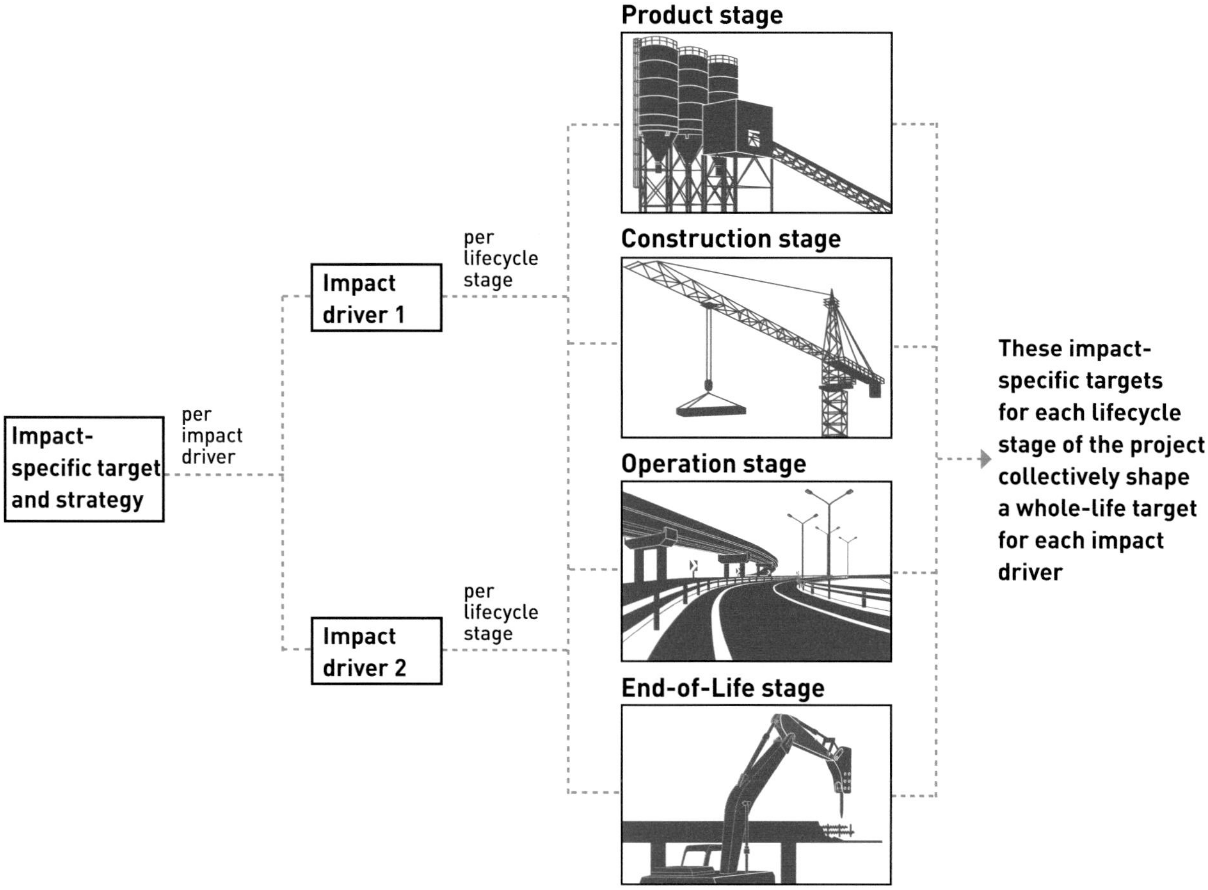

ADHERENCE TO THE MITIGATION HIERARCHY

A key principle that governs the response criteria is adherence to the mitigation hierarchy. The mitigation hierarchy is a set of prioritized, sequential steps that are applied to limit the potential negative impacts and risks of project activities. The mitigation hierarchy prioritizes actions based on four steps:

- Actions to avoid negative impacts
- Actions to minimize negative impacts that were not avoided
- Actions to restore and rehabilitate
- Actions to offset residual impacts

The four-step mitigation hierarchy is a well-established and already mainstreamed approach in environmental impact assessments for many sectors, to achieve the best outcomes for people and nature. At the same time, concepts that are integral to climate and biodiversity action, such as "net-zero," "no net loss," "net positive," or "net gain," are already built into the hierarchy. They are the intended outcomes once all the steps of the hierarchy are effectively implemented.

In the traditional structure of the hierarchy, net gain is achieved through the offset step. However, more recently, organizations such as the International Union for Conservation of Nature (IUCN) and the Science Based Targets Network (SBTN) have emphasized the need to expand the hierarchy to include transformative actions, which go beyond merely eliminating negative impacts to actively providing positive contributions for nature and people. The TNFD has also adopted the SBTN's AR^{3}T framework, an adaptation of the traditional mitigation hierarchy for company action.[22]

The mitigation hierarchy has a central role within the criteria:

- It can be used in its own right or as a mechanism for setting and implementing targets such as no net loss (NNL) or net gain/net positive impact (NPI), to meet regulatory or company internal policy goals[23]
- It links the various response criteria (strategies) with the overall net-zero and net positive targets, assisting projects in exploring how their selected strategies contribute to these targets

While prioritized, the mitigation hierarchy is not a strictly one-way linear process. Its steps can be revisited and iterated to refine mitigation strategies. Thus, it can

NET-ZERO IMPACT AND NET POSITIVE IMPACT AS INHERENT OUTCOMES OF THE MITIGATION HIERARCHY STEPS

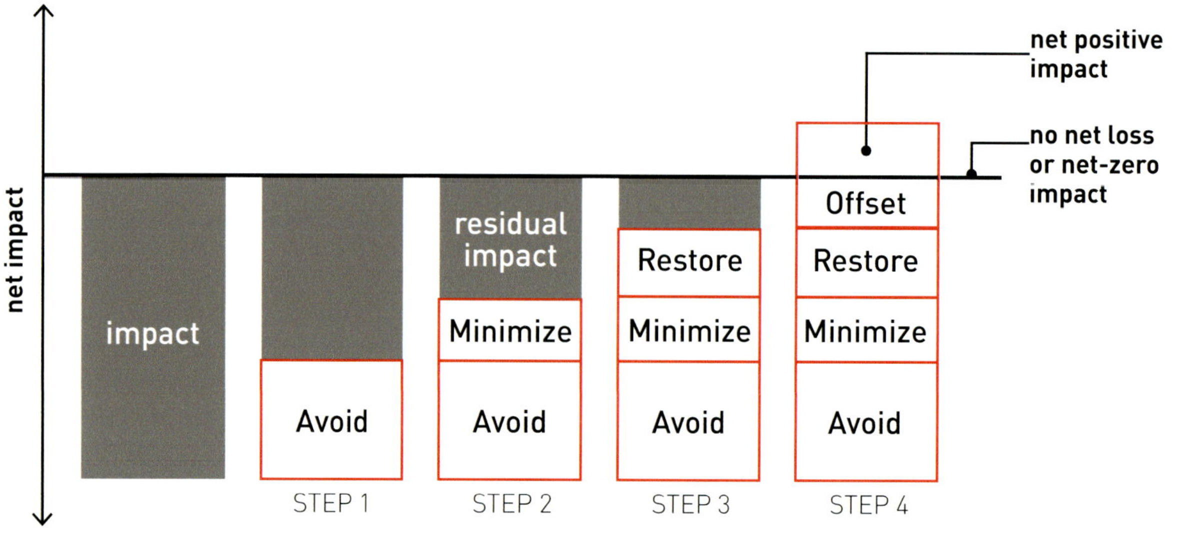

Graph adapted from NPI Alliance (2015). *Net Positive Impact for biodiversity: The conservation case.* Gland, Switzerland: IUCN.

serve as a framing and scoping tool that supports efforts to limit all impacts and risks to the extent possible and on a best-effort basis. It provides a mechanism for making explicit decisions on potential mitigation options and determining the best combination of actions that balance mitigation requirements and project priorities.[24]

Finally, the mitigation hierarchy steps guided the selection of the response criteria to ensure that higher ambition is enabled.

The most straightforward connection of the mitigation hierarchy is with biodiversity. The overall impact that has to be limited or reversed in this case is biodiversity loss, and the mitigation hierarchy can be applied for all relevant impact drivers (or pressures on biodiversity). The biodiversity criteria adopt the United Nations Global Compact (UNGC) and the International Union for Conservation of Nature (IUCN) mitigation hierarchy of 2012[25] and add a new dimension, the "Transform" step, modeled on the SBTN's AR^{3}T framework.

In the case of climate change mitigation, the mitigation hierarchy steps are adapted from the Institute of Environmental Management and Assessment (IEMA) GHG Management hierarchy, first published in 2009 and updated in 2020.[26] The impact that must be mitigated in this case is climate change and its key driver, GHG emissions. The mitigation hierarchy steps had to be adapted to correspond to infrastructure projects, as the 2020 IEMA hierarchy focuses on companies.

In the case of climate change adaptation, the impacts to be mitigated are the effects of climate change. The hierarchy refers to climate risk mitigation with the objective of reducing the probability and/or consequences of a risk event to an acceptable threshold and defining a suitable response. In this case, the "Offset" step of the mitigation hierarchy is not applicable, given that resilience loss in one location cannot be meaningfully offset elsewhere.[27] The hierarchy steps are not based on an existing mitigation hierarchy per se but combine information considered relevant from various sources.

Transformative action should not be seen as an additional step per se to the existing steps of the mitigation hierarchy but rather as an "add-on" to the other steps, serving as a reminder of the need to exceed expectations and drive change.

"Transform" aims to highlight that business-as-usual practice cannot deliver the net-zero emissions and nature-positive ambition. It relates to the level and scale of action needed to achieve ambitious targets and the barriers or enablers. Transformative change and action are already recognized as indispensable components of climate and biodiversity goals, yet it remains unclear how to operationalize these concepts or what types of actions and targets can be considered meaningfully transformative.

Transformative change is explained as "fundamental, system-wide reorganization across technological, economic, and social factors, including paradigms, goals, and values."[28]

There are various working definitions of what constitutes transformative action in the context of both climate and biodiversity (or nature), and no single definition has gained general consensus. The criteria acknowledge this limitation and offer a high-level conceptual definition that still needs to be further

MANAGEMENT RESPONSES IN LINE WITH THE MITIGATION HIERARCHY

	CLIMATE CHANGE MITIGATION	CLIMATE CHANGE ADAPTATION	BIODIVERSITY
	Impact to be mitigated: The climate change and its drivers (GHG emissions)	Impact to be mitigated: The effects of climate change (climate risks)	Impact to be mitigated: Biodiversity loss and its drivers (pressures on biodiversity)
AVOID	Avoid/Eliminate	Avoid/Eliminate threat	Avoid
REDUCE	Reduce	Reduce exposure/ vulnerability to threat Absorb/Minimize impact of failure	Minimize
RESTORE	Substitute/Replace	Restore/Recover	Restore
OFFSET	Compensate/Offset		Offset
	According to the Institute of Environmental Management & Assessment (IEMA) GHG Management Hierarchy (updated 2020)	Adapted from Resilience Management steps	According to the IUCN, UNGC (2012)

\+

Transform

clarified, applied, and tested. This definition builds on examples of how transformative actions are commonly understood in corporate action initiatives and identifies their equivalents for project-level action.

A key characteristic of transformative corporate action is that it exceeds private action:"Private actions alone cannot be considered transformative."[29] Change must be upscaled across entire sectors and transform the underlying economic, social, and political systems within which companies or projects are embedded. This system transformation can be achieved by working together with other companies and stakeholders to develop coordinated and mutually reinforcing actions.

Transformative corporate action is connected to:

- Supply chain and sector-wide efforts to adopt sustainable practices and transform productive systems in line with global climate and biodiversity goals (e.g., ensuring that suppliers provide deforestation- and conversion-free products)
- Proportional contributions toward diffuse, indirect, and historic impacts on top of no net loss for direct operations
- Advocating for profound shifts in behaviors, values, markets, policy ambition, and institutions
- Wider reformation of environmental incentive structures
- Cross-sectoral, collective action for investing in and delivering additional outcomes
- Providing capacity, resources, and promoting knowledge sharing, as well as the collection and sharing of data
- Investment in research and development (R&D)
- Increased accountability and certification requirements
- Public awareness and empowerment initiatives to develop a sense of shared ownership of the vision to address climate and biodiversity challenges and opportunities

- Support for local community rights and social safeguards (e.g., collective action pathways, social protection, and adaptive safety nets)

To a large extent, transformation is linked to efforts to address some of the indirect drivers of climate change and biodiversity loss, including unsustainable production, supply, and consumption patterns.

It may relate to suppliers, specific markets, industry partners, or wider structural issues which require collective action. For example, a project may need to source a certain volume of recycled content or achieve a certain percentage of recyclability to meet mitigation targets; however, barriers such as insufficient supply or technology may hinder progress. Similarly, a project may want to support ecosystem restoration in a particular area; however, cooperation with other stakeholders may be required.[30]

The direct equivalent to transformative project action would be resource circularity strategies. Transformation is inherent to the concept of circular economy, as "it is about transformation of the entire value creation system by decoupling growth from finite resources."[31] It requires collaboration, research and innovation, knowledge building and sharing that surpasses the project boundary. Thus, it is closely related to stakeholder engagement, shared responsibility and ownership, and continuous learning and improvement.

It can be argued that transformative action links corporate and project action.

EXPANSION OF THE ASSESSMENT BOUNDARY BEYOND THE PROJECT SITE

Project activities and the byproducts they generate can have direct, indirect, or cumulative impacts on a wide area that often extends well beyond the immediate physical boundaries of the intervention area. As such, a project's area of influence should not be understood only in biophysical or geographic terms, but rather in a broader context that includes ecological, social, cultural, and economic systems and their dynamics.

Therefore, projects should adopt an expanded boundary for impact and risk identification and response management to understand and account for interactions and interdependencies within the project, connected infrastructure system, and the broader community. This broader perspective can better inform project responses to avoid unintended trade-offs, prevent the exacerbation of existing or reasonably anticipated risks, and enable wider positive outcomes.

The decision on the appropriate assessment and action scope for each project is context-specific and must be grounded in a rigorous, multidisciplinary investigation during the early planning stages of a project.

STAKEHOLDER ENGAGEMENT AND MEANINGFUL PARTICIPATION

This principle is grounded in the recognition that the social impacts of infrastructure projects and their climate and biodiversity actions should be identified, assessed, and mitigated so that they don't compromise the quality of life of the local communities and stakeholders most affected by project activities. Climate and biodiversity action "must be just and equitable." The need to account for stakeholders' concerns and views is embedded in global goals for climate and biodiversity. The Paris Agreement affirms "the importance of education, public awareness, public participation, public access to information and cooperation at all levels"[32] and recognizes its importance for enhancing climate actions. The Global Biodiversity Framework, in turn, includes a specific target calling to "ensure participation in decision-making and access to justice for all."[33] The target recognizes the importance of indigenous people and local communities' insights for context-specific and effective strategies, as well as the need to empower vulnerable groups.

It is directly reflected in the criteria through the concept of "inclusivity," which mandates consultation with the community and all relevant project stakeholders to align project goals with existing broader social goals. "When an inclusive, representative group of stakeholders is engaged throughout the project, the results satisfy the widest possible part of the community that will be affected by and/or use the project."[34]

As part of the criteria, inclusivity is not only a prerequisite for broader sustainable project performance but also a strategy in itself.

Project users and communities "can bring local experience and native expertise into infrastructure planning and development."[35] In parallel with public engagement, inclusivity refers to engaging a diverse, cross-sectoral group of stakeholders with varied backgrounds and skill sets. It is particularly critical for addressing multidisciplinary issues like climate change and biodiversity loss. Inclusivity strategies, if implemented, can act as key enablers for maximizing the potential of the other strategies.

COMPLIANCE WITH MINIMUM SOCIAL SAFEGUARDS

This principle rests on the acknowledgment that integrated climate and biodiversity action has to be just and equitable. It refers to minimum social expectations related to business practices (by the project owner, manager, project team, and supply chain) and project-specific activities. As the criteria focus on addressing environmental issues, it is deemed necessary to prevent situations where certain practices and activities violate fundamental human rights, workers' rights, and principles of good governance, such as fair competition. This principle serves as a reminder that no climate or biodiversity action can justify such violations.

OVERVIEW OF PRESENTED ANALYSIS AS PART OF THE CRITERIA

As part of the research that led to the development of the criteria, a systematic analysis was conducted to ensure their comprehensiveness and to highlight their interconnections or overlaps—and, by extension, the interconnectedness of climate and biodiversity action. Overall, the analysis included:

- Identification of linkages between the response criteria—and more specifically, each of the proposed strategies—and the assessment criteria (impact drivers or risks) identified for each type of action. This step ensured comprehensive coverage of all impact drivers, implementation of mitigation hierarchy steps, and full lifecycle considerations
- Highlighting opportunities for joint action by identifying overlaps between different sets of criteria. To do so, the strategies included in the response criteria of one type of action were linked with the impact drivers or risks of the other sets of criteria

These analyses complement the presentation of the climate and biodiversity action criteria through arrow diagrams that illustrate connections.

In the integrated criteria chapter, the analysis explores the overlaps between the climate and biodiversity action criteria, and a summary—the Opportunity Matrix—is provided for better illustration of the potential for joint action.

The following diagrams aim to illustrate the common key steps of the analysis as presented in the respective chapters.

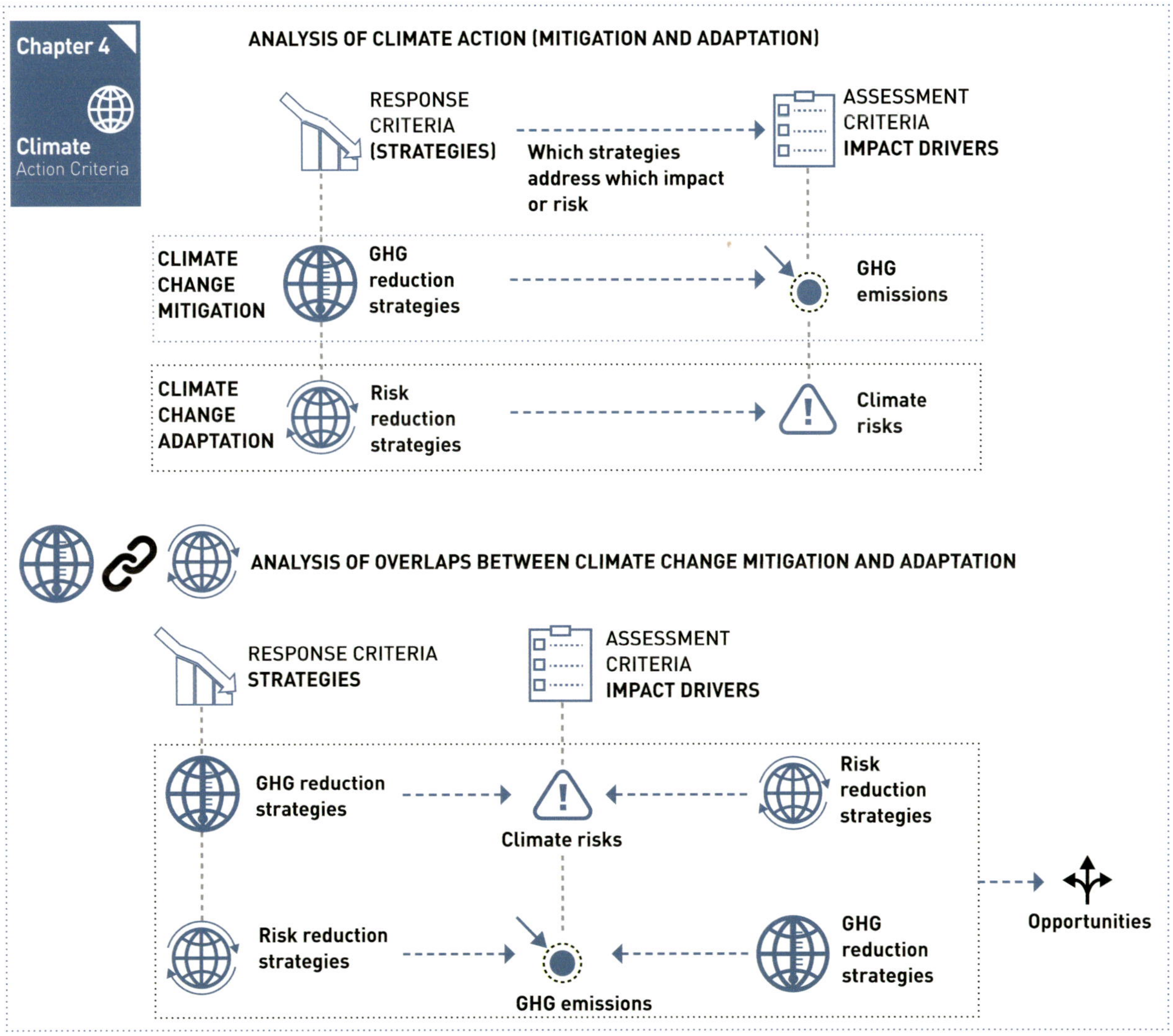

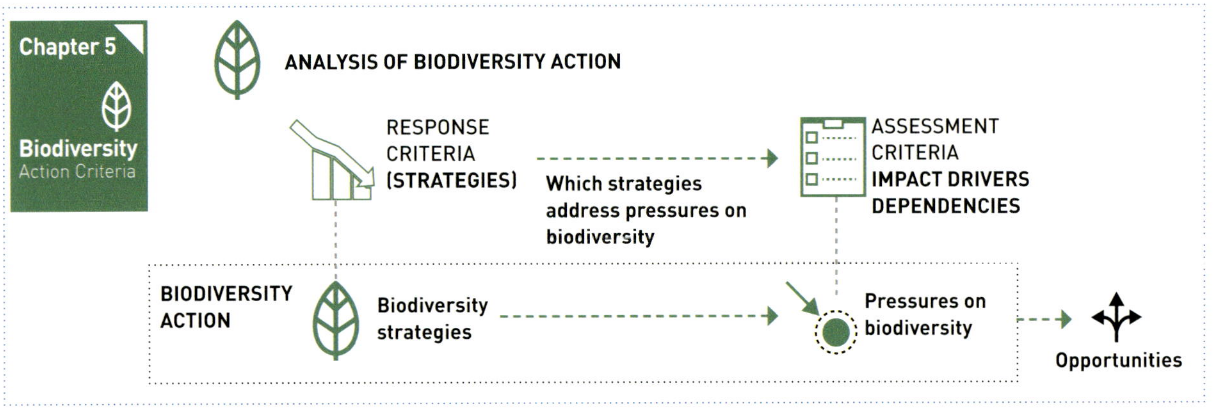
Chapter 5
Biodiversity
Action Criteria
ANALYSIS OF BIODIVERSITY ACTION
RESPONSE
CRITERIA
(STRATEGIES)
Which strategies
address pressures on
biodiversity
ASSESSMENT
CRITERIA
IMPACT DRIVERS
DEPENDENCIES
BIODIVERSITY
ACTION
Biodiversity
strategies
Pressures on
biodiversity
Opportunities

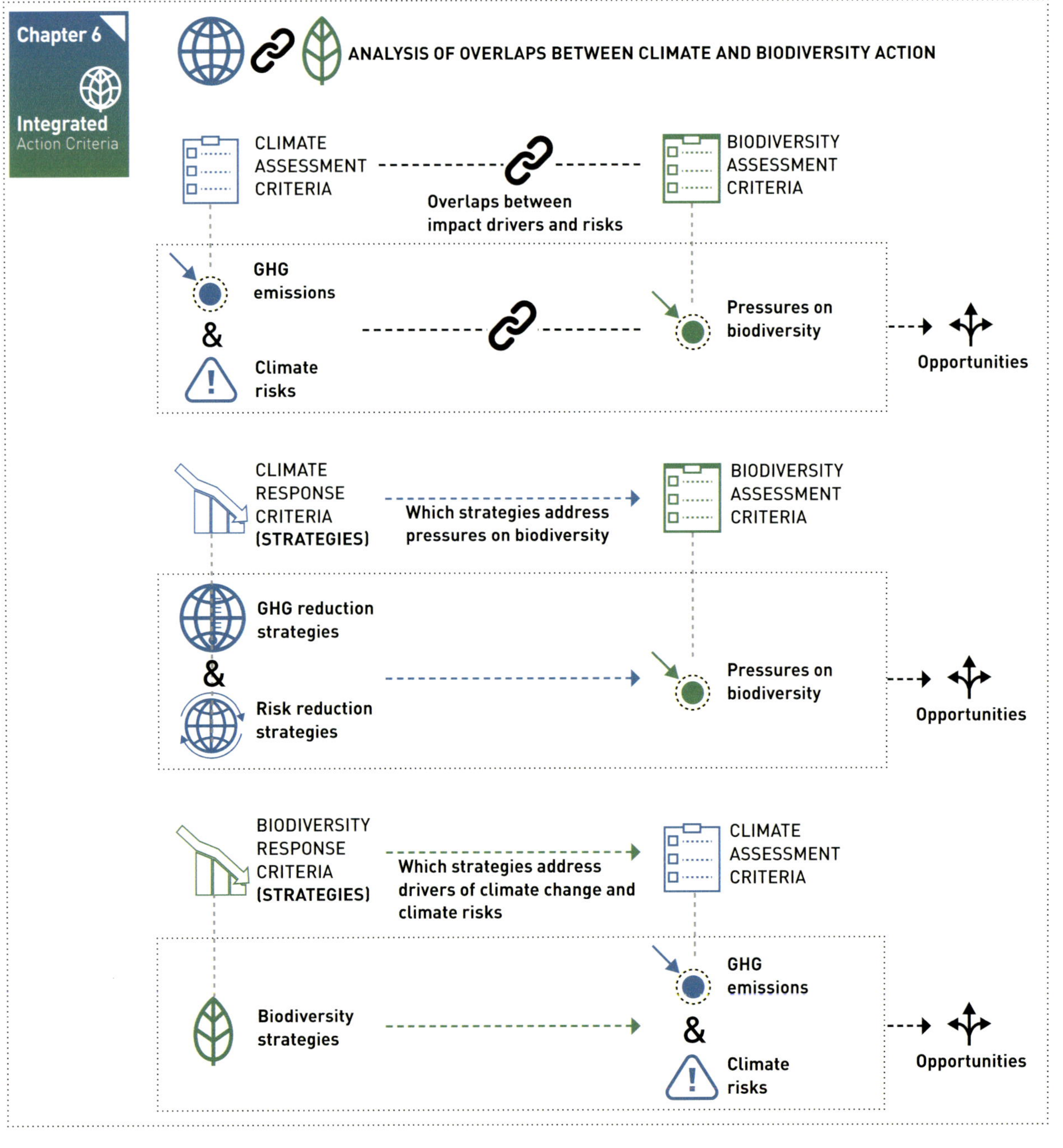
Chapter 6
Integrated
Action Criteria
ANALYSIS OF OVERLAPS BETWEEN CLIMATE AND BIODIVERSITY ACTION
CLIMATE
ASSESSMENT
CRITERIA
Overlaps between
impact drivers and risks
BIODIVERSITY
ASSESSMENT
CRITERIA
GHG
emissions
&
Climate
risks
Pressures on
biodiversity
Opportunities
CLIMATE
RESPONSE
CRITERIA
(STRATEGIES)
Which strategies address
pressures on biodiversity
BIODIVERSITY
ASSESSMENT
CRITERIA
GHG reduction
strategies
&
Risk reduction
strategies
Pressures on
biodiversity
Opportunities
BIODIVERSITY
RESPONSE
CRITERIA
(STRATEGIES)
Which strategies address
drivers of climate change and
climate risks
CLIMATE
ASSESSMENT
CRITERIA
Biodiversity
strategies
GHG
emissions
&
Climate
risks
Opportunities

NOTES

1. L. Bennun et al., *Mitigating Biodiversity Impacts Associated with Solar and Wind Energy Development: Guidelines for Project Developers*, 2021, https://doi.org/10.2305/iucn.ch.2021.04.en.
2. S. Thacker et al., *Infrastructure for Climate Action* (UNOPS, 2021), 19.
3. "About Envision," Institute for Sustainable Infrastructure, accessed November 11, 2025, https://sustainableinfrastructure.org/envision/about/.
4. Taskforce on Climate-Related Financial Disclosures, *Recommendations of the Taskforce on Climate-related Financial Disclosures* (TCFD, 2017), https://assets.bbhub.io/company/sites/60/2021/10/FINAL-2017-TCFD-Report.pdf.
5. Taskforce on Nature-related Financial Disclosures, *Recommendations of the Taskforce on Nature-related Financial Disclosures* (TNFD, 2023), https://tnfd.global/wp-content/uploads/2023/08/Recommendations-of-the-Taskforce-on-Nature-related-Financial-Disclosures.pdf?v=1734112245.
6. Science-based Targets Network (SBTN), *Science-based Targets for Nature: Initial Guidance for Business* (SBTN, 2020). https://sciencebasedtargetsnetwork.org/wp-content/uploads/2020/11/Science-Based-Targets-for-Nature-Initial-Guidance-for-Business.pdf.
7. Climate Disclosure Standard Board (CDSB), *CDSB Framework for Reporting Environmental & Social Information*, January 2022, https://www.cdsb.net/what-we-do/reporting-frameworks.
8. CDSB, *CDSB Framework: Application Guidance for Biodiversity-Related Disclosures*, November 2021, https://www.cdsb.net/biodiversity.
9. Global Sustainability Standards Board (GSSB), *Consolidated Set of the GRI Standards* (GSSB, 2024).
10. GRESB Infrastructure, *2022 Asset Assessment: 2022 GRESB Infrastructure Fund & Asset Assessment Pre-Release* (internal report, 2022).
11. World Economic Forum, *Measuring Stakeholder Capitalism: Towards Common Metrics and Consistent Reporting of Sustainable Value Creation*, accessed November 11, 2025, https://www.weforum.org/publications/measuring-stakeholder-capitalism-towards-common-metrics-and-consistent-reporting-of-sustainable-value-creation/.
12. Joseph Alcamo et al., *Ecosystems and Human Well-Being: A Framework for Assessment* (Island Press, 2003), 53-60.
13. United States Environmental Protection Agency (US EPA), *National Ecosystem Services Classification System (NESCS): Framework Design and Policy Application*, EPA-800-R-15-002 (Washington, DC: US EPA, 2015).
14. Roy Haines-Young and Marion Potschin, *Common International Classification of Ecosystem Services (CICES) V5.1 and Guidance on the Application of the Revised Structure* (Fabis Consulting Ltd., 2018), 22-25, https://cices.eu/content/uploads/sites/8/2018/01/Guidance-V51-01012018.pdf.
15. Lars Hein et al., *SEEA Experimental Ecosystem Accounting: Towards a definition and classification of ecosystem services for SEEA* (Final Report, 2018), https://seea.un.org/sites/seea.un.org/files/lg_24_bg_ecosystem_services_classification.pdf.
16. TEEB, *The Economics of Ecosystems and Biodiversity: Mainstreaming the Economics of Nature: A Synthesis of the Approach, Conclusions and Recommendations of TEEB* (TEEB, 2010), 34, https://www.academia.edu/14081672/.
17. IPBES, *The IPBES regional assessment report on biodiversity and ecosystem services for Asia and the Pacific* (IPBES Secretariat, 2018), 104-107.
18. Institute for Sustainable Infrastructure (ISI), *Envision: Sustainable Infrastructure Framework Version* 3 (ISI, 2018), 73.
19. Peter J. Stephenson and Giulia Carbone, *Guidelines for Planning and Monitoring Corporate Biodiversity Performance*, 6, (IUCN, 2021), https://doi.org/10.2305/iucn.ch.2021.05.en.
20. Ibid., 7.
21. SBTN, *Science-Based Targets for Nature: Initial Guidance for Business* (SBTN, 2020), 6-15, https://sciencebasedtargetsnetwork.org/wp-content/uploads/2020/11/Science-Based-Targets-for-Nature-Initial-Guidance-for-Business.pdf.
22. SBTN, *Step 4: Act* (SBTN, 2023), https://sciencebasedtargetsnetwork.org/companies/take-action/act/.
23. The Biodiversity Consultancy, *A cross-sector guide for implementing the Mitigation Hierarchy* (CSBI, 2015), 8, https://www.thebiodiversityconsultancy.com/fileadmin/user_upload/A_cross-sector_guide_for_implementing_the_Mitigation_Hierarchy.pdf.
24. Ibid.
25. United Nations Global Compact (UNGC) and International Union for Conservation of Nature (IUCN), *A framework for corporate action on biodiversity and ecosystem services* (UNGC, IUCN, 2012), 14, https://portals.iucn.org/library/sites/library/files/documents/2012-036.pdf.
26. Institute of Environmental Management and Assessment, *Pathways to Net Zero: Using the IEMA GHG Management Hierarchy* (IEMA, 2020), https://www.iema.net/resources/reading-room/2020/11/26/pathways-to-net-zero-using-the-iema-ghg-management-hierarchy-november-2020.
27. United Nations Office for Disaster Risk Reduction (UNDRR), *Principles for Resilient Infrastructure* (UNDRR, 2022), 23, https://www.undrr.org/publication/principles-resilient-infrastructure.
28. "Glossary", IPBES, accessed November 25, 2025, https://www.ipbes.net/node/42052#.
29. Hollie Booth et al., "Operationalizing Transformative Change for Business in the Context of Nature Positive," *One Earth* 7, no. 7 (2024): 1235-1249, https://doi.org/10.1016/j.oneear.2024.06.003.
30. Ibid.
31. World Business Council for Sustainable Development, *Circular transition Indicators V4* (WBCSD, 2023), 4, https://www.wbcsd.org/wp-content/uploads/2023/09/Circular_Transition_Indicators_v4.pdf.
32. United Nations, *Paris Agreement*, December 12, 2015, in *United Nations Treaty Collection*, 2, https://unfccc.int/sites/default/files/english_paris_agreement.pdf.
33. Convention on Biological Diversity, *Decision Adopted by the Conference of the Parties to the Convention on Biological Diversity: CBD/COP/DEC/15/4* (December 19, 2022), 13.
34. ISI, *Envision: Sustainable Infrastructure Framework Version 3* (ISI, 2018), 13.
35. UNDRR, *Principles for Resilient Infrastructure* (2022), 45.

FRAMING CLIMATE AND BIODIVERSITY ACTION IN INFRASTRUCTURE PROJECTS

Climate

Action Criteria

Chapter 4

Climate Action Criteria

comprise two sets of criteria for framing climate action in infrastructure projects: the Climate change mitigation criteria and the Climate change adaptation criteria.

KEY CONCEPTS AND APPROACHES SHAPING THE CLIMATE ACTION CRITERIA

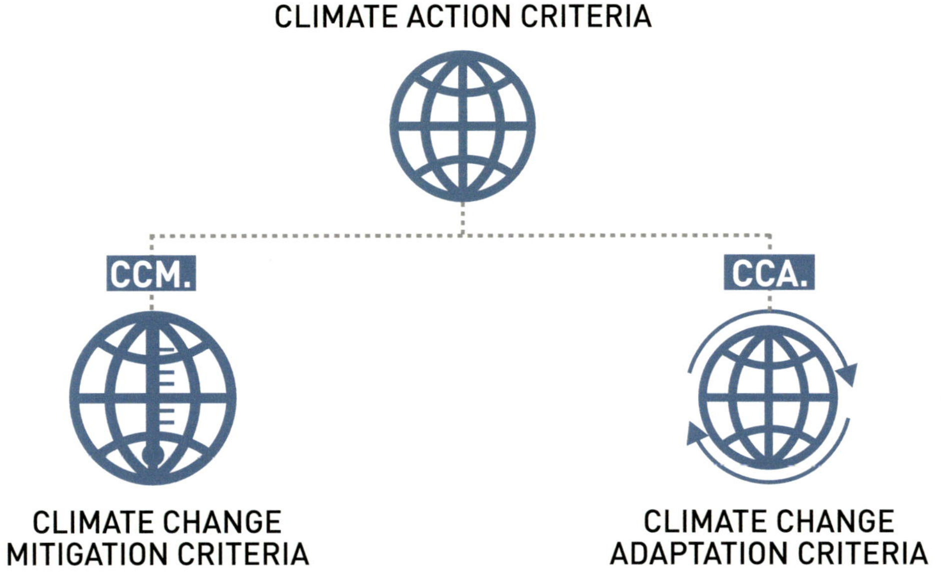

The United Nations Framework Convention on Climate Change (UNFCCC) identifies two responses to climate change: mitigation—reducing greenhouse gas emissions and enhancing sinks—and adaptation to the impacts of climate change. By extension, the climate action criteria comprise two sets: the climate change mitigation criteria and the climate change adaptation criteria.

The Intergovernmental Panel on Climate Change (IPCC), the independent body tasked with assessing the science related to climate change, defines climate change mitigation as "a human intervention to reduce emissions or enhance the sinks of greenhouse gases."[1] One of the primary goals of climate change mitigation is achieving a state of net-zero GHG emissions—or a balance at the global level between emissions by sources and removals by sinks, as explicitly recognized in the Paris Agreement. Since the introduction and widespread endorsement of the Paris Agreement goals, they have become synonymous with what climate change mitigation efforts are expected to achieve.

Climate change adaptation is defined as "the process of adjustment to actual and expected climate change and its effects, in order to moderate harm or exploit beneficial opportunities,"[2] or as "actions that help reduce vulnerability to the current or expected impacts of climate change like weather extremes and hazards, sea-level rise, biodiversity loss, or food and water insecurity."[3]

Mitigation aims to address the project-related drivers of climate change, while adaptation seeks to anticipate and prepare the project to manage the potential impacts of climate change. "Adaptation can be seen as direct damage prevention, while mitigation would be indirect damage prevention."[4]

Both types of responses are urgently needed. Adaptation and mitigation can complement each other and significantly reduce the overall risks of climate change. As recognized in Article 7 of the Paris Agreement, "greater levels of mitigation can reduce the need for greater adaptation efforts."[5] Mitigation actions to limit the increase in the global average temperature are also essential to ensure "the continued availability of the largest possible number of adaptation options,"[6] thereby limiting loss and damage from the adverse impacts of climate change. "Without mitigation, a magnitude of climate change is likely to be reached that makes adaptation impossible for some natural systems, while for most human systems it would involve very high social and economic costs."[7] That is to say, limiting greenhouse gas emissions helps keep more adaptation options within reach for projects and communities.

At the same time, due to the lag times in the global climate system, even the most stringent mitigation actions will not prevent the impacts of climate change in the coming decades. Adaptation to the changes that are already occurring and will continue to affect in the foreseeable future—is therefore unavoidable.

In other words, the outcomes of mitigation both supplement and depend on the outcomes of adaptation. Likewise, decisions and actions for mitigation can have consequences for adaptation, potentially constraining adaptation goals, and vice versa.

These definitions establish the overall scope, objectives, and ambition of the two sets of criteria. The criteria aim to highlight specific synergies between mitigation and adaptation efforts at the infrastructure project level and to demonstrate how the right actions can deliver both required responses to climate change.

GHG EMISSIONS CLASSIFICATION

One key research decision that shaped the climate change mitigation criteria was the selection of an approach to account for the main driver of climate change: Greenhouse gas (GHG) emissions. Two key approaches that have been widely endorsed were considered:

- The Greenhouse Gas Protocol's categorization into Scope 1, 2, and 3 GHG emissions
- The World Green Building Council's whole-life carbon

The Greenhouse Gas Protocol developed a corporate standard for accounting and reporting emissions (first edition published in 2001).[8] Its methodology is primarily designed for a company's GHG inventory. The innovative feature of the Greenhouse Gas Protocol's approach is its classification of emissions into the following categories, or scopes:[9]

- Scope 1 emissions: "Direct GHG emissions from sources that the company owns or controls"
- Scope 2 emissions: "Indirect GHG emissions from the generation of purchased energy (electricity, steam, or heat) consumed by facilities or equipment that the company owns or controls"
- Scope 3 emissions: "Other indirect GHG emissions that occur in a company's value chain, thus are generated from other sources the company does not own or control." These may include emissions from waste disposal, emissions related to business travel, employee commuting, and other activities

The World Green Building Council (WGBC), as part of its Advancing Net Zero project launched in 2016, developed a different methodology to account for emissions from all lifecycle phases of a building or infrastructure project—known as whole-life carbon.[10] "Carbon" here is shorthand for carbon dioxide equivalent (CO_2e), a measure of total greenhouse gases emitted. It encompasses all greenhouse gases and should not be confused with CO_2 emissions alone.

Whole-life carbon includes two main types of carbon: operational carbon and embodied carbon. Operational carbon refers to emissions "associated with the energy used to operate" the building or infrastructure. Embodied carbon refers to "emissions associated with materials and construction processes throughout the whole lifecycle of a building or infrastructure." This includes "material extraction, transport to manufacturer, manufacturing, transport to site, construction, the use phase [...] (excluding operational carbon), maintenance, repair, replacement, refurbishment, deconstruction, transport to end-of-life facilities, processing, and disposal." The WGBC further distinguishes embodied carbon into:

- Upfront carbon: "Emissions caused during the materials production and construction phases," before the infrastructure begins operation
- Use stage embodied carbon: "Emissions associated with materials and processes needed to maintain the infrastructure during use," in addition to operational carbon
- End-of-life carbon: "Emissions associated with deconstruction, [...] transport from the site, and waste processing and disposal" phases that occur after the asset's use phase ends

A key difference between the two approaches is that the Greenhouse Gas Protocol distinguishes GHG emissions based on where they occur along a company's activities and value chain, as well as who controls them. In contrast, the WGBC focuses on an asset and classifies emissions according to the lifecycle stage in which they occur.

Despite the distinct terminologies, both approaches encompass all types of GHG emissions relevant to infrastructure projects across their different lifecycle stages. While the WGBC framework is more directly associated with infrastructure projects—explicitly referring to infrastructure assets—the climate change mitigation criteria opt to follow the Greenhouse Gas Protocol's categorization. This categorization facilitates a clearer and more direct connection between strategies (climate change mitigation response criteria) and impact drivers (climate change mitigation assessment criteria), particularly regarding Scope 1 and 2 GHG emissions. This approach enables emissions to be detached from the lifecycle stage in which they occur, allowing for greater granularity in measurement and the metrics to be pursued. The absence of explicit references to project lifecycle stages in the Scope 1, 2, and 3 categories is

RELATION BETWEEN GREENHOUSE PROTOCOL's SCOPE 1, 2, AND 3 GHG EMISSIONS AND WGBC's WHOLE-LIFE CARBON

	 Product stage	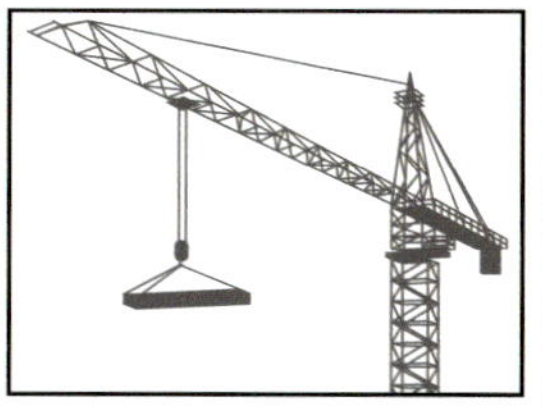 Construction stage	 Operation stage	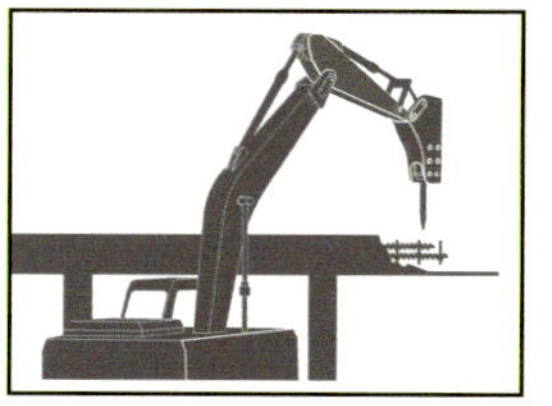 End-of-Life stage
Greenhouse Gas Protocol		**Scope 1 emissions**		
		Related to: • Onsite emissions from fossil fuel use for equipment, construction machinery, etc.	Related to: • Onsite generated energy for project operation	Related to: • Scope 1 emissions generated onsite for the replacement and/or deconstruction of a project
		Scope 2 emissions		
		Related to: • Purchased energy for the construction-installation process, workstations, etc.	Related to: • Purchased energy for operation, maintenance, minor or major rehabilitation	Related to: • Purchased energy consumed onsite for the replacement and/or deconstruction
	Scope 3 emissions (upstream)			
	Related to: • Raw material production • Transport of materials to manufacturer • Manufacturing	Related to: • The transport of materials/products to construction site	Related to: • The transport of materials/ products needed for operations, maintenance, rehabilitation to the site	
		Scope 3 emissions (downstream)		
		Related to: • Construction solid waste, including transport to a waste processing facility and/or landfill, the waste processing, and the disposal to landfill.	Related to: • Generated waste (wastewater and solid waste) during operations, maintenance, minor or major rehabilitation, including transport to a waste processing facility and/or landfill, the waste processing, and the disposal to landfill.	Related to: • Generated waste during end-of-life activities, including transport to a waste processing facility and/or landfill, the waste processing, and the disposal to landfill.
WGBC	**product stage carbon**	**construction stage carbon**	**operational carbon**	**end-of-life carbon**
	← **upfront carbon**	→		
	embodied carbon			

WHOLE-LIFE CARBON VS. SCOPE 1, 2, AND 3 GHG EMISSIONS

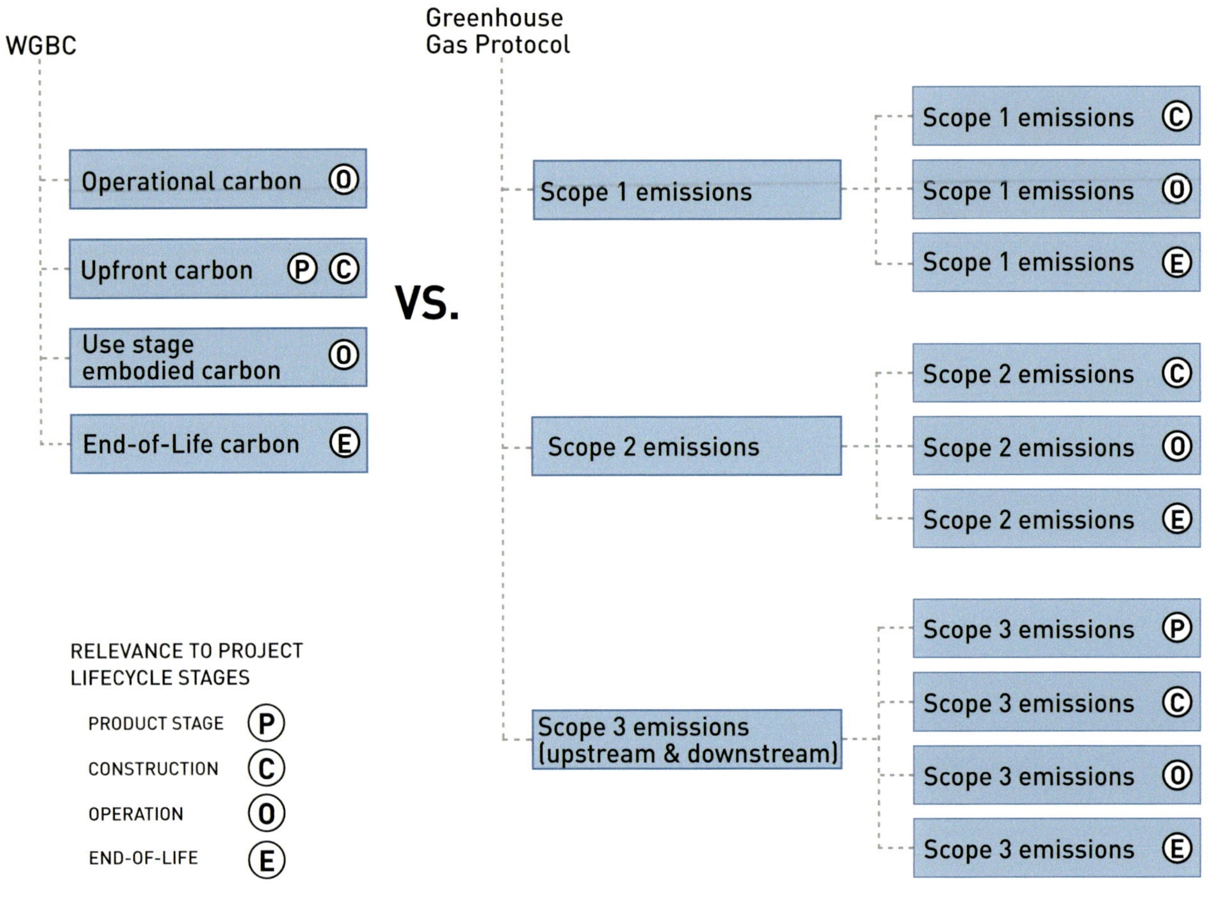

offset by the criteria's additional, specific references to lifecycle stages.

It is important to clarify that this methodological choice should not be interpreted as implying that the use of operational carbon and embodied carbon, rather than Scope 1, 2, and 3 emissions, is invalid for quantifying a project's GHG emissions. Both approaches ultimately yield the same outcome over the project's lifecycle.

RESILIENT INFRASTRUCTURE DEFINITION

As part of the criteria, climate change adaptation is also defined and operationalized through its complementary concept of resilience, described as "the capacity of interconnected social, economic, and ecological systems to cope with a hazardous event, trend, or disturbance, responding or reorganizing in ways that maintain their essential function, identity, and structure."[11] This definition suggests that system-based approaches align more closely with the concept of resilience. Consequently, incorporating resilience thinking into adaptation broadens the scope of assessment and intervention from the project site to encompass entire infrastructure systems and the surrounding community.

Various approaches exist that define what constitutes resilient infrastructure, and they largely converge on the key characteristics that should be inherent to projects or systems to address the effects of climate change.

The UK National Infrastructure Commission identifies six aspects of resilience that together capture the range of possible actions required to deliver resilient infrastructure systems: anticipate, resist, absorb, recover, adapt, and transform.[12]

- Anticipate: Proactive actions taken in advance to prepare for and "respond to shocks and stresses, such as collecting data on the condition of assets"
- Resist: Proactive actions that help "withstand or endure shocks and stresses to prevent an impact on infrastructure services"
- Absorb: Actions that acknowledge an impact on infrastructure services will occur or has occurred, and "aim to lessen" its severity—for example, by building redundancy
- Recover: Actions and procedures that "help quickly restore expected levels of service following an event"

THE CORE RESILIENCE PRINCIPLES AND THE RESILIENCE HIERARCHY STEPS IN RELATION TO EXISTING RESILIENCE FRAMEWORKS

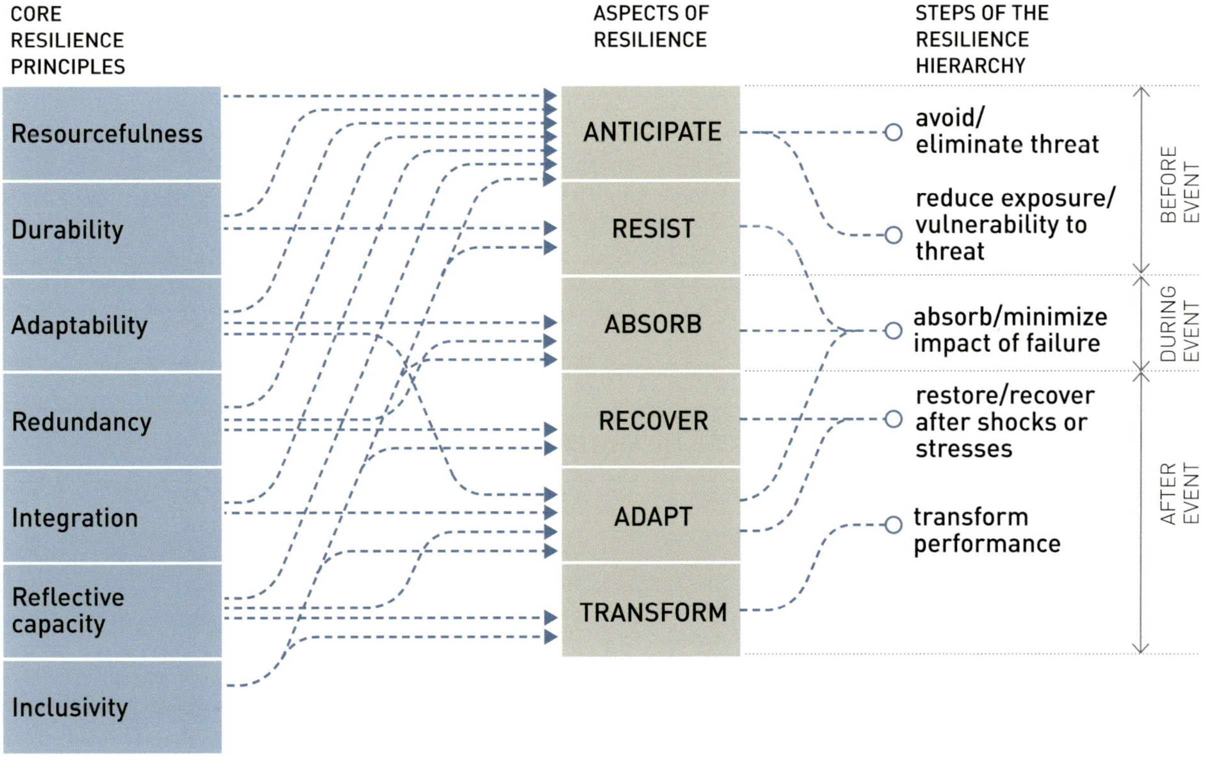

- Adapt: Actions that "modify the system to enable it to continue to deliver services in the face of changes, for example using storage in the electricity system to support renewable generation"
- Transform: Actions that "regenerate and improve infrastructure systems, for example, transforming infrastructure to meet net zero targets"

The criteria adopt the Envision framework's approach to assessing infrastructure project performance in relation to resilience. Envision defines the characteristics of resilient projects based on the Rockefeller Foundation's City Resilience Framework, specifically its "qualities of resilient systems."[13] By referring to these qualities as the "core principles of resilient systems," Envision establishes a connection between the City Resilience Framework and infrastructure project resilience strategies. It requests project teams to explain how their strategies for enhancing resilience address one or more of the following core principles:[14]

- "Reflective (learning and improving)"
- "Resourceful (resource-efficient, creative)"
- "Inclusive (shared action and responsibility)"
- "Integrated"
- "Robust (durable, well-constructed)"
- "Redundant (diverse, fault-tolerant)"
- "Adaptable (flexible)"

This adaptation of system-level and city-level principles to the project scale informs the resilience strategies as included in the climate change adaptation response criteria. These resilience strategies also align with the United Nations Office for Disaster Risk Reduction (UNDRR)'s recently published Principles for Resilient Infrastructure, which apply to nationwide infrastructure systems:[15]

- "Continuously learning"
- "Proactively protected"
- "Environmentally integrated"
- "Socially engaged"
- "Shared responsibility"
- "Adaptively transforming"

All the above approaches were adapted, integrated, and used in developing the proposed resilience hierarchy, as well as determining what constitutes "no resilience loss" and "net resilience gain."

CLIMATE CHANGE MITIGATION CRITERIA

The climate change mitigation criteria consist of 14 criteria, structured around four action steps—Assess, Commit, Respond, and Track—and grouped into five subsets. They are designed to frame key concepts and highlight important considerations that underpin infrastructure project action for climate change mitigation. Each criterion's description conveys its significance for a project's climate mitigation performance and provides a definition adapted to the infrastructure project scale, enhanced where relevant with indicative examples.

CCM. OVERVIEW OF THE CLIMATE CHANGE MITIGATION CRITERIA

ACTION STEPS	SUBSET OF CRITERIA	CRITERIA
1. ASSESS	**Assess the Project's Estimated GHG Inventory**	CCM1.1. Assess the estimated Scope 1 GHG emissions CCM1.2. Assess the estimated Scope 2 GHG emissions CCM1.3. Assess the estimated Scope 3 GHG emissions (upstream and downstream) CCM1.4. Assess the estimated avoided emissions as a result of the project
2. COMMIT	**Commit to and Set Ambitious GHG Reduction Targets**	CCM2.1. Commit to and set targets aligned with the Paris Agreement goals
3. RESPOND	**Prioritize Management Responses Based on the GHG Management Hierarchy**	CCM3.1. Plan for actions to: • Avoid/eliminate GHG emissions • Reduce GHG emissions that were not avoided • Substitute/replace GHG emissions • Compensate/offset residual GHG emissions • Transform performance to achieve net negative emissions
	Implement GHG Emissions Reduction Strategies	CCM3.2. Implement energy efficiency strategies CCM3.3. Implement decarbonization strategies CCM3.4. Implement electrification strategies CCM3.5. Implement carbon capture & storage strategies CCM3.6. Optimize material sourcing practices CCM3.7. Implement resource circularity strategies CCM3.8. Optimize project-related transport and logistics
4. TRACK	**Track Progress against Targets Set**	CCM4.1. Monitor and quantify actual GHG reductions to track progress against targets set

Assess the project's estimated GHG inventory

1. ASSESS

Greenhouse gas emissions are the primary driver of climate change. According to the IPCC, greenhouse gases (GHGs) are the "gaseous constituents of the atmosphere, both natural and anthropogenic, that absorb and emit radiation at a specific wavelength within the spectrum of terrestrial radiation emitted by the Earth's surface, the atmosphere itself, and by clouds. This property causes the greenhouse effect."[16]

The first subset of criteria for framing project action on climate change mitigation concerns the accounting of greenhouse gas emissions estimated to be generated by the project throughout its lifecycle.

In line with the Kyoto Protocol to the United Nations Framework Convention on Climate Change (UNFCCC) of 1997, and the amendment issued by the Greenhouse Gas Protocol in May 2013,[17] greenhouse gases (GHGs) include:

- Carbon dioxide (CO_2)
- Methane (CH_4)
- Nitrous oxide (N_2O)
- Hydrofluorocarbons (HFCs)
- Perfluorocarbons (PFCs)
- Sulfur hexafluoride (SF_6)
- Nitrogen trifluoride (NF_3)

The Greenhouse Gas Protocol classifies emissions into three categories, or scopes:[18]

- Scope 1 GHG emissions
- Scope 2 GHG emissions
- Scope 3 GHG emissions, which are further divided into upstream and downstream emissions

Estimates of the GHG inventory serve as a reference scenario forming the basis for identifying opportunities to reduce emissions. A project may draw upon existing tools such as GHG lifecycle assessments, BIM models, and digital twin technologies to estimate and measure expected GHG emissions and removals, as well as to conduct scenario testing to identify potential GHG reduction opportunities.

CCM1.1.

Assess the estimated Scope 1 GHG emissions

Scope 1 GHG emissions are the "direct GHG emissions" physically released from sources operated by the project, including facilities, processes, or vehicles owned or controlled within the project boundary.[19] For example, they include emissions produced by the combustion of fossil fuels at the project facilities intended for energy production, as well as fugitive emissions such as refrigerants or methane leakage. The assessment should account for the estimated Scope 1 emissions generated over the entire project lifecycle, including those emitted during construction, operation, maintenance and rehabilitation works, and end-of-life activities at the project site.

CCM1.2.

Assess the estimated Scope 2 GHG emissions

Scope 2 GHG emissions are the "indirect GHG emissions" resulting from the generation of purchased energy—such as electricity, steam, or heat—that is generated offsite and consumed by facilities or equipment owned or controlled by the project.[20] Although these emissions are generated offsite, the project maintains direct control over its energy consumption and can therefore quantify the projected energy requirements over the life of the project. Based on this assessment, the project can set targets and implement strategies to reduce Scope 2 emissions to the extent possible. Similar to Scope 1 emissions, Scope 2 emissions occur during all project phases, including construction, operation, maintenance, rehabilitation, and end-of-life

CCM1.3.

Assess the estimated Scope 3 GHG emissions (upstream and downstream)

Scope 3 GHG emissions are "all other indirect emissions" that can be considered consequences of project activities. Upstream emissions are associated with the inputs to the project, such as purchased materials, products (including fuels), equipment, and services. They include emissions stemming from all the activities that occur before a finished product reaches the project site, such as material extraction, production, product manufacturing, and transportation of products to the site. Downstream emissions are associated with outputs of the project activity, including waste generated by the project. They account for emissions generated during the transport of waste to disposal or processing facilities, the processing of waste for reuse, recovery, or recycling, and the final disposal of waste in landfills. Downstream emissions also include those associated with the project's end use or service—for example, private vehicle emissions resulting from the use of road infrastructure.[21]

The Greenhouse Gas Protocol classifies the accounting of Scope 3 emissions as "optional," whereas the accounting of Scope 1 and 2 emissions is "mandatory for all companies." However, it recognizes that "Scope 3 emissions can represent the largest source of emissions for companies and that they present the most significant opportunities to influence GHG reductions and achieve a variety of GHG-related business objectives."[22] The World Green Building Council has also emphasized the urgency for buildings and infrastructure projects to intensify efforts to tackle embodied carbon emissions on a global scale. According to the WGBC, these indirect emissions—historically largely overlooked—"will continue to grow in importance as a proportion of total emissions, particularly as operational emissions are reduced."[23]

CCM1.4.

Assess the estimated avoided emissions as a result of the project

Avoided emissions, also referred to as Scope 4 emissions, quantify the net "positive" climate impact on society—if any—by comparing the GHG impact of a project with that of an alternative reference scenario.[24] They represent emissions avoided when a service is used as a substitute for another service that fulfills the same functions but has a higher GHG density. In other words, avoided emissions are the difference between the GHG emissions that occur as a result of the project and those that would have occurred in its absence. Avoided emissions are accounted for separately from Scope 1, 2, and 3 emissions and do not offset GHG inventory emissions. As such, they are not intended for offsetting claims, but rather serve as an incentive for projects to prioritize the right climate solutions.[25]

Avoided emissions typically result from projects that deliver low- or zero-emission alternatives to higher-emitting options for providing the same service—for example, in energy generation, transportation, or wastewater treatment. Examples include renewable energy projects compared to fossil fuel-based energy generation, or projects that propose a modal shift from higher emitting modes to mass transit transport options. Avoided emissions may also arise from projects that improve road network capacity, thereby reducing emissions associated with congestion and stop-and-go traffic that increase private vehicles' fuel consumption, or from road improvement projects that contribute to a state of good repair. In such cases, the project's contribution lies in the avoided emissions by the end user, making this criterion particularly relevant to the purpose and need of a project.

In summary, the "Assess" step identifies project-driven climate change impact drivers that subsequent action steps will seek to address and reduce throughout the whole life of the project:

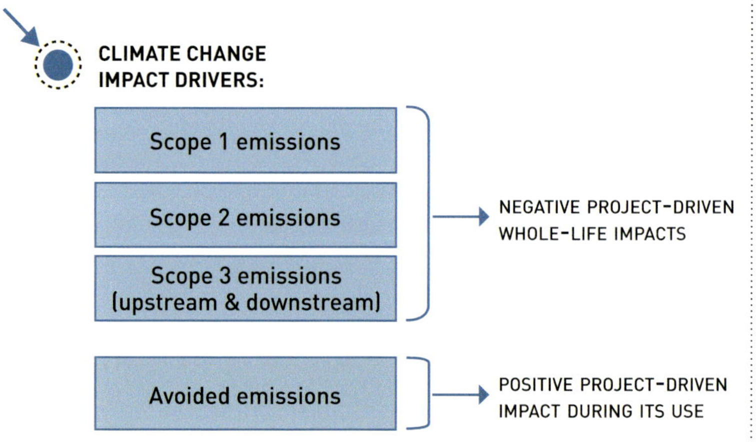

This subset of criteria refers to the commitment to GHG emissions reduction goals and targets established for the project. The level of ambition is defined in relation to the goals set out in the UNFCCC Paris Agreement to "hold the increase in global average temperature to well below 2°C above pre-industrial levels and pursuing efforts to limit the temperature increase to 1.5°C above pre-industrial levels."[26] The targets will shape the project's mitigation pathway, determining the extent of transformation required and guiding how mitigation responses must be tailored.

CCM2.1.

Commit to and set targets aligned with the Paris Agreement goals

As confirmed by the Intergovernmental Panel on Climate Change (IPCC), to limit global warming to 1.5°C by 2100 with no or a limited overshoot,[27] the world needs to halve CO_2 emissions by around 2030 and achieve net-zero CO_2 emissions around the early 2050s, along with deep and sustained reductions in other greenhouse gases.[28]

Net-zero CO_2 emissions (also referred to as carbon neutrality) are achieved "when anthropogenic CO_2 emissions (human-caused emissions such as those from fossil-fueled vehicles and factories) are balanced globally by anthropogenic CO_2 removals over a specified period."[29] Net-zero GHG emissions (also referred to as climate neutrality) are defined as that point when "anthropogenic emissions of greenhouse gases to the atmosphere are balanced by anthropogenic removals over a specified period."[30] The IPCC targets for CO_2 and GHG emissions are grounded in global emission pathways—"modeled trajectories of global anthropogenic emissions over the 21st century."[31] Even the most ambitious pathways do not reach net-zero GHG emissions before 2070, and some never reach net-zero.[32]

While the IPCC sets different timelines for CO_2 emissions (carbon) and GHG emissions targets, many high-profile initiatives refer to net-zero GHG emissions before 2050. This is partly because the difference between CO_2 and greenhouse gas targets is rather insignificant in sectors or countries where CO_2 emissions dominate and/or deep reductions in non-CO_2 are technically feasible.[33] Since infrastructure projects—such as energy supply and transport—pertain to those sectors where CO_2 emissions are dominant, this criterion calls for a commitment to align project targets with net-zero GHG emissions goals by 2050. For water, wastewater, and solid waste projects that also produce other direct GHG emissions such as methane (CH_4) and nitrous oxide (N_2O), the time-bound target may be adjusted to reflect realistic and achievable targets, such as achieving deep reductions by 2050 and net-zero by 2070.

Net-zero targets should cover all relevant GHG emission types/classes, as well as all relevant greenhouse gases. Projects must set emission reduction targets that are meaningful, achievable within specific timeframes, and science-based. Science-based targets reflect the scale of reductions that "the latest climate science deems necessary" to limit warming to 1.5°C above pre-industrial levels, as derived from objective scientific methods.[34]

Targets for 2050 represent long-term targets, while near-term targets for rapid, deep emission cuts must also be set for around 2030. Science-based targets call for roughly halving emissions before 2030, as this interim target is considered the most effective and scientifically sound way to limit global temperature rise to 1.5°C.

As part of the criteria, measurable targets may be either impact driver-specific (e.g., 100% reduction in Scope 2 GHG emissions) or strategy-specific (e.g., 100% renewable energy use by a specified date). In cases where it is not possible to define metrics for addressing an impact driver, a strategy-specific target could be set. Therefore, these two types of targets can be used complementarily by project teams to ensure measurable and verifiable outcomes.

The mitigation hierarchy, presented in the next criterion, can be used to test the feasibility of targets established for each impact driver.

Examples of targets for each driver of climate change could include:

For Scope 1 emissions:

- Percentage of onsite renewable energy (e.g., 100% renewable energy or zero-ready projects)
- Percentage of electrified equipment, fleet, etc.
- Percentage of direct emissions captured by onsite carbon sinks

For Scope 2 emissions:

- Percentage of renewable energy purchased
- For Scope 3 upstream emissions:
- Percentage of recycled-content materials used
- Percentage of locally sourced materials
- Percentage of materials sourced from suppliers implementing sustainable practices

For Scope 3 downstream emissions:

- Percentage of solid waste diverted from landfills during construction, operation, maintenance, and rehabilitation
- Percentage of recyclable or reusable components
- Targets could also be defined per lifecycle stage, such as:
- Zero-emission construction site
- Zero-emission operations facility
- Performance targets during the construction stage may be established as contractual requirements for construction contractors, subcontractors, manufacturers, and suppliers.

Prioritize management responses using the GHG management hierarchy

3. RESPOND

This subset of criteria calls on projects to prioritize management responses based on the principles of the mitigation hierarchy. Adherence to the mitigation hierarchy serves as a cross-cutting principle, tailored to the specific nature of the impact drivers addressed within each set of criteria. In the case of the climate change mitigation criteria, the primary driver is GHG emissions.

CCM3.1.

Plan for actions to:

- **Avoid/eliminate GHG emissions**
- **Reduce GHG emissions that were not avoided**
- **Substitute/replace GHG emissions**
- **Compensate/offset residual GHG emissions**
- **Transform performance to achieve net negative outcomes**

The GHG emissions mitigation hierarchy is adapted from IEMA's GHG Management Hierarchy, originally developed for businesses.[35] An additional step has been added to emphasize the need to go beyond carbon-neutral projects toward achieving positive climate outcomes.

Avoid/Eliminate: Actions to prevent GHG emissions wherever possible across the project lifecycle. This step is most relevant during early planning decisions, such as questioning the necessity of a new project and exploring alternative approaches—for example, maximizing the use of existing assets. The greatest opportunities to influence GHG reductions typically occur during the early planning and design stages of a project.[36]

Reduce: Actions to minimize GHG emissions that cannot be avoided, through improved efficiency in operations, processes, fleet and energy management, and the use of innovative or emerging technologies.

Substitute/Replace: Actions to substitute high-carbon sources by adopting renewable or low-carbon technologies (e.g., onsite generation), reducing the GHG intensity of energy use and purchased energy, and procuring materials, equipment, and services with lower embodied carbon.

Compensate/Offset: Actions to compensate for unavoidable residual emissions through carbon removals, carbon insetting, and certified carbon offsetting.

Transform: Actions aimed at driving higher ambition and supporting climate action beyond carbon neutrality toward carbon positivity (e.g., potentially addressing historic and diffuse emissions).[37] Achieving the goal of limiting warming to 1.5°C requires deep emission reductions, necessitating major and immediate transformations to make this technically feasible. This transformation entails actions and initiatives that extend beyond project boundaries to drive a shift across the entire infrastructure sector and construction industry practices. Such a shift can be enabled through leadership from the full project team—including the infrastructure owner, project manager, design and construction teams, suppliers, and manufacturers. Transformative actions also include sector-wide and cross-sectoral collective actions that strengthen capacity, mobilize resources, promote knowledge sharing, and facilitate investment in research and development (R&D).

Implement GHG Emission Reduction Strategies

3. RESPOND

This subset of the response criteria aims to secure that all drivers of climate change are addressed and that, collectively, projects implement all steps of the GHG management hierarchy to enable net-zero and net positive outcomes. Accordingly, it combines energy-related strategies—which address Scope 1 and 2 GHG emissions—with material- or product-related strategies that address Scope 3 GHG emissions.

Four of the listed energy-related strategies are common across various deep decarbonization pathways and transition scenarios that outline emissions trajectories consistent with the Paris Agreement commitments. These strategies correspond to key technological shifts identified as critical drivers for achieving the transition toward net-zero projects:[38]

- Energy efficiency
- Electricity decarbonization through the use of renewable energy sources
- Electrification, the process of replacing fossil fuel use with electricity as the primary energy source
- Carbon capture and storage for the hard-to-electrify segments of systems

To address a project's embodied emissions (Scope 3 GHG emissions, or embodied carbon), three additional material-related strategies are included. Each of these GHG reduction strategies complements the others, and efforts across all seven are necessary to effectively reach a net-zero outcome.[39]

CCM3.2.

Implement energy efficiency strategies

"Energy efficiency is called the 'first fuel' in clean energy transitions, as it provides some of the quickest and most cost-effective CO_2 mitigation options while lowering energy bills and strengthening energy security."[40]

This criterion refers to the reduction of energy demand—the required electricity and fuel input for a project—through initiatives such as process redesign to avoid energy wastage and loss, use of high-performance equipment and best-available energy-efficient technologies, optimized energy management, deployment of automated control systems, and the elimination of unnecessary energy use through behavioral changes. It is applicable across all stages of the project lifecycle—construction, operation, maintenance, and end-of-life—to address the project's Scope 2 GHG emissions.

In energy generation projects, operational energy efficiency is measured as energy conversion efficiency, with the goal of maximizing the capture of electrical, mechanical, or thermal energy output from the system. Similarly, in energy distribution projects, energy efficiency is assessed by reductions in energy losses during transmission, storage, and delivery.

IMPACT DRIVERS ADDRESSED:
Scope 2 GHG emissions
Scope 1 GHG emissions (in the case of energy generation)

CCM3.3.

Implement decarbonization strategies

This criterion refers to the decarbonization of electricity or fuels used by the project through:

- Onsite renewable energy generation
- Purchase of renewable fuels
- Purchase of renewable electricity from the grid via clean energy procurement mechanisms, such as direct renewable power purchase agreements or Renewable Energy Certificates (RECs)

By generating the electricity it requires to function from onsite renewable sources and using renewable fuels for equipment and fleet, a project addresses its Scope 1 GHG emissions.

A project can also use RECs to address its Scope 2 GHG emissions by "verifying use of zero- or low-emissions renewable source of electricity."[41] RECs are instruments for "accounting and assigning ownership to the attributes of renewable electricity generation and use," including the emissions profile of that generation. They can therefore "serve as the basis for renewable electricity consumption claims" by the project.[42]

IMPACT DRIVERS ADDRESSED:
Scope 1 GHG emissions
Scope 2 GHG emissions

CCM3.4.
Implement electrification strategies

Electrification is the process of "replacing technologies or processes that use fossil fuels with electrically-powered equivalents"[43] (e.g., electrification of fleets, construction equipment, etc.). It is a key strategy for reducing Scope 1 GHG emissions associated with fossil fuel combustion. To maximize emission reduction impacts, electrification should be implemented in combination with energy demand reduction through energy efficiency and the decarbonization of electricity generation.[44]

IMPACT DRIVERS ADDRESSED:
Scope 1 GHG emissions

CCM3.5.
Implement carbon capture & storage strategies

All 1.5°C emission pathways rely on carbon removal to some extent,[45] to compensate for emissions from the hard-to-electrify portions of energy systems, as well as from other project-related activities whose decarbonization is not currently technically possible. Carbon removals refer to the carbon sinks associated with project activities that remove carbon from the atmosphere and permanently store it through natural or mechanical carbon sequestration methods.

"Existing and potential carbon dioxide removal (CDR) measures include afforestation and reforestation, land restoration and soil carbon sequestration, bioenergy with carbon capture and storage (BECCS), direct air carbon capture and storage (DACCS), enhanced weathering, and ocean alkalinization. These measures differ widely in terms of maturity, potential costs, risks, co-benefits, and trade-offs."[46] Among these, afforestation and BECCS are the most commonly used measures.

Carbon removals should be distinguished between those that occur within the project's boundary and supply chain and those that occur externally, including purchased carbon credits.

A project can purchase carbon offsets to address its Scope 1, 2, and 3 GHG emissions by verifying global emissions reductions from additional external projects "intended to reduce GHG emissions, increase the storage of carbon, or enhance GHG removals from the atmosphere."[47] Carbon offsets must meet specific requirements to be valid and effective, such as being "additional," "permanent," and "verified."[48] Credits are issued for verified emissions reductions and can be subtracted from a project's generated emissions to determine its net emissions.[49]

While compensation efforts are necessary in the short term to enable net-zero projects, it is important to prioritize direct emissions reductions and protecting existing carbon pools over enhancing carbon dioxide removals.

IMPACT DRIVERS ADDRESSED:
Scope 1, 2, and 3 GHG emissions

CCM3.6.
Optimize material sourcing practices

Scope 3 upstream emissions—arising during material extraction, processing, and manufacturing—are largely beyond the direct control of the project. However, projects can control these emissions through the selection criteria for the suppliers and/or manufacturers of materials, products, and equipment used as project inputs. Optimizing material sourcing practices involves prioritizing suppliers that implement sustainable strategies. When choosing materials and products, projects should evaluate different options using third-party verified Environmental Product Declarations (EPDs) to gain a clearer understanding of market performance. EPDs, along with other environmental properties of building materials, provide manufacturers' information on the whole-life carbon intensity of materials.

IMPACT DRIVERS ADDRESSED:
Scope 3 GHG emissions (upstream)

CCM3.7.

Implement resource circularity strategies

Resource circularity refers to the flow of materials through a project, including inflows—such as resources, materials, products, and parts sourced—and outflows, such as byproducts and waste streams. Its objective is to maintain a circular flow of resources, recovering, retaining, and enhancing their value to minimize the need for primary material inputs and reduce waste outputs.[50] Resource circularity thus has a positive impact on both upstream and downstream Scope 3 emissions of a project.

Circular inflows are "determined by the non-virgin content of project inflows,"[51] such as recycled-content, renewable content (e.g., sustainable bio-based sources), or the substitutability of materials. Circular outflows are "determined by the recovery potential and actual recovery of project outflows."[52] Recovery involves "the technical recovery of components and materials at a functional equivalence through reuse, repair, refurbishment, repurposing, remanufacturing, recycling, or biodegradation (including composting)."[53] Achieving this requires seizing opportunities to optimize design, such as adopting modular configurations, enabling disassembly, reparability, and maximizing recyclability.[54] Longer product lifetimes reduce the rate at which resources are consumed, leading to less environmental impacts and less waste. A product's lifetime is "the period starting with the product's release for use after manufacturing of recovery and ends when the product becomes obsolete."[55] Durability—defined as "the ability to function as required, under specified conditions of use, maintenance, and repair"—supports longer lifetimes than the industry average and is an integral component of resource circularity.[56]

It is widely recognized that "circularity must be accelerated to optimize resource use and enable the systems' transformation required for sustainable production and consumption. A net-zero, nature-positive world can't be delivered without significantly growing the circular economy."[57] By applying the regenerative model that circularity represents, projects can design out waste, increase product productivity, and maintain sustainable resource use within boundaries.[58]

Projects should identify all potential opportunities to minimize—to the extent possible—resource use and waste and maximize product productivity. One example of a tool to support circularity strategies is the material passports. Material passports allow a project or company to identify the value of its own excess materials and/or leverage excess materials from other companies. They are digital datasets that link data to physical products, optimizing value recovery and facilitating efficient material flows across the supply chain. Digital passports provide information on:

- Circularity characteristics (such as durability, reparability, spare parts, recycled content, reuse potential, and design for disassembly)
- Lifecycle assessment (LCA), footprinting, and emissions data
- Supporting information for green claims

Additional tools include material reuse libraries, digital banks, and reuse search engines, which match available materials for reuse with market demand, offering an alternative to primary material consumption and reducing construction waste.

It is worth highlighting that there is an active area of research on substitute materials and technologies, focusing on solutions that increase durability, reparability, upgradability, and reusability while reducing embodied carbon. This research is driven by the need to reduce the overall carbon footprint of the construction sector. Examples include:

- Recovered materials as high-quality secondary raw materials, e.g., ultra-high-performance concrete produced from excavation waste

- Graphene-enhanced concrete for downsizing structures without compromising structural performance
- Nanotechnology-based construction materials for improved mechanical properties and durability
- Composites, such as carbon- and glass-fiber reinforced polymers, for structure reinforcement to improve mechanical properties and durability[59]
- CO_2 reuse in cement production or carbon fiber manufacturing with high strength and low weight-to-strength ratios, as well as high-temperature resistance[60]
- Green cement produced from steel waste generated during iron-to-steel conversion

Circularity is a broad term that encompasses materials, products, water, and in some cases energy. Water circularity is an important element of circular practices but is not accounted for in this criterion.

IMPACT DRIVERS ADDRESSED:
Scope 3 GHG emissions (upstream & downstream)

CCM3.8.

Optimize project-related transport and logistics

This criterion aims to address Scope 3 upstream emissions generated during the transportation of materials to and from the project site. Opportunities to reduce transport-related emissions depend on the proximity of material source sites and the modes of transportation used. Additionally, workforce commuting to and from the project site represents a source of Scope 3 emissions that should be accounted for. Examples of strategies to reduce transport-related emissions include:[61]

- Sourcing materials locally or regionally to reduce transportation distances
- Proper logistics planning to reduce the number of deliveries
- Using waterborne or rail transport instead of trucking where feasible
- Establishing onsite plants (e.g., concrete or asphalt plants) instead of trucking material to the site
- Onsite reuse of soils or other materials
- Employee commuting programs with incentives, such as shuttles to transit stations, ride-share programs, or bicycle facilities

IMPACT DRIVERS ADDRESSED:
Scope 3 GHG emissions (upstream)

OVERVIEW OF HOW GHG REDUCTION STRATEGIES ADDRESS ALL IMPACT DRIVERS OF CLIMATE CHANGE AND IMPLEMENT ALL STEPS OF THE GHG MANAGEMENT HIERARCHY

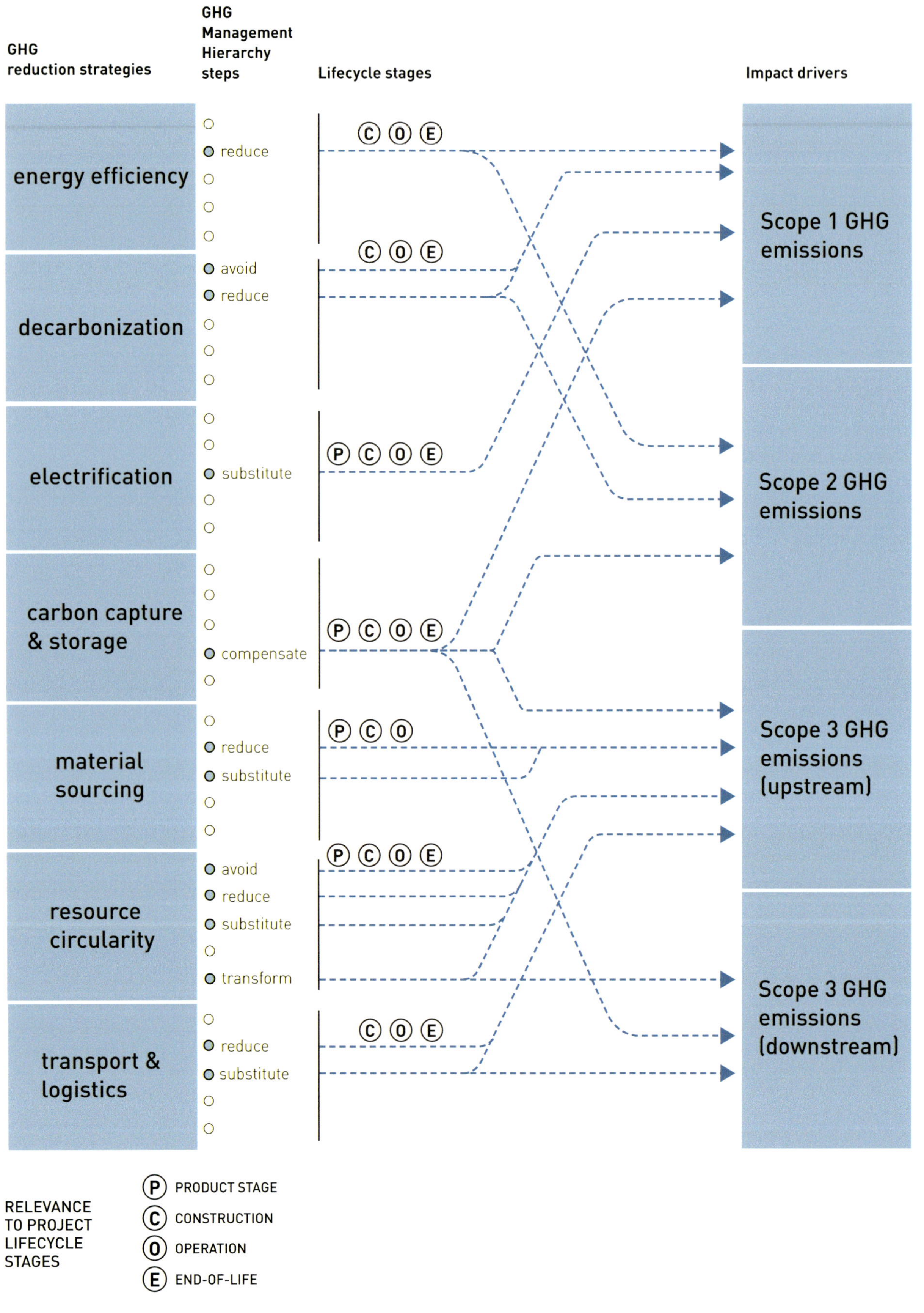

4. TRACK

Track criteria aim to monitor the progress against impact driver- and strategy-specific targets. They hold a special place within the criteria, as revising and adapting initial plans is critical in the context of climate change and its associated long-term targets. For climate change mitigation, ongoing research continues to advance the transition to climate neutral projects. Emerging technologies that are currently at the pilot or near-commercial stage may become available in the near future, enabling projects to increase ambition and improve outcomes. Similarly, lessons learned from pilot projects can inform new best practices.

CCM4.1.

Monitor and quantify actual GHG reductions to track progress against targets set

This criterion calls for projects to establish mechanisms to monitor and verify progress toward both their long-term and near-term targets. Projects can demonstrate improvements in GHG emissions through direct, real-time measurements and by tracking, collecting, and analyzing data related to their implemented strategies to verify their effectiveness. Monitoring can enable readjustments to near-term targets and strategies to keep on track with longer-term targets.

As recommended, overarching climate goals should be disaggregated into specific, measurable targets for all relevant climate change impact drivers. These specific targets can form a set of performance indicators for monitoring project performance.

Performance tracking during the construction stage may be included as a contractual requirement for construction contractors and suppliers.

CLIMATE CHANGE ADAPTATION CRITERIA

The climate change adaptation criteria consist of 17 criteria structured around four action steps and grouped into six subsets. They are designed to frame key concepts and important considerations that underpin climate change adaptation action. Each criterion's description outlines its importance to a project's climate adaptation performance and provides a definition tailored to the infrastructure project scale, enhanced where relevant with indicative examples.

CCA. OVERVIEW OF THE CLIMATE CHANGE ADAPTATION CRITERIA

ACTION STEPS	SUBSET OF CRITERIA	CRITERIA
1. ASSESS	**Assess Exposure to Climate-Related Hazards/Threats**	CCA1.1. Assess exposure to all location-relevant climate-related hazards/threats
	Assess Vulnerability to Climate-Related Risks	CCA1.2. Assess climate-related service continuity risk CCA1.3. Assess climate-related physical asset risk CCA1.4. Assess climate-related resource availability risk (water, materials, land, workforce) CCA1.5. Assess climate-related supply chain continuity risk
2. COMMIT	**Commit to and Set Ambitious Resilience Targets**	CCA2.1. Commit to and set targets for no resilience loss and/or net resilience gain
3. RESPOND	**Prioritize Management Responses Based on the Resilience Hierarchy**	CCA3.1. Plan for actions to: • Avoid/eliminate threat • Reduce exposure/vulnerability to threat • Absorb/minimize impact of failure • Restore/recover after shocks or stresses • Transform performance
	Implement Risk Reduction Strategies	CCA3.2. Optimize project siting and site design to avoid threats CCA3.3. Implement strategies to maximize resourcefulness CCA3.4. Implement strategies to maximize durability CCA3.5. Implement strategies to maximize adaptability CCA3.6. Implement strategies to maximize redundancy CCA3.7. Implement strategies to maximize integration CCA3.8. Implement strategies to optimize system recovery CCA3.9. Develop local skills and capabilities CCA3.10. Promote inclusivity
4. TRACK	**Track Progress against Targets Set**	CCA4.1. Implement strategies to promote reflective capacity

A series of extreme weather events—from hurricanes and wildfires to heat waves and floods—are already being experienced globally and are projected to increase in frequency and intensity due to climate change. Climate-related extreme weather events rank as the top global risk for the next decade, according to the World Economic Forum's 2025 Global Risks Report, and they have consistently been among the top-rated risks over the past ten years.[62] "The severity of the impacts of climate extremes depends strongly on the level of vulnerability and exposure to these events"[63] of a given site. As noted, "Exposure and vulnerability are key determinants of disaster risk and of impacts when the risk is realized."[64]

Similar to disaster risk management, climate change adaptation focuses on reducing exposure and vulnerability while increasing resilience to the potential adverse impacts of climate extremes. According to the Sendai Framework for Disaster Risk Reduction, the first priority is to understand risk: "Disaster risk management needs to be based on an understanding of disaster risk in all its dimensions of vulnerability, capacity, exposure of persons and assets, hazard characteristics and the environment."[65]

Therefore, "understanding the multi-faceted, dynamic nature of both vulnerability and exposure is a prerequisite for determining how weather and climate events contribute to the occurrence of disasters, and for designing and implementing effective adaptation strategies."[66]

CCA1.1.

Assess exposure to all location-relevant climate-related hazards/threats

Hazards or threats are events that have the potential to cause damage or harm, whether naturally occurring (hazards) or human-induced (threats). The IPCC defines a hazard as "the potential occurrence of a natural or human-induced physical event or trend that may cause loss of life, injury, or other health impacts, as well as damage and loss to property, infrastructure, livelihoods, service provision, ecosystems, and environmental resources."[67]

Threats/ and hazards can be classified by:[68]

- Duration: acute shocks—extreme weather events over hours and days—or chronic stressors—slow-onset events over years or decades. Examples of climate change-driven acute shocks include cyclones, wildfires, heat waves, and flooding. Examples of chronic stressors include sea level rise, droughts, water scarcity, temperature variability, changing precipitation patterns, environmental degradation, and species extinction. A key difference between acute shocks and chronic stressors—apart from their duration—is their predictability. Chronic stressors tend to have higher predictability and confidence levels than acute shocks
- Extent of effects: project site level (e.g., localized stormwater overflow), infrastructure system-wide level, or community-wide level (e.g., regional changes in climate)

The assessment begins with the physical hazards observed in a given location and should account for climate change's impact on each hazard's characteristics, frequency, duration, and severity.[69]

Climate scenarios are central to assessing exposure to potential future climate-related hazards and threats. The climate modeling community has settled on several plausible scenarios; however, it is important to note that these are not forecasts. There is inherent uncertainty in climate modeling and weather projections. This means there is uncertainty in estimates of risk under current conditions, and even greater uncertainty for future conditions. To account for this uncertainty, a range of plausible physical climate-related scenarios should be considered when assessing how resilient a site or location is against potential future hazard trends. Moreover, the TCFD's recommendations call for considering "challenging" high-risk scenarios or worst-case scenarios to account for potential significant increases in average global temperatures due to "no climate action" or "limited action," which would pose greater threats. Building a strong knowledge base on climate risks is an ongoing process in many countries, and the availability of up-to-date climate risk information is critical.

Assess vulnerability to climate-related risks

1. ASSESS

According to the IPCC, the core definition of risk is "the potential for adverse consequences."[70] The word "potential" underscores that uncertainty is a key element of the concept of risk. Risk is increasingly unpredictable and cannot be fully eliminated. In the context of climate change, risks can arise from the potential physical impacts of climate change as well as from human responses to it. Risk results from the interaction between the vulnerability of the affected system or asset, its exposure over time to a hazard, the nature of the climate-related hazard itself, and the likelihood of its occurrence.[71]

Climate risks are particularly relevant to long-lived, fixed assets such as infrastructure projects. Projects should assess which climate-related risks may arise from identified potential hazards and threats, and how vulnerable the project and its connected infrastructure systems are to these risks.

The climate-related risks listed as part of the criteria build upon the Taskforce on Climate-related Financial Disclosures (TCFD)'s "climate-related physical risks."[72] The TCFD provides examples of such risks and their potential financial impacts, translating physical risks into reduced revenues, increased operating or capital costs, and other financial consequences. The selection of criteria focuses on the potential sources of these financial impacts, including:

- Decreased production capacity or output
- Supply chain disruption
- Negative impacts on the workforce
- Damage to property and assets in high-risk locations
- Changes in water availability, sourcing, and quality
- Raw material availability

This subset of the assessment criteria groups the potential sources of project impact into four types of climate-related risks relevant to infrastructure projects:

- Service continuity risk
- Physical asset risk
- Resource availability risk (water, raw materials, workforce, land)
- Supply chain continuity risk

These climate-related risks are interconnected. Realized physical asset risk can lead to service continuity risk and to resource availability risk due to increased demand for materials for repair or replacement. Likewise, a supply chain interruption could generate resource availability risk and, by extension, affect service continuity. Despite their interconnections, these risks are treated under separate criteria to enable visible connections with risk-specific actions.

CCA1.2.

Assess the project's vulnerability to climate-related service continuity risk

Infrastructure service disruptions can have serious consequences for public safety, health, the economy, and the environment. Projects must identify risks of potential service disruptions or productivity losses resulting from failures associated with extreme weather events. Asset design and system configuration should be stress-tested under different future climate scenarios to identify potential failure points in critical project components, as well as potential barriers to recovery. Furthermore, the assessment should expand the project boundary to identify interdependencies within the broader infrastructure system, network, or across other infrastructure systems. To this end, infrastructure projects can conduct systems mapping to examine how different decisions can affect the level of service delivered to the community, and to understand the interconnections among various infrastructure systems—such as energy, digital communications, transport, and water—and the potential cross-system impacts of those decisions.

CCA1.3.

Assess the project's vulnerability to climate-related physical asset risk

The impacts of climate change are experienced locally. This criterion refers to the potential for damage to assets located in high-risk areas, such as coastal zones or flood-prone regions, that are exposed to climate change-driven physical risks. Depending on the location, projects may face increased physical risks arising from acute shocks—such as cyclones, heat waves, floods, and wildfires—or from chronic stressors, including sea level rise and prolonged droughts. This assessment should ideally be performed during the early planning phases of a project, as part of the screening of alternatives, to enable the selection of the most suitable project location. This step is critical, as unsuitable site selection may not only lock a project into high-risk conditions but also exacerbate existing risks in a given location by spreading the effects of a hazard and extending its impacts to interconnected infrastructure or the surrounding community.

CCA1.4.

Assess project's vulnerability to resource availability risk

This criterion refers to the assessment of a project's long-term dependencies on resources essential for its continued and efficient function throughout its service life. The key resources a project depends on include water, equipment for operations, raw materials for maintenance and rehabilitation, and land for future expansion. In addition, a crucial resource is human capital—the workforce (its capacity and skills) on which the project's long-term operations rely. A project must have a clear understanding of its long-term input needs, how these could be affected by climate change, and how their continued sourcing can be secured in the context of a changing climate. Resource security should be managed in a way to ensure that the project does not compete with the surrounding community for limited resources.

Water availability represents a key risk in the context of climate change, which alters both the distribution and availability of water resources, making it increasingly difficult and energy-intensive to meet demand. Currently, about 40 percent of the world's population experiences water scarcity, and by 2050, more than half is expected to face water stress.[73] Water stress refers to the "inability to meet human and ecological demands for water. Compared to scarcity, water stress is a more inclusive and broader concept," considering not only quantity but also "water quality, environmental flows, and accessibility of water."[74] Growing water stress and competing demands are adding pressure on freshwater resources.

The long-term availability of raw materials, equipment, and other inputs is also critical for continued project operation, maintenance in a state of good repair, and planned rehabilitations. Resource availability is closely linked to the supply chain continuity risk, described under another criterion.

Land availability is another factor projects should consider to ensure long-term resilience. It is tied to the project's capacity to fulfill its mission over time. Project design should therefore factor in the projected capacity needs and any potential future expansion of operations.

Finally, projects must assess the local and regional workforce to identify potential specific skill or capability gaps. The planning and management of human capacity should consider potential climate-related impacts such as employee health and safety, or potential increased absenteeism due to climate-driven disasters. Climate change can also affect the ability to attract and retain a skilled workforce, indirectly through what the TCFD terms "reputational risk."[75] According to the TCFD, climate change can be a "potential source of reputational risk tied to changing community perceptions of an organization's contribution to—or detraction from—the transition to a lower-carbon economy."[76] This risk is related to the project's level of climate ambition and its mandatory or voluntary actions to combat climate change and its effects.

CCA1.5.

Assess the project's vulnerability to supply chain continuity risk

Climate-driven natural disasters can create ripple effects across global supply chains. As a result, projects are exposed to potential future climate-related supply chain disruptions. Such disruptions may arise from direct damage to suppliers' production sites, damage to transport routes that impede product delivery, or indirectly from reliance on resources affected by chronic stressors such as water shortages. Mapping supplier locations throughout the supply chain, assessing their exposure to climate-driven risks, and tracking their performance can help safeguard long-term supply chain reliability. At the same time, projects should anticipate risks and enhance their ability to operate under varying conditions by diversifying their supply base, implementing dual sourcing for critical materials, identifying viable resource substitutes (replacing products or components with similar alternatives when necessary), and exploring alternative transport routes or modes. Regionalizing sourcing can further safeguard against global-scale shortages and disruptions.

In summary, the "Assess" step identifies climate change-related hazards/threats and risks that are relevant to the project and that subsequent action steps will seek to address and reduce throughout the project's life:

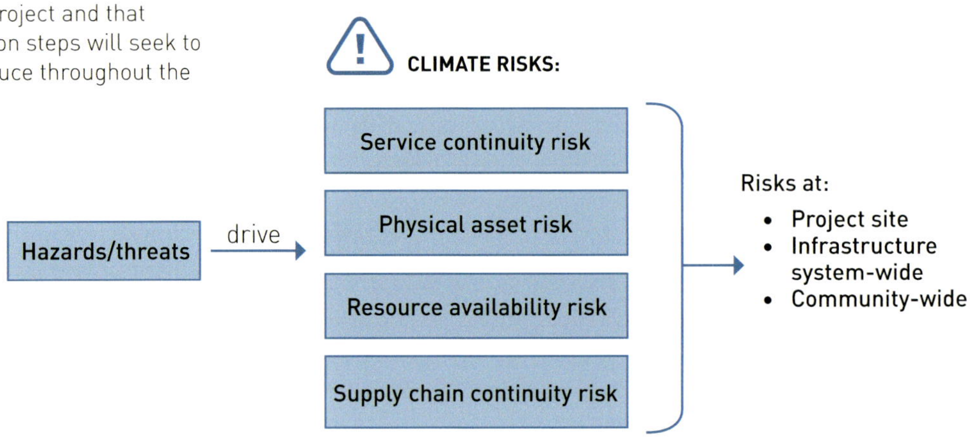

Commit to and set ambitious resilience targets

2. COMMIT

Unlike climate change mitigation, there is no quantified global goal for climate change adaptation. Although adaptation is acknowledged as a shared global challenge, similar to mitigation, it presents significant difficulties due to the local and context-specific nature of adaptation interventions. The Paris Agreement recognized adaptation as a key component of the long-term global response to climate change, establishing the goal of "enhancing adaptive capacity, strengthening resilience, and reducing vulnerability to climate change, with a view to contributing to sustainable development and ensuring an adequate adaptation response in the context of the agreement's temperature goal of limiting warming to 1.5°C or 2°C."[77]

At COP26 in 2021, the Glasgow–Sharm el-Sheikh work program on the Global Goal on Adaptation was initiated to develop a framework for the Global Goal on Adaptation, which was subsequently adopted at COP28. The framework identified key targets critical for climate adaptation that are closely linked with the UN SDGs, including water management, food security, health, infrastructure development, ecosystem conservation, poverty alleviation, and preservation of cultural heritage. Adaptation efforts should comprehensively consider all target areas. Alignment with the UN SDGs underscores the interconnected nature of climate adaptation and broader sustainable development objectives, emphasizing the need to integrate adaptation into broader development agendas to achieve holistic and sustainable outcomes.[78]

The commitment criterion also adopts the UN Office for Disaster Risk Reduction (UNDRR)'s concept of "net resilience gain," an innovative idea introduced in its Principles for Resilient Infrastructure.[79] Net resilience gain is defined as "a long-term collaborative commitment to (a) address systemic resilience loss by minimizing or eliminating actions that erode, reduce, or undermine systemic resilience, and (b) enhance systemic resilience by prioritizing actions that create systems intrinsically resilient to potential disruptions."[80]

The UNDRR concept of net resilience gain is defined at the national infrastructure system level. To apply it at the infrastructure project level, the steps of the resilience mitigation hierarchy were adapted to be more specific to project-level action.

CCA2.1.

Commit to and set targets for no resilience loss and/or net resilience gain as a result of the project

This criterion aims to clarify the resilience outcomes and levels of resilience ambition that projects are expected to deliver as they anticipate potential climate-driven events.

As part of the criteria, "no resilience loss" and "net resilience gain" are defined in relation to the infrastructure system that the project is part of or connected to, and the broader community that would experience the adverse impacts if climate-related risks materialize, potentially resulting in project failure. "No resilience loss" refers to a state in which the project does not exacerbate existing or projected risks to the infrastructure system or the community. This can be achieved through increased preparedness and anticipation in line with the resilience hierarchy. "Net resilience gain" refers to positive contributions at the system and community levels, such as eliminating existing, observed threats. It also encompasses transformative action that extends the project's boundaries, for example, resilience knowledge sharing and resilience literacy initiatives.

The project must establish methods for measuring or quantifying resilience goals. These may include, but are not limited to:

- Reduced magnitude of losses at a project, infrastructure system level, and community levels
- Reduced probability of occurrence of failure
- Reduced recovery time during a potential event of failure
- Minimum service levels maintained during a potential failure

The targets should be continuously reviewed, challenged, and reformulated as needed.

Prioritize management responses based on the resilience hierarchy

3. RESPOND

As with all similar action steps, the first subset refers to adherence to the mitigation hierarchy. It is adapted from the six aspects of resilience defined by the UK National Infrastructure Commission. The hierarchy prioritizes actions before, during, and after an event or shock.

CCA3.1.

Plan for actions to:

- **Avoid/eliminate threat**
- **Reduce exposure/ vulnerability to threat**
- **Absorb/minimize impact of failure**
- **Restore/recover after shocks or stresses**
- **Transform performance to achieve a net resilience gain**

Avoid/eliminate threat: Proactive actions taken before an event to avoid or eliminate potential threats altogether, such as proper siting of infrastructure outside hazard-prone areas, flood-prone zones, or water-stressed regions.

Reduce exposure/vulnerability to threat: Actions taken in advance (before an event) to help resist or endure shocks and stresses, thereby preventing impacts on infrastructure services. Reducing vulnerability is a core element of climate adaptation.[81]

Absorb/minimize impact of failure: Actions taken before and during an event that acknowledge that an impact on infrastructure services may occur, aiming to absorb or lessen that impact and adapt in the face of changes through preparedness and anticipatory measures.[82]

Restore/recover after shocks or stresses: Actions that reduce recovery time and quickly return service levels to normal following a disruption event, such as processes for restarting operations after a power outage.

Transform performance: Actions that "regenerate and improve infrastructure."[83] According to the Organisation for Economic Co-operation and Development (OECD), transformative action in the context of resilience involves continuous learning applied to a project to enhance its performance. This is closely linked to criterion CCA4.1, which emphasizes the need to continually assess, build knowledge, learn, and improve to inform future decision-making. Transformation also encompasses community behavioral change and an understanding of how to act before, during, and after a disaster.[84]

Another interpretation of transformation is provided by the UK National Infrastructure Commission, which links adaptive transformation with high climate ambition, such as transforming infrastructure to meet net-zero targets.[85]

Implement risk reduction strategies

3. RESPOND

"Resilience can be said to be threat agnostic."[86] It assumes that "at some stage, some threat or combination of threats will materialize and disrupt the system. The goal is to be ready for whatever happens, even if it cannot be anticipated and has never happened before."[87] This is why resilience frameworks provide conceptual or process-based approaches, describing resilience through principles or qualities that can be replicated or applied in different contexts.

The selection of specific risk reduction strategies that form the adaptation response criteria aims to:

- Address and reduce project vulnerability to all climate-related risks
- Ensure project preparedness before, during, and after a potentially disruptive event
- Collectively implement all steps of the resilience hierarchy

As part of the criteria, infrastructure project responses to potential climate change hazards and risks adopt the Rockefeller Foundation's qualities of resilient systems. These qualities, referred to as "core resilience principles" throughout this publication, form the central component of the risk reduction strategies. Three additional strategies supplement the core resilience principles:

- Proper siting and site design to better avoid hazards in the first place
- Optimized system recovery to improve performance during a failure
- Development of local skills and capabilities to highlight the role of the human factor in enabling resilience

CCA3.2.

Optimize project siting and site design to avoid threats

This criterion aims to address project exposure and vulnerability to potential physical asset risks arising from siting. It represents a critical first step in assessing and identifying the physical asset risks relevant to a given location during planning and the screening of alternative siting options. Site-specific hazards determine both project vulnerability and the potential for the project to exacerbate the hazard and spread its effects (e.g., creating impervious surfaces in a floodplain).[88]

Siting can eliminate certain known or high-confidence hazards and threats at the outset, thereby reducing the scope and intensity of subsequent adaptation efforts. Once a suitable location has been selected, proper site design and configuration can further enhance project resilience. The project site should be designed to adapt to the specific attributes of its location and surroundings. As noted, "Altering the characteristics of the built and natural environment changes the future risk profiles, the likelihood and magnitude of hazard events, and the vulnerability of society to those hazards."[89] Environmental degradation and species extinction are among the potential climate change-driven chronic stressors that projects should consider when building their resilience strategies, and they represent threats that interventions should not exacerbate. Environmental degradation can significantly raise vulnerability to natural hazards, including extreme climate events.

Preservation, "management, and maintenance of the surrounding natural environment are 'no-regrets' options for boosting the resilience of infrastructure assets."[90] Integrating infrastructure with the natural environment employs natural capital to complement grey infrastructure, using natural buffers to attenuate the impacts of hazards.

Incorporating ecosystem services and nature-based solutions for disaster risk reduction supports the resilience of infrastructure systems by increasing their protection against hazards.

CLIMATE RISKS ADDRESSED:
Physical asset risk

CCA3.3.

Implement strategies to maximize resourcefulness

Resourcefulness is the ability to deliver greater value with less input, thereby reducing pressure on limited natural resources, such as water and primary materials. It encompasses alternative practices that treat process byproducts as valuable resources. Resource efficiency, one of the climate-related opportunities identified by the TCFD, is an inherent component of resourcefulness. At the same time, resourcefulness is a broader and more inclusive concept. It involves experimentation and innovation—testing of alternative ideas and resourceful solutions to achieve project goals or meet project needs during a shock or under stress. "This may include investing in capacity to anticipate future conditions, set priorities, and respond, for example, by mobilizing and coordinating wider human, financial and physical resources."[91] Resourcefulness is also "instrumental to the ability to restore the functionality of critical systems under severely constrained conditions."[92]

Examples of resourcefulness strategies include:

- Sizing the project to require less material
- Designing the project to use less material, for example through the use of high performance materials such as advanced concrete, steel, or prefabricated components with high strength and enhanced durability
- Reducing material requirements for repair and maintenance
- Using recycled materials, including onsite reuse of materials and soil or materials with recycled content
- Using substitute materials and technologies that increase durability, reparability, upgradability, or reusability of products
- Implementing water conservation measures
- Diversifying water supply sources, for example by using wastewater as an additional source
- Rainwater harvesting and storage

Resourcefulness shares many principles with resource circularity, as described under criterion CCM.3.7, and additionally addresses water circularity. Circularity principles "offer an opportunity to recognize and capture the full value of water (as a service, input to processes, source of energy, and carrier of nutrients and other materials)."[93]

CLIMATE RISKS ADDRESSED:
Resource availability risk (materials, water, land)
Supply continuity risk
Physical asset risk

CCA3.4.

Implement strategies to maximize durability

Durability refers to "well-conceived, constructed, and managed physical assets so that they can resist the impacts of hazard events without significant damage or loss of function."[94] Durability encompasses not only the ability to withstand or absorb shocks and stresses but also the capacity to resist long-term wear and decay associated with project operations. This implies a longer useful life and reduces the need for maintenance and replacement. Material degradation is often accentuated by exposure to chronic stressors, such as extreme heat, precipitation, or flooding, resulting from climate change.

Robust design identifies potential failures in critical project systems or components and incorporates safeguards to ensure that "failure is predictable, safe, and not disproportionate to the cause."[95] This requires actively avoiding "design thresholds that might lead to catastrophic collapse if exceeded."[96] The ability to fail safely during a shock requires safe-to-fail features and strategic design to ensure failure is contained and minimized.

In the event of infrastructure disruption, systems must be designed to fail safely to protect occupants by maintaining critical life-support conditions, or passive survivability, for example through reliable backup power systems and effective hazardous substances controls. Incorporating safe-to-fail mechanisms is equally important to avoid cascading failures that could impact other assets and the broader community.[97]

Examples of durability strategies include:

- Exceeding minimum specifications, regulations, or industry norms for project durability regarding materials, structures, and construction quality
- Maximizing durability for critical project components where failures or degradation are most likely, considering anticipated impacts from future demand and loads
- Committing to maintaining the project in a state of good repair (e.g., through predictive and proactive maintenance). Combining routine maintenance with periodic condition-based maintenance improves reliability, extends infrastructure lifespan, and enhances performance over time[98]
- Improving construction quality through increased use of prefabrication, modular assembly, and offsite construction (manufacturing in a controlled environment enhances quality compared with onsite construction)
- Improving construction quality through intelligent construction systems, machines, and technologies with real-time monitoring of material placement and compaction (e.g., using electromagnetic technologies such as infrared)[99]
- Using materials with crack-healing properties, such as self-healing asphalt and self-healing bioconcrete, to improve mechanical properties

CLIMATE RISKS ADDRESSED:
Resource availability risk (materials, water, land)
Physical asset risk

CCA3.5.

Implement strategies to maximize adaptability

This resilience principle recognizes that, due to the changing nature of climate change-related hazards, "the requirements placed on a project in the future may look different from those placed today,"[100] both in functional and spatial terms.

Adaptability is the ability of a project for increased operational tolerance to long-term changes. Increased operational tolerance requires expanding the range of conditions under which a project can function, grow, or be configured. It entails manual override capacity and flexible management structures to cope better with external stresses under various future scenarios and combination of hazards. Adaptability is highly connected with redundancy—for example, through backup systems or distributed networks that maintain functionality—and reflective capability, systems that learn or change over time to improve preparedness for altered conditions. Adaptive capacity is a continuous process that allows projects to redefine operational best practices as new evidence becomes available. This enables not only the management of unexpected risks but also the leveraging of emerging opportunities.

Adaptability is further enabled by the introduction of new knowledge and technologies, which can supplement or reinterpret indigenous or traditional knowledge and practices.[101] Plans, actions, and project components should be flexible enough to allow rapid adjustment during shocks or stresses, as well as the ability to evolve over time.

This ability also increases the possibilities for repurposing infrastructure to alternative future uses—for example, through modular approaches that allow easy reconfiguration or refurbishment to accommodate shifts in supply or user demand. As a result, adaptability can extend the useful life of a project.

Examples of strategies that support adaptability include:

- Use of modular structures and materials that are easily adaptable for changing configurations
- Use of prefabricated components that can be separated easily for recycling or reuse
- Use of decentralized systems, distributed networks
- Use of intelligent systems capable of evolving over time
- Provision of alternative supply options (methods or locations) for critical resources
- Use of natural systems

CLIMATE RISKS ADDRESSED:
Service continuity risk
Resource availability risk (materials, land)
Supply chain continuity risk

CCA3.6.

Implement strategies to maximize redundancy

Redundancy refers to "spare capacity purposely created within built-in project systems so they can accommodate disruption, extreme pressures, or surges in demand."[102] It enables core project systems to continue functioning even when disrupted. Redundancy includes diversification—the availability of multiple ways to achieve a given need or fulfill a particular function—thereby "spreading risk" from a single point of reliance to multiple options.

Redundancy should be intentional, cost-efficient, and not the result of inefficient design, such as oversizing of structures or systems. In the case of transportation projects, redundancy can also include the provision of multiple transport mode options as alternatives to private vehicle use, increased system capacity to reduce congestion, and sufficient capacity to accommodate projected growth in demand.

Examples of strategies that support redundancy include:

- Use of decentralized systems and distributed networks
- Establishment of resource reserves
- Provision of alternative supply options (methods and locations) for resources critical to the project's continued function
- System capacity planning to address projected growth in commercial, industrial, or residential demand
- Availability of multiple transportation modes options
- Increased system capacity to reduce congestion
- Increased pedestrian proximity and accessibility to active, shared, or mass transportation
- Extended active transportation networks

CLIMATE RISKS ADDRESSED:
Service continuity risk

CCA3.7.

Implement strategies to maximize integration

Integration is a quality within and between systems, and across different scales of operation, that improves overall resilience and system performance. The first level involves the integration of internal systems within the project, the next level involves integration with external infrastructure systems, and optimal performance is achieved through integration at the community level. A project's functional integration into connected systems, where beneficial and appropriate, helps identify conflicts, reduces the risk of systemic and cascading failures, and promotes consistent decision-making and efficiency by leveraging co-benefits. Integrated systems can operate in multiple ways, under various conditions, or in different configurations. Multi-functional components can help avoid duplication and provide system diversity without relying solely on redundant backups. As noted, "The exchange of information between systems enables them to function collectively and respond rapidly through shorter feedback loops."[103]

CLIMATE RISKS ADDRESSED:
Service continuity risk
Resource availability risk (materials, land)

CCA3.8.

Implement strategies to optimize system recovery

Risks cannot be fully eliminated. This criterion complements the core resilience principles by emphasizing the need to address disaster relief, recovery, and the minimization of loss during failure scenarios.

Minimizing the downtime of critical systems is a key component of infrastructure resilience. Infrastructure provides essential services that communities rely on for survival. Depending on the project type, emergency management and disaster recovery planning may already be a critical consideration. However, all projects can benefit from increased focus on resilience and system recovery. Infrastructure owners and project teams should consider how improved hazard scenario planning and reduced recovery time can provide economic benefits both to the organization and to the broader community.

Projects should maximize efforts to restore project functionality and speed recovery by establishing target recovery times and agreed minimum service levels in the event that identified hazards materialize.

The development of optimal recovery strategies and system response can largely benefit from lessons learned through past experience and is therefore closely linked to the reflective capacity principle.

CLIMATE RISKS ADDRESSED:
Service continuity risk

CCA3.9.

Develop local skills and capabilities

This criterion complements the core resilience principles by recognizing the value of human capital for long-term project resilience. It refers to the need to invest in local workforce skill development and capacity building through training programs that address specific skill and capability gaps relevant to long-term resilience targets. The operation of infrastructure projects may involve new or uncommon methodologies and technologies, and a lack of a skilled workforce may pose significant risks to the project.[104] These risks can be exacerbated during and after climate-related disruption events, potentially hindering the project's ability to respond and jeopardizing the health and safety of personnel. As climate risks evolve, training should be ongoing to incorporate new knowledge on risk and vulnerability, as well as up-to-date risk management best practices. "Skills and capacity building within a workforce can create systemic change that carries over to future projects"[105] and the industry as a whole. This principle also applies to the project's supply chain. Training programs can be targeted to economically depressed, underemployed, or disadvantaged communities. Education and training initiatives may also extend beyond the workforce to build broader community awareness, as well as climate and resilience literacy.

CLIMATE RISKS ADDRESSED:
Resource availability risk (workforce)

CCA3.10.

Implement strategies to maximize inclusivity

This core resilience principle aims to foster active engagement, involvement, and participation of people and communities in the resilience planning and development of an infrastructure project. Inclusivity refers to the establishment of shared actions, responsibilities, and ownership in climate risk management. Consultation with the community is important to understand and determine the acceptable levels of risk, expectations for service base levels, and to align the project's goals with existing broader community resilience objectives. Projects should support the achievement of these goals and may even help initiate processes for developing community resilience goals if they do not already exist. Knowledge sharing on climate risks with local communities can help build better coordination in responses, build resilience literacy, and drive community behavioral change—for example, reducing demand during disruptive events. Project users and "communities can bring local experience and native expertise into infrastructure planning and development."[106]

In parallel, inclusivity involves engaging a diverse, cross-sectoral group of stakeholders with varied backgrounds and skill sets. It is particularly important when dealing with multidisciplinary issues such as climate change, disaster risk reduction, or emergency response. Technical cooperation can add value by bringing attention to threats and vulnerabilities that might otherwise be overlooked, harmonizing and standardizing risk data, and providing more accurate resilience models. Shared responsibility and the exchange of information and expertise enable coordinated responses to shared hazards and vulnerabilities.[107,108]

When implemented, inclusivity strategies can act as a key enabler for maximizing the potential of the other resilience strategies described in the criteria. More importantly, stakeholder engagement and participation are recognized as a cross-cutting principle that should govern all sets of criteria and will be highlighted as such in the integrated set of criteria.

OVERVIEW OF HOW RESILIENCE STRATEGIES ADDRESS ALL CLIMATE RISKS AND IMPLEMENT ALL STEPS OF THE RESILIENCE HIERARCHY

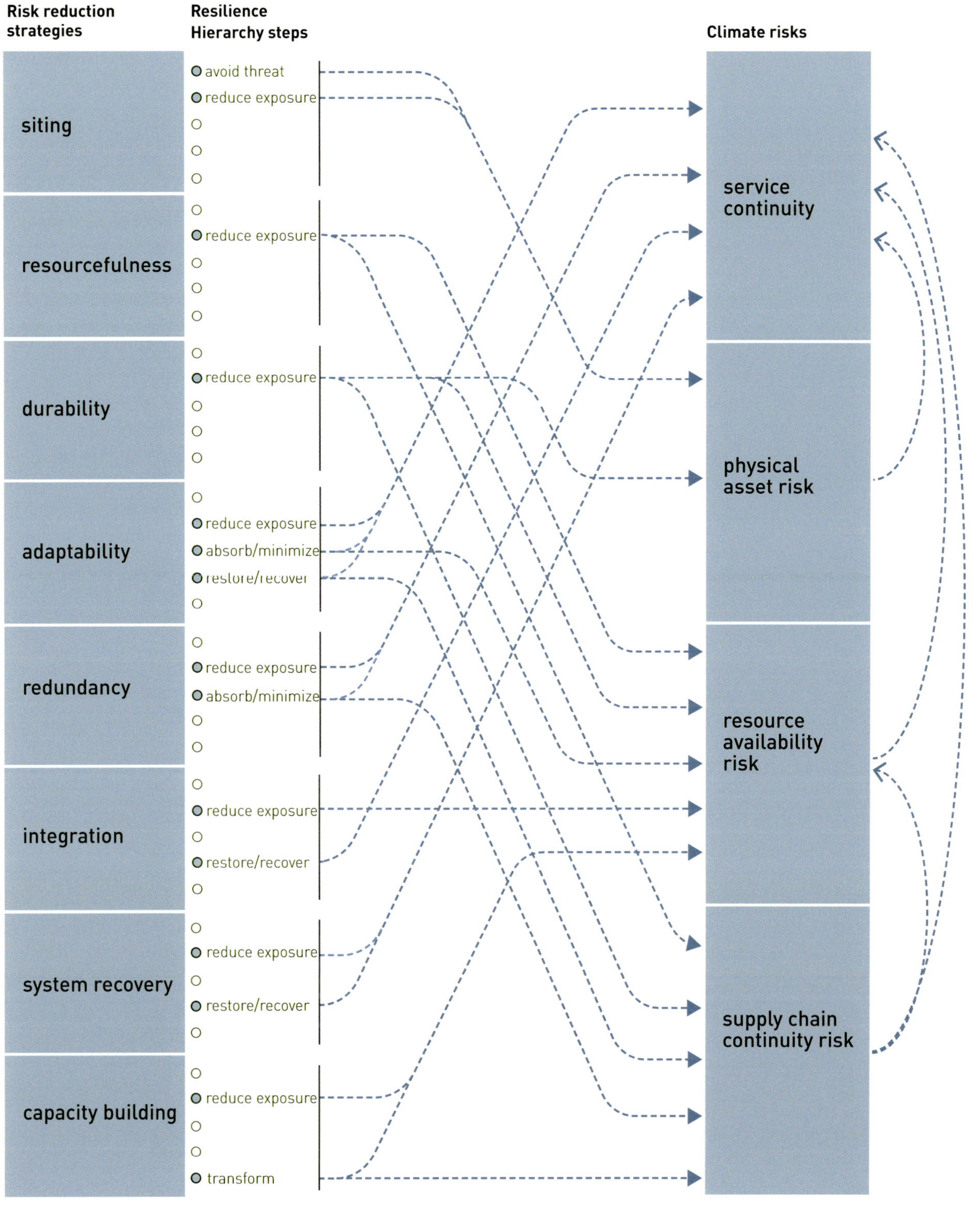

Track progress against targets set

4. TRACK

CCA4.1.

Implement strategies to promote reflective capacity

"Adaptation is a process rather than an endpoint."[109] The goal of resilience strategies is to enable, rather than hinder, this ongoing process. Risk—and its measurement, perception, and understanding—evolves over time. Future systems may be subject to requirements that differ from those applied today. Reflective projects "are accepting of the inherent and ever-increasing uncertainty and change in today's conditions,"[110] which is particularly relevant for long-lived infrastructure. Such projects have tools and methods in place to continuously evolve through plan-do-check-act systems, revisit plans, and "modify standards or norms based on emerging evidence, rather than seeking permanent solutions."[111] "Small-scale failures or near-failure events should be analyzed as learning opportunities and used to revise management strategies."[112] It should be expected that operational best practices will change as new evidence becomes available.[113]

Due to deep uncertainty, dynamic complexity, and the long timeframe associated with climate change, robust adaptation efforts require iterative risk management strategies.[114] Climate-related risks and opportunities are highly dynamic, depending on changes in complex climate systems as well as economic and societal contexts. The characteristics of climate-related risks and opportunities for projects are likely to change over time, whether gradually or abruptly. Updating risk and opportunity identification systems and refining or reformulating climate-related strategies and targets is therefore critical. Incorporating new learning into strategies and ambitions is a key means of effective and efficient management.

"Progress toward resilience in the context of changing climate extremes can benefit from questioning assumptions and paradigms and stimulating innovation to encourage new patterns of response."[115]

Examples of strategies include:[116]

- Developing a robust plan-do-check methodology to identify priorities, evaluate progress, and make adjustments to continually improve performance
- Creating an operations and management plan that establishes plan-do-check-act systems to learn and continually improve resilience capabilities
- Integration data or monitoring systems with reporting or preparedness systems to learn and enhance performance over time
- Incorporating equipment or software in the design for detailed or real-time monitoring of all primary project functions
- Developing a plan to use monitoring data to improve water quality and efficiency, reduce leakage, and/ or conserve water
- Establishing a schedule for future re-evaluation and modification of the maintenance plan based on monitored data

This criterion aligns with the IPCC's statement that "resilience is a positive attribute when it maintains capacity for adaptation, learning and/or transformation."[117]

It is important to highlight that the core resilience principles are inherently interconnected. Durability facilitates resourcefulness, particularly efficiency in material use, enabling the downsizing of structures while maintaining structural performance through high-quality materials. Integration enables redundancy, eliminating the risk of oversized systems or duplicated components and minimizing resource input requirements. Redundancy, in turn, supports adaptability by reducing over-reliance on a single asset. Finally, reflective capacity and inclusivity act as enablers for the rest of the core principles through the generation of new knowledge and shared actions and responsibilities.

The high interdependence of the resilience principles is evident in their overlapping or shared strategies. Many of these overlaps relate to the extension of a project's useful life, an objective addressed collectively by all principles. Extending the useful life of assets translates into longer maintenance or replacement cycles, which significantly affects GHG emissions, resource consumption, and costs.

MITIGATION AND ADAPTATION CRITERIA OVERLAPS

Climate action requires the use of both climate change mitigation and adaptation criteria. The Paris Agreement recognizes adaptation as a key component of the long-term global response to climate change, establishing the goal of "enhancing adaptive capacity, strengthening resilience, and reducing vulnerability to climate change,"[118] to ensure an adequate adaptation response in the context of global warming of 1.5°C or 2°C. Accordingly, a climate action project must implement strategies to: (1) address the drivers of climate change (mitigation) and, in parallel, (2) address potential future climate risks (adaptation). This twofold action does not necessarily mean doubling efforts; rather mitigation and adaptation actions should converge. Overlaps between GHG reduction strategies and resilience strategies create opportunities for projects to simultaneously tackle drivers of climate change and climate risks, complementing each other to maximize outcomes.

Two characteristic examples of overlapping strategies are resourcefulness (an adaptation criterion) and resource circularity (a mitigation criterion). Resource efficiency and resource diversification are core components of both concepts. Additionally, innovation and collaboration are essential for maximizing the outcomes of both approaches. Resource circularity is a more inclusive concept that encompasses various resilience principles, such as durability and adaptability. Thus, by implementing a strategy for reducing Scope 3 upstream and downstream emissions, a project can enhance its resilience against risks related to physical assets, resource availability, and supply chain continuity.

OVERLAPS BETWEEN THE RESOURCEFULNESS AND RESOURCE CIRCULARITY CONCEPTS

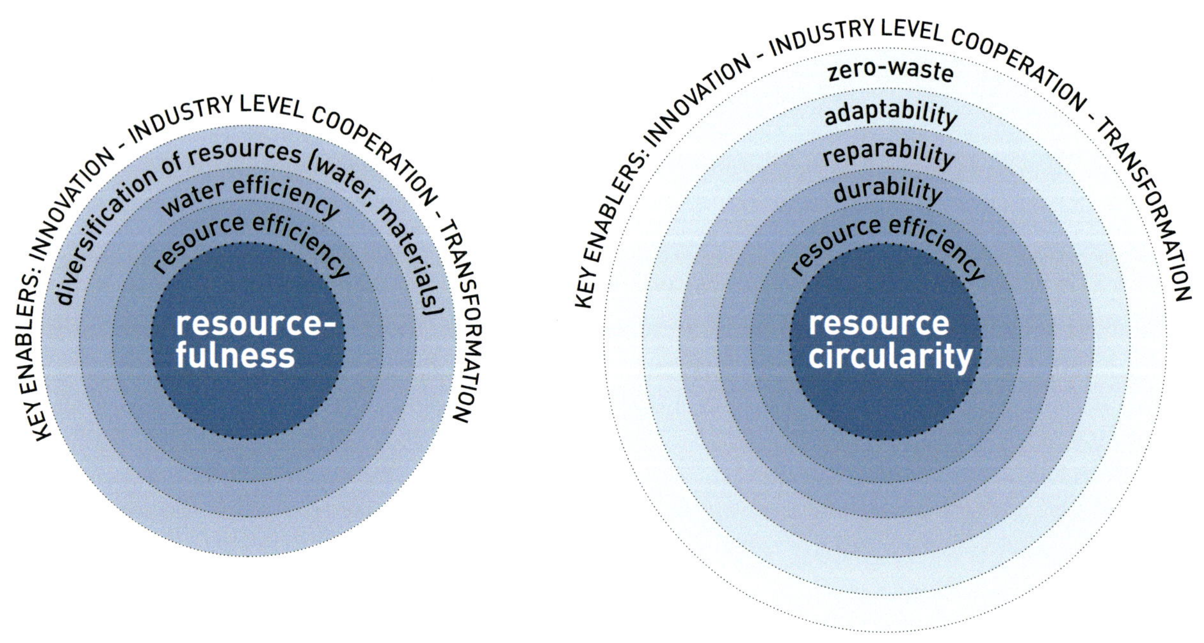

Where the two concepts overlap:

- Resource efficiency (materials)
- Diversification of resources
- Innovation
- Industry-level cooperation
- Transformation

The resilience principles play a critical role within climate change action criteria. Concepts such as resourcefulness, durability, adaptability, and integration can be translated into GHG reduction potential, thereby complementing mitigation strategies. The diagram below highlights how GHG emissions can be addressed through both GHG reduction strategies and resilience strategies.

However, it is important to balance the resilience principles based on project needs and goals to deliver optimal outcomes. For example, if an adaptability strategy aims to upcycle end-of-life project components and materials, it may require the use of additional materials to maintain usability. Consequently, designing for material reuse and upcycling could conflict with objectives aimed at reducing material intensity. Projects should balance their overall climate mitigation and adaptation objectives, explore various combinations of strategies, and select the best-fit set of responses for both climate change mitigation and adaptation.

GHG reduction strategies, beyond their high-level contribution to reducing the need for greater adaptation efforts, can also directly reduce project vulnerability to climate-related risks.

GHG EMISSIONS ADDRESSED BY BOTH GHG REDUCTION STRATEGIES AND RESILIENCE STRATEGIES

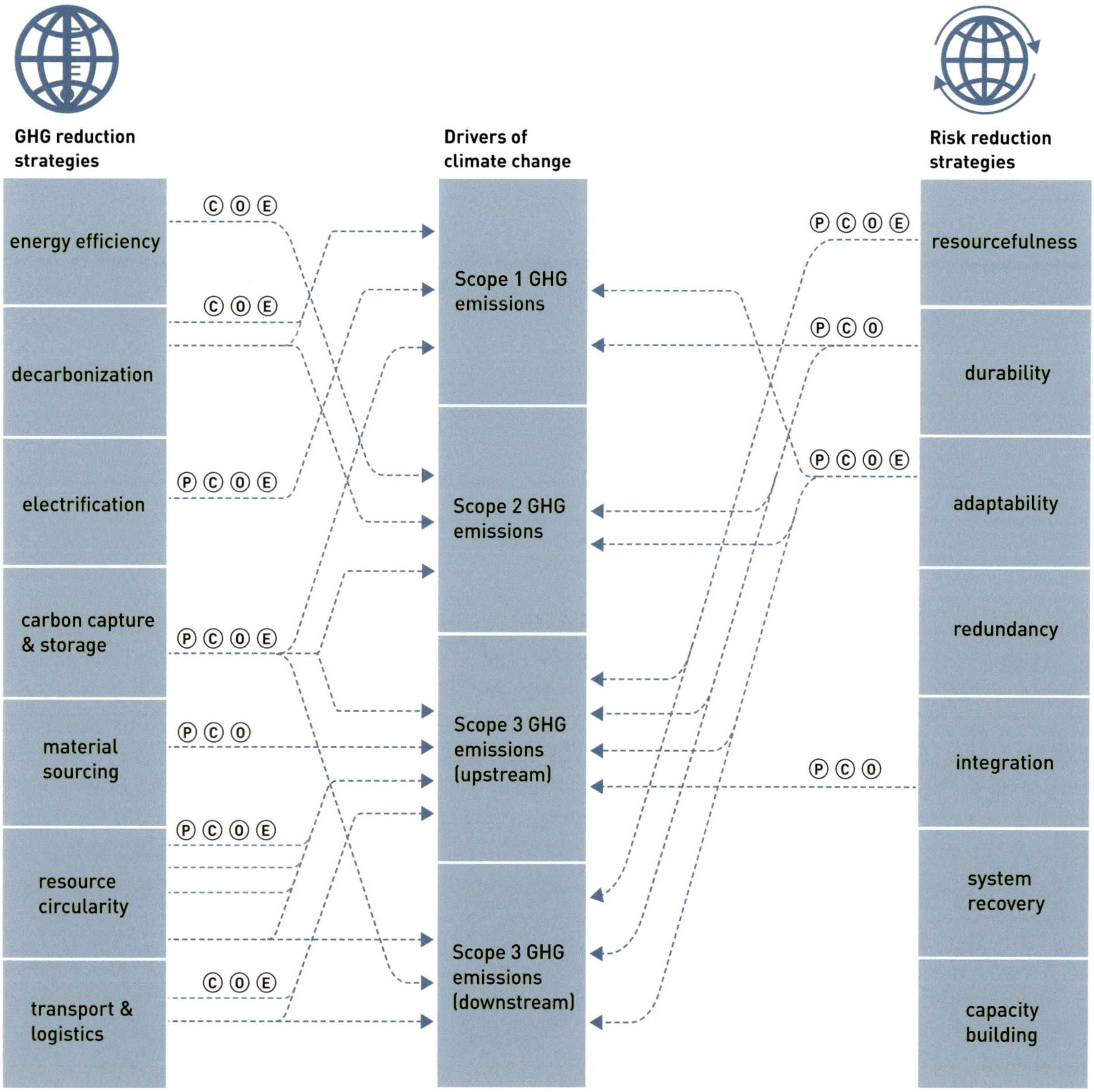

Opportunity:
Joint benefits of core resilience principles for climate mitigation and adaptation

CLIMATE RISKS ADDRESSED BY BOTH RISK REDUCTION AND GHG REDUCTION STRATEGIES

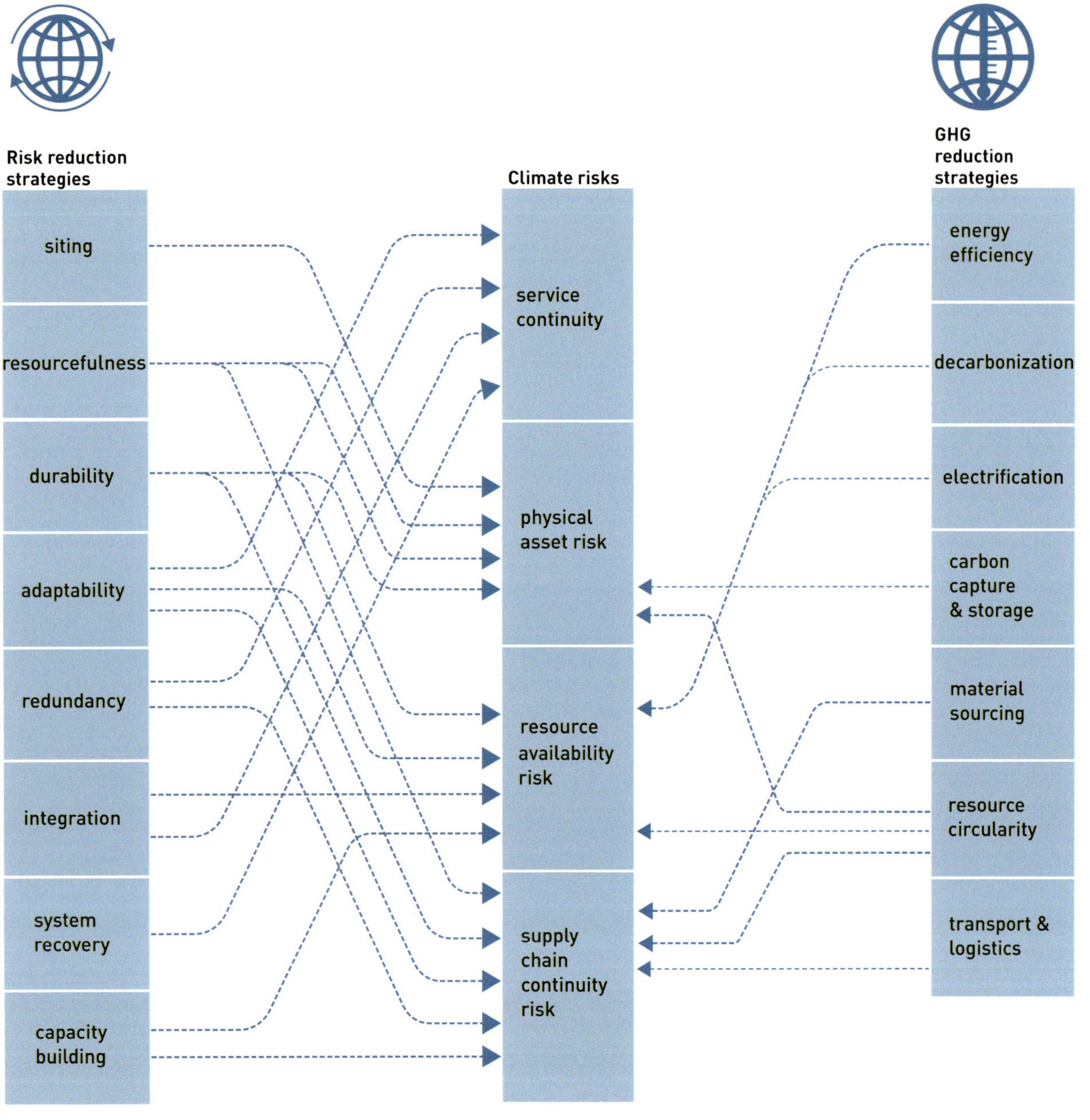

Opportunity:
Climate mitigation strategies contributing to adaptation (resilience)

For example, energy efficiency, decarbonization, and electrification strategies reduce dependency on fuels, decreasing vulnerability to future fuel shortages. Carbon storage and capture strategies through onsite natural sinks can act as buffers, mitigating the effects of potential extreme events. Sustainable material sourcing can reduce vulnerability to climate-related reputational risks that may impact future supply chain continuity.

Finally, maximizing redundancy illustrates how GHG reduction and resilience strategies can work together. Redundancy entails the risk of adding or oversizing project components, which can increase resource use and emissions. If GHG reduction strategies are considered when designing redundancy, this risk can be minimized or eliminated, resulting in more positive outcomes.

OPPORTUNITIES WHEN COMBINING GHG REDUCTION STRATEGIES WITH REDUNDANCY STRATEGIES

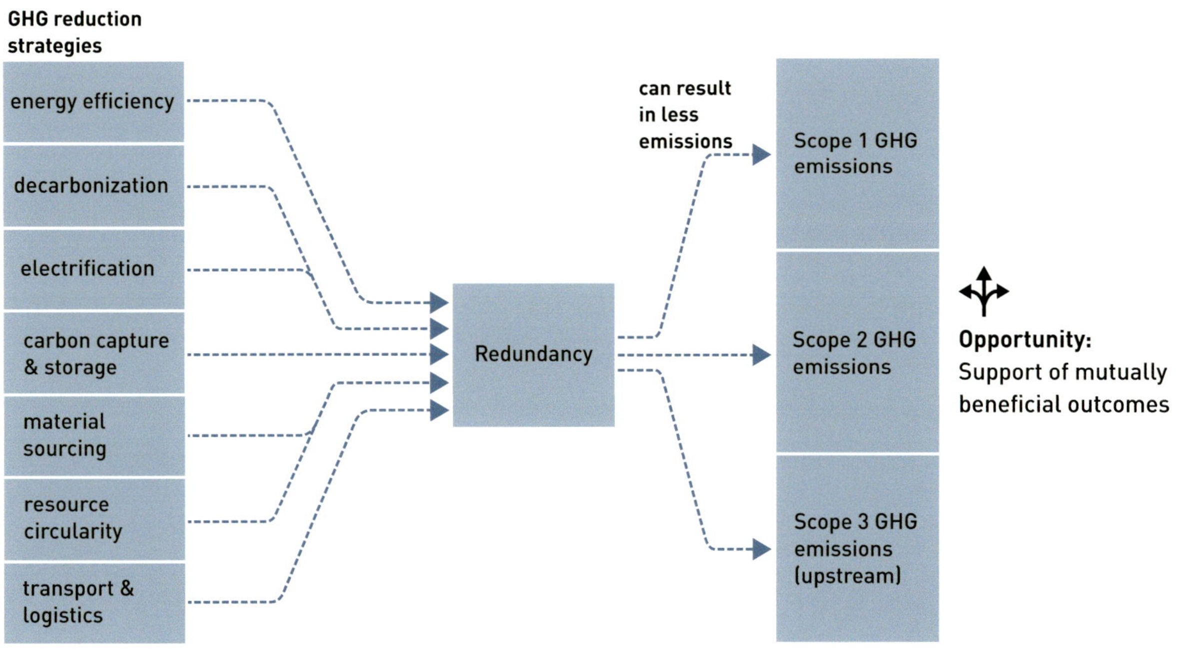

NOTES

1. "IPCC Glossary Search," Intergovernmental Panel on Climate Change (IPCC), accessed November 12, 2025, https://apps.ipcc.ch/glossary/.
2. Ibid.
3. "What Is Climate Change Adaptation and Why Is It Crucial?" United Nations Development Programme (UNDP), January 2024, https://climatepromise.undp.org/news-and-stories/what-climate-change-adaptation-and-why-it-crucial.
4. IPCC, *Climate Change 2007: Impacts, Adaptation and Vulnerability*. Contribution of Working Group II to the Fourth Assessment Report of the Intergovernmental Panel on Climate Change, ed. M.L. Parry et al., (Cambridge University Press, 2007), 748.
5. United Nations, *Paris Agreement*, December 12, 2015, in *United Nations Treaty Collection*, 9, https://unfccc.int/sites/default/files/english_paris_agreement.pdf
6. United Nations Framework Convention on Climate Change (UNFCCC), *Matters relating to adaptation: Glasgow–Sharm el-Sheikh work programme on the global goal on adaptation referred to in decision 7/CMA.3* (December 2023).
7. IPCC, Climate Change 2007, 747.
8. World Business Council for Sustainable Development (WBCSD) and World Resources Institute (WRI), *The Greenhouse Protocol: A Corporate Accounting and Reporting Standard* (WBCSD, WRI, 2004), 5-7, http://pdf.wri.org/ghg_protocol_2004.pdf.
9. Ibid., 25.
10. World Green Building Council (WGBC), *Bringing Embodied Carbon Upfront: Coordinated Action for the Building and Construction Sector to Tackle Embodied Carbon* (WGBC, 2019).
11. IPCC, *Glossary*.
12. UK National Infrastructure Commission, *Anticipate, React, Recover: Resilient Infrastructure Systems* (May 2020), 13, https://nic.org.uk/app/uploads/Anticipate-React-Recover-28-May-2020.pdf.
13. The Rockefeller Foundation and ARUP. *City Resilience Framework* (April 2014; updated December, 2015), 5, https://www.rockefellerfoundation.org/wp-content/uploads/City-Resilience-Framework-2015.pdf.
14. Institute for Sustainable Infrastructure (ISI), *Envision: Sustainable Infrastructure Framework Version 3* (ISI, 2018), 179.
15. United Nations Office for Disaster Risk Reduction, *Principles for Resilient Infrastructure* (UNDRR, 2022), 23, https://www.undrr.org/publication/principles-resilient-infrastructure.
16. IPCC, Glossary.
17. WBCSD and WRI, *Greenhouse Gas (GHG) Protocol: Required Greenhouse Gases in Inventories: Accounting and Reporting Standard Amendment* (WBCSD and WRI, 2013), 5.
18. WBCSD and WRI, *The Greenhouse Gas Protocol: A Corporate Accounting and Reporting Standard* (2004), 25.
19. Adapted from WBCSD and WRI, *The Greenhouse Gas Protocol: A Corporate Accounting and Reporting Standard* (2004), 25.
20. Ibid.
21. Ibid.
22. GHG Protocol, *Scope 3 Frequently Asked Questions* (June 2022), https://ghgprotocol.org/sites/default/files/2022-12/Scope%203%20Detailed%20FAQ.pdf.

23. WGBC, *Bringing embodied carbon upfront* (2019), 7.
24. WBCSD and Net Zero Initiative, *Guidance on Avoided Emissions: Helping business drive innovations and scale solutions toward Net Zero* (March 2023), https://www.wbcsd.org/wp-content/uploads/2023/09/Climate-Avoided-Emissions-guidance_WBCSD.pdf.
25. Ibid.
26. United Nations, *Paris Agreement*, 3.
27. IPCC, *Climate Change 2022: Impacts, Adaptation and Vulnerability.* Contribution of Working Group II to the Sixth Assessment Report of the Intergovernmental Panel on Climate Change, ed. H.O. Pörtner et al. (Cambridge University Press, 2022), 2484, https://doi.org/10.1017/9781009325844. A temperature overshoot is "exceedance of a specified global warming level followed by a decline to or below that level during a specified period." Pathways with "no or limited overshoot" are exceeding 1.5°C by no more than 0.1°C.
28. "The Evidence Is Clear: The Time for Action Is Now. We Can Halve Emissions by 2030," IPCC press release, April 4, 2022, https://www.ipcc.ch/2022/04/04/ipcc-ar6-wgiii-pressrelease/.
29. IPCC, *Glossary*.
30. Ibid.
31. IPCC, *Annex I: Glossary*, ed. R. van Diemen. In *Climate Change and Land: An IPCC Special Report on Climate Change, Desertification, Land Degradation, Sustainable Land Management, Food Security, and Greenhouse Gas Fluxes in Terrestrial Ecosystems*, eds. P. R. Shukla et al., 2019, https://doi.org/10.1017/9781009157988.010.
32. Andy Reisinger et al, "Science-Based Targets Miss the Mark," *Communications Earth & Environment* 5, no 383 (July 23, 2024), https://doi.org/10.1038/s43247-024-01535-z.
33. Ibid.
34. "What are Science-Based Targets?," Science-based Targets Initiative, accessed October 30, 2024, https://sciencebasedtargets.org/how-it-works.
35. Institute of Environmental Management and Assessment (IEMA), *Pathways to Net Zero: Using the IEMA GHG Management Hierarchy* (IEMA, 2020), 4, https://www.isepglobal.org/media/bn3lgz1c/iema_ghg_hierarchy_nov_2020.pdf.
36. International Federation of Consulting Engineers, *Infrastructure and Climate Change: Sustainable Infrastructure Series – A Global Leadership Forum discussion paper* (FIDIC, 2023), 13, https://issuu.com/fidic/docs/infrastructure_and_climate_change_july_2023_?fr=xKAE9_zU1NQ.
37. IEMA, *Pathways to Net Zero* (November 2020).
38. Task Force on Climate-related Financial Disclosures (TCFD), *Technical Supplement: The Use of Scenario Analysis in Disclosure of Climate-Related Risks and Opportunities* (June 2017), 22-23, https://assets.bbhub.io/company/sites/60/2020/10/FINAL-TCFD-Technical-Supplement-062917.pdf.
39. Sarah Ladislaw and Stephen J. Naimoli, "Climate Solutions Series: Deep Decarbonization Pathways," *Center for Strategic and International Studies (CSIS)*, March 10, 2020, https://www.csis.org/analysis/climate-solutions-series-deep-decarbonization-pathways.
40. "Energy Efficiency," International Energy Agency (IEA), accessed November 21, 2023, https://www.iea.org/energy-system/energy-efficiency-and-demand/energy-efficiency.
41. United States Environmental Protection Agency (US EPA) and Green Power Partnership, *Offsets and RECs: What's the difference?* (February 2018), 3, https://www.epa.gov/sites/default/files/2018-03/documents/gpp_guide_recs_offsets.pdf.
42. Ibid., 4.
43. "Electrification," IEA, accessed October 30, 2024, https://www.iea.org/energy-system/electricity/electrification.
44. Ibid.
45. Kelly Levin et al., "What Does 'Net-Zero Emissions' Mean?" *World Resources Institute*, March 20, 2023, https://www.wri.org/insights/net-zero-ghg-emissions-questions-answered#.
46. Intergovernmental Panel on Climate Change (IPCC), *Global Warming of 1.5°C: An IPCC Special Report on the Impacts of Global Warming of 1.5°C above Pre-industrial Levels and Related Global Greenhouse Gas Emission Pathways, in the Context of Strengthening the Global Response to the Threat of Climate Change, Sustainable Development, and Efforts to Eradicate Poverty*, ed. V. Masson-Delmotte et al. (Cambridge University Press, 2018), https://doi.org/10.1017/9781009157940.
47. US EPA and Green Power Partnership, *Offsets and RECs: What's the Difference?*, 3.
48. Ibid., 3.
49. Ibid.
50. WBCSD, *Circular transition Indicators V4* (May 2023), 14, https://www.wbcsd.org/wp-content/uploads/2023/09/Circular_Transition_Indicators_v4.pdf.
51. Ibid., 16.
52. Ibid.
53. Ibid., 14.
54. Ibid., 76.
55. United Nations Environment Programme, *The Long View: Exploring Product Lifetime Extension* (2017): 14, https://wedocs.unep.org/20.500.11822/22394.
56. WBCSD, *Circular Transition Indicators V4* (May 2023): 19.
57. Ibid., 6.
58. Adapted from WBCSD, *Circular Transition Indicators V4* (May 2023).
59. Matt Alderton, "Could carbon fiber be the superhero of composite building materials?," *Autodesk*, December 10, 2021, https://www.autodesk.com/design-make/articles/carbon-fiber-building?
60. Tereza Pultarova, "Pollutant or Resource? [Carbon Dioxide]", *Engineering and Technology* 14, no 2 (March 2019), https://doi.org/10.1049/et.2019.0204.
61. ISI, *Envision: Sustainable Infrastructure Framework Version 3* (2018), 107.
62. World Economic Forum, *The Global Risks Report 2025: Insight Report* (WEF, 2025), 67.
63. IPCC, *Managing the Risks of Extreme Events and Disasters to Advance Climate Change Adaptation. A Special Report of Working Groups I and II of the Intergovernmental Panel on Climate Change*, ed. C.B. Field et al. (Cambridge University Press, 2012), 10.

64. Ibid., 6.
65. UNDRR, *Sendai Framework for Disaster Risk Reduction 2015-2030* (2015), 36, https://www.undrr.org/media/16176/.
66. IPCC, *Managing the Risks of Extreme Events and Disasters to Advance Climate Change Adaptation* (2012), 10.
67. IPCC, *Glossary.*
68. ISI, *Envision: Sustainable Infrastructure Framework Version 3* (2018), 170.
69. Ibid., 170.
70. IPCC, *Climate Change 2022: Impacts, Adaptation and Vulnerability*, 5.
71. Ibid., 253.
72. TCFD, *Recommendations of the Task Force on Climate-related Financial Disclosures: Final Report* (June 2017), 6, https://assets.bbhub.io/company/sites/60/2021/10/FINAL-2017-TCFD-Report.pdf.
73. World Water Council. *International High Level Panel on Water announces outcomes to alert the world to the severity of global water issues* (March 2018), https://www.worldwatercouncil.org/sites/default/files/2018-03/International_High_Level_Panel_on_Water_announces_outcomes_to_alert_the_world_to_the_severity_of_global_water_issues.pdf.
74. The CEO Water Mandate, *Driving Harmonization of Water Stress, Scarcity, and Risk Terminology: Discussion Paper* (January 2014), https://ceowatermandate.org/files/Driving_Harmonization_of_Water_Terminology_draft.pdf.
75. TCFD, *Recommendations of the Task Force on Climate-related Financial Disclosures: Final Report* (June 2017), 6.
76. Ibid.
77. United Nations, *Paris Agreement*, 9.
78. Wubet, Selamawit, "Adapting to the Inevitable: The Global Goal on Adaptation and the Urgency of Action," *Global Center on Adaptation*, July 9, 2024, accessed October 30, 2024, https://gca.org/adapting-to-the-inevitable-the-global-goal-on-adaptation-and-the-urgency-of-action/.
79. UNDRR, *Principles for Resilient Infrastructure* (2022).
80. Ibid., 16.
81. UK National Infrastructure Commission, *Anticipate, React, Recover: Resilient Infrastructure Systems*, 13.
82. Ibid.
83. Ibid.
84. Infrastructure Australia, *Resilience Principles: Infrastructure Australia's approach to resilience* (June 2022), https://www.infrastructureaustralia.gov.au/sites/default/files/2022-06/Resilience%20Principles%20-%20Infrastructure%20Australia%27s%20approach%20to%20resilience.pdf.
85. UK National Infrastructure Commission, *Anticipate, React, Recover: Resilient Infrastructure Systems* (May 2020), 13.
86. Organisation for Economic Co-operation and Development (OECD), *Resilience Strategies and Approaches to Contain Systemic Threats* (September 2019), 3, https://one.oecd.org/document/SG/NAEC(2019)5/en/pdf.
87. Ibid.
88. ISI, *Envision: Sustainable Infrastructure Framework Version 3* (2018), 168.
89. UNDRR, *Principles for Resilient Infrastructure* (2022), 39.
90. Ibid., 42.
91. The Rockefeller Foundation and ARUP, *City Resilience Framework*, 5.
92. Ibid.
93. "Water in Circular Economy and Resilience (WICER)," World Bank Group, September 15, 2021, https://www.worldbank.org/en/topic/water/publication/wicer.
94. The Rockefeller Foundation and ARUP, *City Resilience Framework*, 5.
95. Ibid.
96. Ibid.
97. UNDRR, *How to make infrastructure resilient: The Handbook for Implementing the Principles for Resilient Infrastructure* (UNDRR, 2023), 29, https://www.undrr.org/media/87213/download?startDownload=20251205.
98. UNDRR, *Principles for Resilient Infrastructure* (2022), 36.
99. National Academies of Sciences, Engineering, and Medicine, *NCHRP Report 750: Strategic Issues Facing Transportation, Volume 7: Preservation, Maintenance, and Renewal of Highway Infrastructure* (The National Academies Press, 2020), https://doi.org/10.17226/25757.
100. UNDRR, *Principles for Resilient Infrastructure* (2022), 55.
101. The Rockefeller Foundation and ARUP, *City Resilience Framework*, 5.
102. Ibid.
103. Ibid.
104. ISI, *Envision: Sustainable Infrastructure Framework Version 3* (2018), 82.
105. Ibid.
106. UNDRR, *Principles for Resilient Infrastructure* (2022), 45.
107. The Rockefeller Foundation and ARUP, *City Resilience Framework*, 5.
108. UNDRR, *Principles for Resilient Infrastructure* (2022).
109. K. Vincent, *Climate Adaptation and Its Measurement: Challenges and Opportunities* (REAL, 2024), 6.
110. The Rockefeller Foundation and ARUP, *City Resilience Framework*, 5.
111. Ibid.
112. UNDRR, *Principles for Resilient Infrastructure* (2022), 58.
113. Ibid.
114. IPCC, *Managing the Risks of Extreme Events and Disasters to Advance Climate Change Adaptation*, 27-8.
115. Ibid., 20.
116. ISI, *Envision: Sustainable Infrastructure Framework Version 3* (2018), 178-9.
117. IPCC, *Glossary.*
118. United Nations, *Paris Agreement*, 9.

Biodiversity

Action Criteria

Chapter 5

Biodiversity Action Criteria

comprise one set of criteria for framing action in infrastructure projects against biodiversity loss and towards net biodiversity gain.

KEY CONCEPTS AND APPROACHES SHAPING THE BIODIVERSITY ACTION CRITERIA

BIODIVERSITY ACTION CRITERIA

The Methodology chapter outlines the common characteristics, structural components, overarching concepts, and cross-cutting principles that shape the climate and biodiversity criteria. This section aims to further explain the distinctive concepts specific to the biodiversity action criteria, which required a more tailored approach compared to the climate criteria.

The Convention on Biological Diversity (CBD) defines biodiversity—a contraction of "biological diversity"—as "the variability among living organisms from all sources including, among other things, terrestrial, marine, and other aquatic ecosystems and the ecological complexes of which they are part; it includes diversity within species, between species, and between ecosystems."[1] Biodiversity is the living component of natural capital and plays a key role in maintaining and supporting ecosystem structure, processes, and functions.[2]

DEPENDENCIES ON BIODIVERSITY OR ECOSYSTEM SERVICES

In the case of climate change mitigation and adaptation criteria, the first action step focuses on project-driven impacts and existing risks that may be exacerbated by the project. These potential impacts are estimated and quantified to establish a baseline for measuring improvements in project performance. In the case of biodiversity, apart from assessing potential or actual impacts (pressures) on biodiversity, the criteria also include an evaluation of the project's and the community's dependencies on biodiversity.

Human activities depend on and impact biodiversity. The Taskforce on Nature-related Financial Disclosures (TNFD) defines dependencies as "aspects of environmental assets and ecosystem services that an organization relies on to function."[3] The concept of dependency is particularly relevant for infrastructure projects, as many project-related activities are highly dependent on biodiversity—for example, maintaining a clean and regular water supply. Moreover, biodiversity and healthy ecosystems play a critical role in enhancing project resilience and reducing the risks and damages posed by the adverse impacts of climate change. "They can serve as natural buffers against extreme climate and weather events."[4] At the same time, ecosystems such as forests, rangelands, croplands, peatlands, and wetlands are important global carbon stores. Their conservation, restoration, and sustainable use are critical to achieving the Paris Agreement climate goals.[5] Consequently, the decline of biodiversity can undermine climate change mitigation and adaptation efforts, preventing climate projects from realizing their intended outcomes. Project and community dependencies and impacts on biodiversity therefore represent both potential future risks that must be addressed and opportunities that should be considered during project planning.

Project and community dependencies on biodiversity are articulated through the concept of ecosystem services, which refer to the flows of ecosystems and the "contributions from biodiversity to people's well-being or quality of life."[6] These provide a starting point for addressing the question: How can biodiversity-related performance be assessed?

Biodiversity can be viewed as the "stock" of ecosystems, while ecosystem services represent the "flows" from ecosystems to human activities.[7] "Biodiversity influences the functioning and productivity of ecosystems, acting as an enabling asset that is essential for and underpins ecosystem services. Greater biodiversity generally results in higher quality, quantity, and resilience of ecosystems and their services."[8] For example, species abundance, diversity, or the presence of key species within a specific ecosystem can help maintain ecosystem functioning and resilience, thereby supporting the continued provision of ecosystem services. Therefore, biodiversity decline or loss is coupled with a decline or loss of ecosystem services.

In other words, there is a connection between the condition of ecosystems and their capacity to deliver services, with the ecosystem's condition often serving as a proxy for this capacity.[9] "In a narrow sense, the sustainability of a specific ecosystem service may refer simply to the biological potential of the ecosystem to maintain its yield" (e.g., crop production or soil carbon storage capacity).[10]

Understanding the concept of ecosystem services is essential for recognizing nature's relationship with climate change mitigation and adaptation, as well as with overall human well-being. The Millennium Ecosystem Assessment, a global initiative established in 1999, sought to assess how changes in ecosystems affect human well-being and helped bring this concept into widespread use.[11]

Ecosystem services—also referred to as nature's contributions to people by the Intergovernmental Science-Policy Platform on Biodiversity and Ecosystem Services (IPBES)—were integral to the CBD's Aichi Biodiversity Targets under the 2011-2020 Strategic Plan and continue to play a role in the post-2020 Kunming-Montreal Global Biodiversity Framework, particularly under Target 11: Restore, Maintain and Enhance Nature's Contributions to People.[12]

"Decision-makers need clear information on how biodiversity underpins these services, the demand for them, the capacity of ecosystems to provide them, and the pressures impairing that capacity."[13]

Various approaches exist for classifying ecosystem services, often embedded within conceptual frameworks that underpin global and national ecosystem assessment and accounting systems. These frameworks aim to systematically record data on the stocks and flows of ecosystems. Unlike standard environmental impact assessments, which focus on the impacts of human actions on the environment, ecosystem assessments place ecosystems at the center of development goals. Conceptual frameworks organize and structure the work required to assess complex ecosystems, social dynamics, and human-environment interactions.

For the biodiversity criteria, the United Nations System of Environmental Economic Accounting—Ecosystem Accounting (SEEA-EA, 2021) classification[14] was selected as the reference list for ecosystem services. It is among the most recently updated ecosystem assessment and accounting systems and is also used by the TNFD in its Recommendations.[15] The reference list provides common labels and descriptions for key ecosystem services.

The UN SEEA categorizes ecosystem services into three main groups:

- Provisioning services: "Representing the contributions to benefits that are extracted or harvested from ecosystems," such as food, fiber, fuel, and water[16]
- Regulating and maintenance services: "Resulting from the ability of ecosystems to regulate biological processes and to influence climate, hydrological and biochemical cycles," such as filtration, purification, regulation.[17]
- Cultural services: "Experiential and intangible services related to the perceived or actual qualities of ecosystems"[18]

The UN SEEA further distinguishes ecosystem services into final and intermediate services. "Final ecosystem services are those in which the user of the service is an economic unit."[19] Intermediate are those "in which the user of the services is an ecosystem asset, and where there is a connection to the supply of final ecosystem services."[20] In the case of regulating and maintenance services, a single ecosystem service may be considered "final or intermediate depending on the context."[21]

The UN SEEA reference list of ecosystem services can be used as a checklist by projects to identify:

- The supply of ecosystem services available to the project and the community, which is location-specific and depends on the ecosystem type (e.g., terrestrial, freshwater, or marine)
- The project's demand for available services necessary to fulfill specific needs, which is activity-specific
- The contributions to well-being of both internal and external project stakeholders, which are location-, culture-, and perception-based

This mapping can highlight potential risks to the project arising from project-driven or pre-existing impacts on biodiversity. It can also inform the selection of project strategies to address these risks and leverage opportunities. Additionally, it helps ensure that the project does not negatively affect or create conflicting demands on communities' dependencies on biodiversity.

Given the climate-biodiversity and infrastructure focus of the criteria, project dependencies on biodiversity will focus only on those ecosystem services that are relevant to both climate and infrastructure projects. The full list of ecosystem services, however, will be used to assess community dependencies, which should be enhanced and not undermined by the project.

LIST OF ECOSYSTEM SERVICES (ADAPTED FROM UN SEEA)

PROVISIONING SERVICES

- Crop provisioning (food, energy crops)
- Wood provisioning
- Water supply: potable water
- Grazed biomass provisioning
- Terrestrial and aquatic plants as energy source
- Water supply: non-potable water for processes
- Livestock provisioning
- Other raw material provisioning
- Water supply: non-potable water for irrigation
- Aquaculture provisioning
- Genetic material

REGULATING AND MAINTENANCE SERVICES

- Global climate regulation
- Landslide mitigation
- Storm mitigation (other than water-related events)
- Rainfall pattern regulation
- Solid waste remediation
- Noise attenuation
- Local climate regulation
- Water purification: (nutrients and other pollutants)
- Pollination
- Air filtration
- Water flow regulation: Baseline & peak flow
- Biological control: Pest control
- Soil quality regulation
- Flood mitigation: Coastal protection
- Biological control: Disease control
- Soil erosion control
- Flood mitigation: River flood mitigation
- Nursery population and habitat maintenance

CULTURAL SERVICES

- Recreation-related
- Education, scientific and research
- Ecosystem and species appreciation
- Visual amenity
- Spiritual, artistic and symbolic

DEFINITIONS OF CLIMATE- AND INFRASTRUCTURE PROJECT-RELEVANT ECOSYSTEM SERVICES (ADAPTED FROM UN SEEA)[22]

REGULATING AND MAINTENANCE SERVICES

Global climate regulation
Reduction of GHG concentrations in the atmosphere through the sequestration and retention (storage) of carbon in ecosystems. (final service)

Rainfall pattern regulation
Maintenance of rainfall patterns at the sub-continental scale through vegetation-driven evapotranspiration. (final or intermediate service)

Local climate regulation
"Regulation of ambient atmospheric conditions—including micro and mesoscale climates"—through evaporative cooling by vegetation and water bodies, as well as shading provided by trees. (final or intermediate service)

Air filtration
"Filtration of airborne pollutants through their deposition, uptake, fixation, and storage" by plants. (final service)

Soil quality regulation
"Decomposition of organic and inorganic materials," contributing to soil fertility and characteristics. (intermediate service)

Soil erosion control
Stabilization provided by vegetation that reduces soil and sediment loss. (final or intermediate service)

Landslide mitigation
Stabilization provided by vegetation against the damaging effects of soil and rock mass movement (wasting). (final ecosystem service)

Solid waste remediation
"Transformation of organic or inorganic substances through the action of micro-organisms, algae, plants, and animals." (final or intermediate service)

Water purification (nutrients and other pollutants)
"Restoration and maintenance of the chemical condition of surface water and groundwater through the breakdown or removal of nutrients and other pollutants." (final or intermediate service)

Water flow regulation: Baseline & peak flow
"Regulation of river flows, groundwater, and lake water levels, through the capacity of ecosystems to absorb, store water and gradually release it during dry seasons or periods via evapotranspiration." This also includes the ability to mitigate the effects of floods and other extreme water-related events. (final or intermediate service)

Flood mitigation: Coastal protection
Protection of seashore from tidal surges provided by linear elements, such as, coral reefs, sand banks, dunes, or mangroves. (final service)

Flood mitigation: River flood mitigation
Provision of structural support and a physical barrier against high water levels through riparian vegetation. (final service)

Storm mitigation (other than water-related events)
Mitigation of the impacts of wind, sand, and other non-water storms through vegetation, including linear elements. (final service)

Noise attenuation
Reduction of noise impacts on people through vegetation, which absorbs, scatters, and partially blocks sound, mitigating its harmful or stressful effects. (final service)

Pollination
"Fertilization of crops by wild pollinators," supporting crop production and maintaining or enhancing the abundance and diversity of other species. (final or intermediate service)

Biological control: Pest control
Prevention or reduction of pest impacts on biomass production or other ecosystem processes. (final or intermediate service)

Nursery population and habitat maintenance
Sustaining species populations either "the maintenance of habitats (e.g., for nurseries or migration or protection of natural gene pools." (intermediate service)

PROVISIONING SERVICES

Crop provisioning (energy crops)
Cultivation of plants "harvested by economic units for various uses," including energy production. (final service)

Wood provisioning
"Growth of trees and other woody biomass in cultivated and uncultivated contexts" for harvesting (timber production and energy). (final service)

Terrestrial and aquatic plants as energy source
Growth of plants e.g., algae for use as energy source. (final service)

Other raw material provisioning
Provisioning of materials such as soil, sand, rock, gravel, and other raw resources.

Water supply: Potable water
"Combined ecosystem contributions of water flow regulation, and purification" to the supply of potable water of appropriate quality to users (final service)

Water supply: Non-potable water for processes
Provision of water for uses less dependent on water quality, such as cooling, where water purification services are not required. (final service)

Water supply: Non-potable water for irrigation
Provision of water for uses that are less dependent on water quality, such as landscape irrigation. (final service)

NATURE-BASED SOLUTIONS (NbS)

As part of the criteria, ecosystem services are also used to illustrate the benefits of nature-based solutions (NbS). Nature-based solutions are increasingly and widely recognized for their significant potential to generate climate-biodiversity co-benefits. The need for NbS is explicitly mentioned in the Global Biodiversity Framework, in the work on the Global Goal on Adaptation, and within the context of the UN SDGs. Therefore, nature-based solutions are a key component of the biodiversity response criteria.

The term nature-based solutions was first introduced during the UNFCCC negotiations in 2009[23] and was formally defined by the International Union for Conservation of Nature (IUCN) in 2016 as "actions to protect, sustainably use, manage and restore natural or modified ecosystems, which address societal challenges, effectively and adaptively, providing human well-being and biodiversity benefits."[24] The overarching goal of NbS is "to support the achievement of society's development goals and safeguard human well-being in ways that reflect cultural and societal values and enhance the resilience of ecosystems, their capacity for renewal and the provision of services; nature-based solutions are designed to address major societal challenges, such as food security, climate change, water security, human health, disaster risk, social and economic development."[25] By definition, NbS are integral to climate change mitigation and adaptation.

The NbS concept builds on ecosystem services and shares common principles with related approaches, including the ecosystem approach, ecosystem-based adaptation and mitigation, and green and blue infrastructure. A growing body of literature supports the view that the assessment of performance of NbS should be ecosystem service-based.

However, concerns have been raised that, without clear criteria, almost any action could be labeled as an NbS. To address this, the IUCN published the Global Standard for Nature-based Solutions in 2020. The Standard identifies eight criteria to clarify what the NbS concept entails and what is required for successful deployment. It defines NbS as solutions that:[26]

- "Effectively address societal challenges"
- Are "sustainable"
- Are "adaptively managed based on evidence"
- Are "economically viable"
- Are "based on inclusive, transparent and empowering governance processes"
- "Equitably balance trade-offs"
- Have a design "informed by scale"
- "Result in a net gain to biodiversity and ecosystem integrity"

The last criterion is particularly important within the criteria, as it connects NbS with net positive targets. Achieving biodiversity net gain means leaving biodiversity "in a measurably better state than before the development took place."[27]

NbS can be implemented independently or integrated with other types of solutions, such as technological or engineering solutions.

In the context of integrated climate-biodiversity action, three main objectives are identified for NbS:

- Need to maintain ecological function and ecosystem services
- Maximize carbon sequestration by natural ecosystems
- Adapt to the impacts of climate change

Nature conservation is embedded within the concept of NbS. Conservation is a top priority and an important adaptation strategy for biodiversity, as "genetically-diverse populations and species-rich ecosystems have a greater potential to adapt to climate change."[28]

Ecosystem restoration is defined as "the process of assisting in (initiating or accelerating) the recovery of an ecosystem that has been degraded, damaged, or destroyed by human activity."[29] However, restoration cannot replace conservation or be used to justify destruction or unsustainable use. While restoration can successfully reestablish biodiversity, structure and function in a degraded ecosystem, it may not always fully restore the original ecosystem's characteristics or functionality.[30]

Ecosystem restoration also enhances the resilience of biodiversity in the face of climate change. For example, "restoration with a variety of native species ensures ecosystem resilience in the face of climate change and has benefits for biodiversity, but also relies on novel species assemblages to match future climatic conditions."[31] Moreover, "measures narrowly focusing on the protection and restoration of biodiversity generally yield important co-benefits for climate change mitigation, though these may be suboptimal compared to measures that account for both biodiversity and climate."[32]

According to the IPBES-IPCC Joint Report, "the term 'rehabilitation' may be more appropriate than 'restoration,' in the context of climate change, where re-establishing the pre-existing conditions may not be possible, but an enhanced state and functions appropriate to shifting conditions is feasible."[33] Increasingly, under climate change, restoration is viewed for carbon sequestration, e.g., rebuilding carbon stocks."[34] However, adaptive management that re-evaluates results through and readjusts management strategies to achieve goals should be integral of any solution.

As part of the criteria NbS are grouped into four main types:[35]

- Solutions that preserve existing ecosystems and, by extension, their supply of ecosystem services

FOUR KEY TYPES OF NATURE-BASED SOLUTIONS

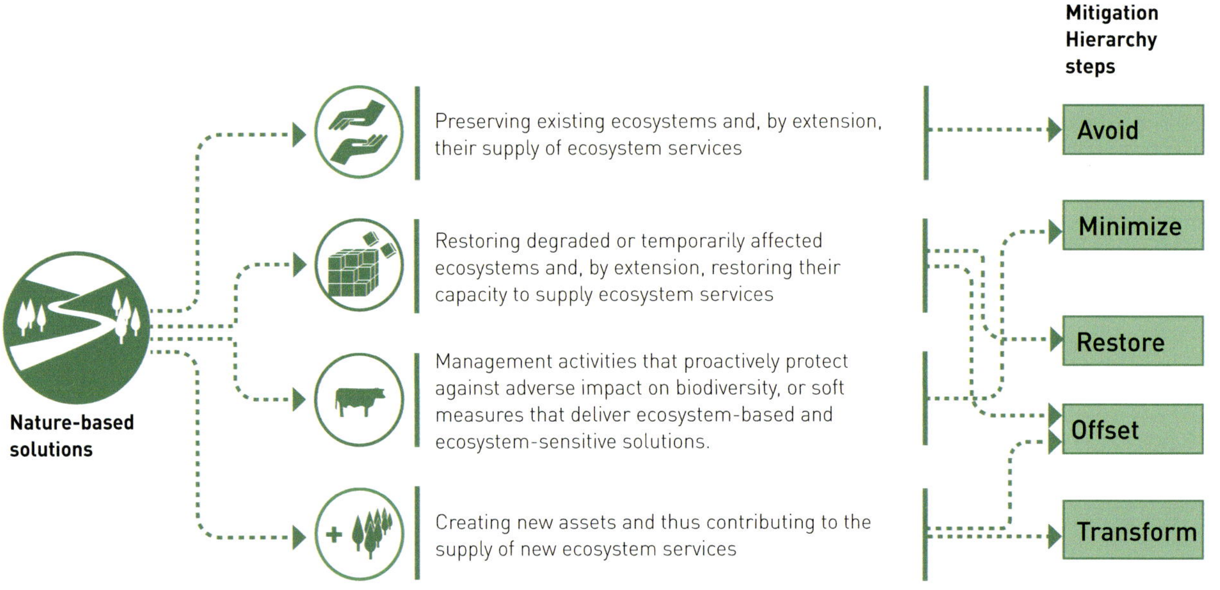

- Solutions that restore degraded or temporarily affected ecosystems and thereby restore their capacity to supply ecosystem services
- Management activities that proactively protect against adverse impacts on biodiversity (ecosystems and species), or soft measures that deliver ecosystem-based and ecosystem-sensitive solutions
- Solutions that create new ecosystem assets, thereby contributing to the supply of new ecosystem services

Net biodiversity gain is more closely associated with the creation of new ecosystem assets, although it can also result from restoration efforts that aim to improve biodiversity relative to its prior state. Overall, the four types of interventions are not mutually exclusive and can be combined.

The biodiversity criteria are designed to provide a flexible approach to guide project teams in selecting strategies appropriate for a specific project type and location. They do not prescribe specific solutions for particular projects, locations, or ecosystem types, nor do they directly link specific NbS solutions to outcomes. Instead, the criteria describe NbS through the full set of ecosystem services—the benefits that ecosystems can potentially contribute to a project or the broader community. This approach, combined with an assessment of project dependencies, allows teams to first identify the ecosystem services they aim to achieve and then determine which nature-based solutions are compatible with the project's specific context and can support the delivery of the required services. Importantly, NbS can deliver multiple services simultaneously.

A key challenge with biodiversity, unlike climate, is that the location of project activities is as critical as the activities themselves for defining project-driven impacts and appropriate management responses, including NbS.

This makes location-specific data essential for identifying best-fit solutions. The geographical context and ecosystem type ultimately determine which types of NbS are applicable, implementable, and compatible with the project's objectives and environmental setting. Moreover, compatibility with the socio-cultural context should be also pursued, as it is a key factor for the success of NbS and depends on how different stakeholder groups value nature and its goods and services. All forms of relevant information, including indigenous and local knowledge and practices, should be considered.

LINKED CRITERIA

The biodiversity action criteria follow the IUCN's framework of linked indicators. The International Union for Conservation of Nature's (IUCN) Guidelines for Planning and Monitoring Corporate Biodiversity Performance illustrate the links between pressures, the state of biodiversity, benefits (ecosystem services/dependencies), and responses. The guidelines employ what they term a framework of linked indicators. "A change in response is expected to lead to a change in pressure which leads to a change in state of biodiversity which provides more benefits to people, encouraging more responses."[36]

When applying this framework using the steps of the mitigation hierarchy for project responses to pressures on biodiversity, it provides a proxy for the state of biodiversity and the associated benefits (ecosystem services) for each type of response, such as avoid, minimize, restore, or offset, as illustrated in the diagram below.

However, it is important to note that the pressure-state-benefit-response framework has its limitations, as reality is far more complex than linear causal relationships. "There are no one-to-one linkages between indirect drivers, direct drivers, biodiversity, and the actions on these drivers needed to halt and reverse biodiversity loss; rather, a 'many-to-many' relationship exists between these elements."[37]

DOING EVERYTHING ELSE BETTER

The uncertainty associated with the high complexity of both biodiversity and climate change—and their interactions—makes the prediction of the outcomes of management responses even more challenging. As noted, in the context of climate change, a key part of biodiversity action is enhancing the persistence of ecosystems so they adapt. In this regard, a key near-term measure is reducing non-climatic pressures on biodiversity caused by habitat conversion, fragmentation, overharvesting, pollution, and invasive alien species. It is what the IPBES-IPCC Report refers to as "doing everything else better,"[38] identifying it as a crucial measure to maximize opportunities for species and ecosystems to adapt to and survive climate change, as well as an important climate change adaptation action.

As part of the biodiversity action criteria, measures addressing other pressures on biodiversity constitute the second set of responses to biodiversity loss that complement nature-based solutions.

These measures comprise various sustainable strategies, including siting, micro-siting and physical measures, as well as prevention, abatement, and control measures, which should be integrated into climate change-focused actions and prioritized and designed in accordance with the mitigation hierarchy.

NATURE-POSITIVE INFRASTRUCTURE

Finally, it is important to clarify the criteria's relationship to "nature-positive," an emerging concept that is increasingly used in recent global, business, and project goal-setting for biodiversity. Leading organizations and high-profile industry initiatives are currently working to operationalize the concept; however, definitions range from conceptual to process-based and target-based approaches.[39]

The criteria adopt a conceptual definition of nature-positive infrastructure from the Playbook for Nature-Positive Infrastructure Development by FIDIC, WWF and AECOM. As the Playbook explains, their definition closely mirrors that of the UK Council's for Sustainable Business (2022): "A nature-positive approach puts nature and biodiversity gain at the heart of decision making and design. It goes beyond reducing and mitigating negative impacts on nature as it is a proactive and restorative approach focused on conservation, regeneration and growth."[40] The Playbook also emphasizes the need to consider climate change, given the interlinkages between biodiversity and climate. In general, the need to integrate climate change is a principle on which most definitions of nature-positive—despite variations in terminology—converge. In short, nature-positive reflects the intended outcomes of integrated climate and biodiversity action.

LINKAGES BETWEEN PROJECT RESPONSES (IN LINE WITH THE MITIGATION HIERARCHY) AND ECOSYSTEM SERVICES

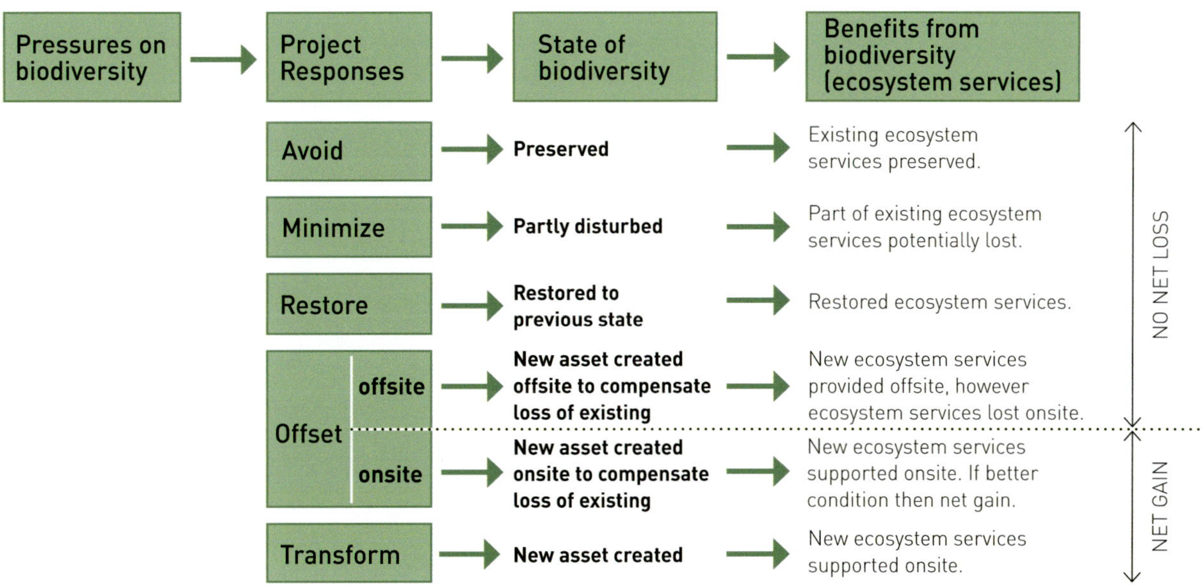

BIODIVERSITY ACTION CRITERIA

The biodiversity criteria comprise 14 criteria structured around four action steps and grouped into seven subsets. They are designed to frame key concepts and highlight important considerations that underpin action to halt and reverse biodiversity loss.

BIO. OVERVIEW OF BIODIVERSITY ACTION CRITERIA

ACTION STEPS	SUBSET OF CRITERIA	CRITERIA
1. ASSESS	**Assess the State of Biodiversity**	BIO1.1. Assess the state of ecosystems BIO1.2. Assess the state of species
	Assess Project-Driven Pressures on Biodiversity	BIO1.3. Assess land/freshwater/sea change BIO1.4. Assess resource exploitation BIO1.5. Assess climate change and its related impacts BIO1.6. Assess pollution (water, air, soil, waste, noise, light) BIO1.7. Assess introduction of invasive species
	Assess Dependencies on Biodiversity	BIO1.8. Assess project's dependency on ecosystem services BIO1.9. Assess community's dependency on ecosystem services
2. COMMIT	**Commit to and Set Ambitious Biodiversity Targets**	BIO2.1. Commit to and set ambitious targets aligned with global policy goals to both halt and reverse biodiversity loss
3. RESPOND	**Prioritize Management Responses Based on the Mitigation Hierarchy**	BIO3.1. Plan for actions to: • Avoid impacts on biodiversity • Minimize impacts on biodiversity • Restore impacts on biodiversity • Offset impacts on biodiversity • Transform performance
	Implement Biodiversity Strategies	BIO3.2. Implement nature-based solutions (NbS) BIO3.3. Use prevention, abatement, and control measures against pressures on biodiversity (other than NbS)
4. TRACK	**Track Progress against Targets Set**	BIO4.1. Monitor pressures and responses and track progress against targets set to verify selection of strategies or adapt them as needed

Assess the state of biodiversity

1. ASSESS

The state of biodiversity refers to the overall quality of its components—genes, species, and ecosystems—and depends on the condition and extent of these components. The present recommended criteria focus specifically on ecosystems and species, as they are more measurable components of biodiversity than genetic diversity.

This assessment establishes the baseline/reference state against which any potential change in the state of biodiversity resulting from project activities will be assessed. This information supports the identification of potential pressures on biodiversity (impact drivers of biodiversity loss) and informs the management of these pressures.

BIO1.1.

Assess the state of ecosystems

An ecosystem is "a dynamic complex of plant, animal and microorganism communities and the non-living environment, interacting as a functional unit."[41] The state of an ecosystem can be measured in terms of extent—the area covered by a particular ecosystem that could potentially be affected by project activities (measured in spatial terms)—and its condition, defined as "the quality of an ecosystem measured by its abiotic and biotic characteristics."[42]

Biotic and abiotic characteristics include:[43]

- "Ecosystem's composition, function, and structure"
- "Landscape characteristics, such as connectivity" to facilitate the movement of organisms and the flow of resources (e.g., nutrients, moisture) between similar habitats
- "Physical and chemical characteristics, such as soil structure and nutrient levels"

Ecosystem condition "underpins the ecological integrity of the ecosystem and supports its capacity to supply ecosystem services on an ongoing basis."[44]

BIO1.2

Assess the state of species

Species are "a group of individuals that actually or potentially interbreed and produce fertile offspring."[45] The state or status of species refers to the diversity of species in an area, species population and abundance, the presence of threatened and endangered species listed on the IUCN Red List of Threatened Species and national conservation lists, and the risk of species extinction.[46] Infrastructure projects can affect species directly, for example through mortality caused by collisions or electrocution on electricity transmission and distribution lines, and indirectly through the loss or degradation of their habitat.

Assess existing and project-driven pressures on biodiversity

1. ASSESS

This subset of criteria uses the Intergovernmental Panel on Biodiversity and Ecosystem Services (IPBES) framework on pressures affecting biodiversity to define potential project-driven impacts that must be addressed. It also includes an assessment of existing (historic) impact drivers of biodiversity loss that the project aims to mitigate. According to IPBES, the main anthropogenic pressures on biodiversity, or direct drivers of biodiversity change, are:[47]

- Land, freshwater, and sea change
- Direct resource exploitation
- Climate change and its related effects
- Pollution
- Introduction of invasive species

Potential impacts arising from these pressures affect both ecosystems and species. Impacts can be measured in terms of extent (e.g., area of ecosystems, species population size) and quality (e.g., ecosystem condition, species abundance and diversity), including:[48]

- "Decrease in habitat cover"
- "Reduction in native vegetation cover"
- "Change in distribution and behavior" of species dependent on the habitat
- "Decrease in population sizes of species"
- "Decrease in the abundance and diversity (or health)" of plant and animal species

The criteria are structured around these pressures, which are particularly relevant to infrastructure projects throughout their lifecycle.

BIO1.3.

Assess land, freshwater and sea change

Land, freshwater, and sea change (extent/area) causes habitat and ecosystem loss, degradation, and fragmentation, which can lead to species extinction and the loss of ecosystem functions and associated ecosystem services.[49] "Land-use change is the leading cause of terrestrial biodiversity loss."[50] Land change resulting from a project should not be understood solely as land clearance, or long-term loss of natural vegetative cover; it also encompasses the introduction of new elements or structures that may act as barriers to species movement (fauna), increase accident risks or fatalities, and lead to greater human intrusion. Similarly, freshwater change refers to modifications of natural systems, such as alterations to surface water flow, which can impact aquatic habitats and the species they support.

BIO1.4.

Assess resource exploitation

This criterion addresses the exploitation of plants and other organisms, as well as the extraction of natural resources such as timber and water. The rate of resource exploitation frequently exceeds the capacity of ecosystems to regenerate, resulting in impacts such as species extinction, genetic drift, and habitat degradation. Similarly, unsustainable water use can result in habitat loss, fragmentation, and degradation, reduce food and water availability for species, and disrupt overall ecosystem functioning.

Projects should assess their estimated resource use and adopt strategies to minimize it. This criterion overlaps with the climate change adaptation criterion on resource availability risk.

BIO1.5.

Assess climate change and its related effects

Climate change is a direct driver of biodiversity decline and is expected to become the leading threat to biodiversity in the coming decades.[51] Climate change and its related effects, such as shifts in temperature, precipitation, and sea level rise, have both direct and indirect impacts on species distribution, physiology, behavior, and habitat modification. For example, ocean acidification resulting from increased atmospheric carbon dioxide absorption has profound implications for marine ecosystems, diminishing their capacity to sequester carbon and creating feedback loops that further exacerbate climate change. As noted, "climate threats interact with one another, and their impacts accumulate; e.g., ocean temperature, acidification and hypoxia interact to produce complex biotic responses."[52]

Beyond its direct pressure on biodiversity, climate change has a "multiplier effect," intensifying the impacts of the other direct drivers of biodiversity decline and loss. It interacts with and increasingly exacerbates non-climatic stressors such as habitat loss, invasive species, pollution, disease, and overexploitation through compounding effects. These effects include habitat degradation, increased disease susceptibility, and altered movement patterns of species that cause ecological damage.

Erratic, extreme, or altered rainfall patterns and storms increase runoff carrying pollutants into water bodies. Rising water temperatures from global warming can enhance the proliferation of harmful microorganisms, negatively impacting aquatic life. Melting ice can release pollutants accumulated over decades, and sea level rise can lead to saltwater intrusion, affecting freshwater supplies. In turn, water pollution reduces the ability of water bodies to act as carbon sinks; for example, eutrophication alters the composition and quality of sediments, affecting the ability to store carbon.

The non-linear effects and complex interrelations between climate change, biodiversity, and non-climatic stressors make it hard to attribute observed changes in biodiversity to specific causes. Climate change vulnerability assessments of species and ecosystems are increasingly applied; however, real-world evaluations of their effectiveness remain limited due to their recent implementation.[53]

BIO1.6.

Assess pollution (water, air, soil, waste, noise, light)

Pollution refers to "contaminants introduced to the environment that cause instability or harm. Pollution can take numerous forms, as a variety of chemical compounds, types of light and sound, and products can cause environmental damage depending on their properties and concentrations."[54] The types of pollutants affecting ecosystems range from human-made chemical products, nutrients, and microplastics to ambient sources, such as noise and artificial light.

The Kunming-Montreal Global Biodiversity Framework (GBF) includes a target to reduce pollution from all sources to levels that are not harmful to biodiversity. Due to the particular harmful impacts of nutrient pollution—such as nitrogen and phosphorus—pesticides, and highly hazardous chemicals, the GBF calls for reducing nutrients by at least 50% by 2030.[55]

Project activities, including the use of fertilizers and pesticides, industrial emissions, byproducts of processes, wastewater, and solid waste, can modify the physical and chemical state of soil, air, and water, leading to degradation of ecosystem quality and threats to plant and animal species. Air pollutants emitted by infrastructure projects may include NOx, SOx, persistent organic pollutants (POPs), volatile organic compounds (VOCs), hazardous air pollutants (HAP), particulate matter (PM), and others.[56]

Development activities can temporarily or permanently disturb land. Construction activities, such as excavation, temporary staging, access for construction equipment, and material storage, can alter the soil's ability to retain water, support infiltration, store carbon, retain nutrients, or sustain healthy plant growth. The most common impact on soil is compaction.

Finally, artificial light emissions and noise pollution resulting from project operations and construction activities can also affect biodiversity by altering species behavior and distribution, including migration and breeding patterns, for example, through disruption of foraging, breeding, or social behaviors.

BIO1.7.

Assess introduction of invasive species

Invasive alien species—plants, animals, or microorganisms introduced to a project site either deliberately or accidentally—pose a major and growing threat to ecosystems, habitats, and native species through their establishment and propagation. "Invasive alien species affect biodiversity by competing with native species for resources, by direct predation or by introducing pathogens. They also modify the composition and structure of ecosystems, reducing the services they provide."[57] One of the Global Biodiversity Framework targets is dedicated to invasive species, urging to: "Eliminate, minimize, reduce and or mitigate the impacts of invasive alien species on biodiversity and ecosystem services by identifying and managing pathways of the introduction of alien species, preventing the introduction and establishment of priority invasive alien species, reducing the rates of introduction and establishment of other known or potential invasive alien species by at least 50 per cent by 2030, and eradicating or controlling invasive alien species, especially in priority sites, such as islands."[58]

Infrastructure project construction or maintenance activities can serve as pathways for the introduction of invasive species.

A 2023 IPBES Assessment Report[59] confirms the current and increasing threat posed by invasive species, which can cause irreversible changes to biodiversity and ecosystems, including local and global species extinctions, biological homogenization, and alterations to ecosystem structure and functioning, such as soil and water characteristics. According to the joint IPCC-IPBES report, "interactions between climate change and invasive species are a particular concern," as invasive species "are projected to benefit from climate change, which accelerates their rates of colonization."[60]

Managing invasive alien species can help to mitigate the negative effects of other drivers of change of biodiversity.

INTERACTIONS BETWEEN CLIMATE CHANGE, ITS ASSOCIATED EFFECTS, AND OTHER PRESSURES ON BIODIVERSITY

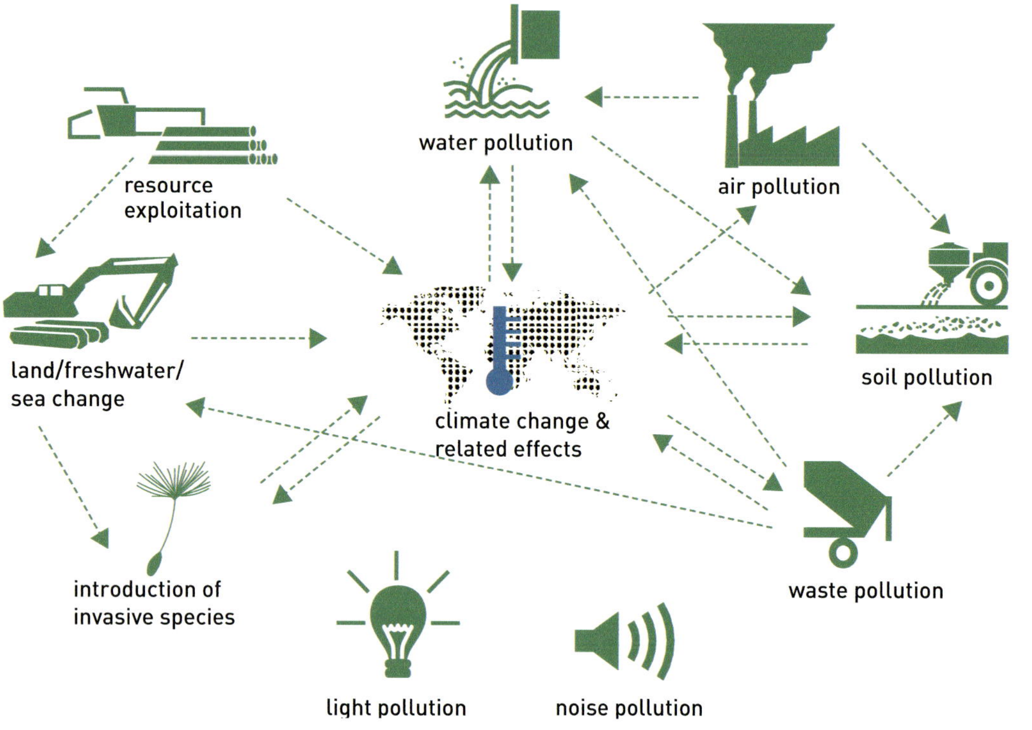

This subset of criteria aims to encourage projects to integrate biodiversity and associated ecosystem services into project decision-making processes to leverage their significant potential.

BIO1.8.

Assess project's dependencies on ecosystem services (climate- and infrastructure-relevant)

Project dependencies on biodiversity are defined as the ecosystem services that a project relies on to function, either as inputs or by enabling, enhancing, or influencing the environmental conditions required for successful project performance.[61] Dependencies may include, for example, water flow and quality regulation, regulation of hazards such as flood control, and carbon sequestration, among others. Impacts on biodiversity can reduce ecosystems' capacity to deliver these services and may pose significant risk to a project that relies on them.

The UN SEEA reference list of ecosystem services is used to define dependencies on biodiversity. Climate change-relevant and infrastructure project-relevant services have been identified to define a project's potential dependencies on biodiversity. These

LIST OF CLIMATE- AND INFRASTRUCTURE-RELATED ECOSYSTEM SERVICES

- Crop provisioning (energy crops)
- Wood provisioning
- Terrestrial and aquatic plants as energy source
- Other raw material provisioning
- Water supply: potable water
- Water supply: non-potable water for processes
- Water supply: non-potable water for irrigation
- Global climate regulation
- Rainfall pattern regulation
- Local climate regulation
- Air filtration
- Soil quality regulation
- Soil erosion control
- Landslide mitigation
- Solid waste remediation
- Water purification: (nutrients and other pollutants)
- Water flow regulation: Baseline & peak flow
- Flood mitigation: Coastal protection
- Flood mitigation: River flood mitigation
- Storm mitigation (other than water-related events)
- Noise attenuation
- Pollination
- Biological control: Pest control
- Nursery population and habitat maintenance

primarily include regulating services, as well as provisioning services. A clear definition and comprehensive classification of ecosystem services are essential for better understanding of what biodiversity can deliver. Projects can use the narrowed-down UN SEEA classification as a checklist to identify:

- The supply of ecosystem services available at the specific project location, which depends on ecosystem type (e.g., terrestrial, freshwater, or marine) and is location-specific
- The project's demand for these services, based on what is planned to be used to fulfill specific project needs (project-activity specific)
- Potential additional ecosystem system services that can be introduced to supplement available services and support the project in achieving its biodiversity targets, with identification of those services essential for biodiversity goals (location- and activity-specific)

BIO1.9.

Assess community's dependencies on ecosystem services

As part of this criterion, a project can use the full UN SEEA reference list of ecosystem services as a checklist to identify:

- Which of the available ecosystem services are used by the community and other external stakeholders, in order to highlight potential unintended trade-offs in the provision of these services due to project-driven impacts or conflicts and pressures on demand
- Which ecosystem services introduced by the project could be used by the community
- Which additional ecosystem services could contribute to the well-being of the community
- Existing or potential project-driven inequalities in the distribution of benefits

This assessment should reflect what people value most about an ecosystem—issues perceived as most important—which can vary among different stakeholders' groups.[62] Achieving this requires the engagement, active dialogue, and meaningful involvement of all affected stakeholders.

In summary, the "Assess" step identifies project-driven pressures on biodiversity that are relevant to the project, as well as existing (historical) pressures that the project aims to address, and that subsequent action steps will seek to reduce across the whole life of the project. It also includes an assessment of the project's and community's dependencies on biodiversity—ecosystem services—that represent both risks and opportunities for project performance:

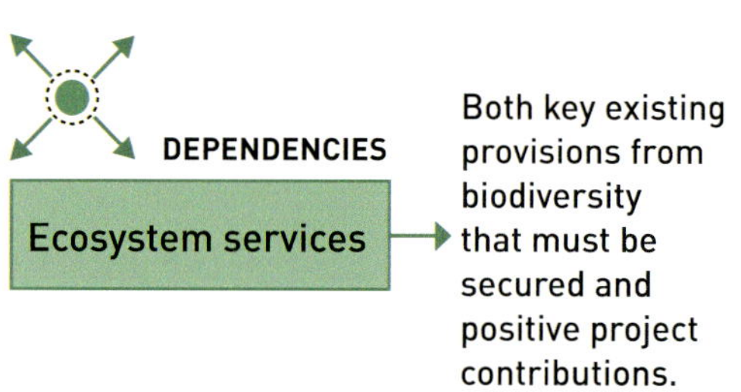

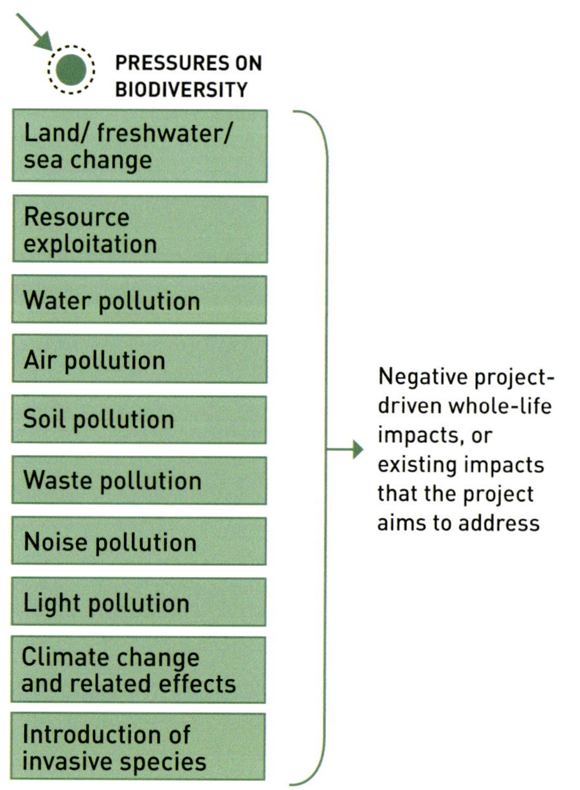

This subset of criteria refers to project commitments to halt and reverse biodiversity loss, and how these commitments are informed by the 2050 Goals and 2030 Targets in the Kunming-Montreal Global Biodiversity Framework. Adopted in 2022 by the Convention on Biological Diversity (CBD), the Framework sets out the CBD's 2050 vision of a world "living in harmony with nature," where "biodiversity is valued, conserved, restored and wisely used—maintaining ecosystem services, sustaining a healthy planet and delivering benefits essential for all people."[63]

BIO 2.1.

Commit to and set ambitious targets aligned with global policy goals to both halt and reverse biodiversity loss

The Kunming-Montreal Global Biodiversity Framework sets "whole-of-government and whole-of-society"[64] goals for 2050:[65]

- "The integrity, connectivity and resilience of all ecosystems are maintained, enhanced, or restored, substantially increasing the area of natural ecosystems by 2050"
- "Human induced extinction of known threatened species is halted, and, by 2050, the extinction rate and risk of all species are reduced tenfold and the abundance of native wild species is increased to healthy and resilient levels"
- "Biodiversity is sustainably used and managed, and nature's contributions to people, including ecosystem functions and services, are valued, maintained and enhanced—with those currently in decline being restored—supporting the achievement of sustainable development for the benefit of present and future generations by 2050"

The Framework also establishes interim global targets for 2030 to conserve biodiversity, reduce pressures, and mainstream biodiversity into decision-making at all levels. Emerging approaches summarize these goals and targets under the concept of "nature positive." However, given the lack of consensus on how to accurately define "nature positive" outcomes, this criterion instead refers to a commitment to contribute to no net loss (halt) and net biodiversity gain (reverse).

Building on existing commitments and actions related to nature, projects should aim to:

- Halt biodiversity loss by 2030 by minimizing their contributions to negative outcomes for nature and reducing pressure on ecosystems
- Reverse biodiversity loss by 2050 by maximizing their contributions to positive outcomes in terms of "the health, abundance, diversity, resilience of species, ecosystems, and natural processes"[66]

A project goal to halt biodiversity loss requires addressing all relevant pressures on biodiversity to collectively secure no net loss of biodiversity.

To reverse biodiversity loss, projects must achieve "biodiversity net gain (BNG)." BNG is an approach to development and land or marine management that "leaves biodiversity in a measurably better state than before the development took place."[67] BNG goes beyond existing habitat and species protection. Intended to reinforce the mitigation hierarchy, it seeks to create and establish new habitat as well as enhance existing ones, safeguarding and improving the ecological connectivity essential for wildlife.

Projects must apply the mitigation hierarchy to all relevant throughout the entire project lifecycle to understand the level of effort required to achieve the established targets.

It is important to highlight that alignment with these goals is not limited to the achievement of final outcomes but also includes adherence to principles such as ensuring "full, equitable, inclusive, effective and gender-responsive representation and participation in decision-making, and access to justice and information related to biodiversity by indigenous peoples and local communities."[68]

Prioritize management responses based on the mitigation hierarchy

3. RESPOND

Efforts to reduce impacts should be based on a best effort basis, recognizing that it is not possible to reduce all potential impacts to zero. The criterion adopts the United Nations Global Compact (UNGC) and International Union for Conservation of Nature (IUCN) mitigation hierarchy steps[69] and adds a new dimension modeled on the SBTN's AR³T framework.[70]

BIO 3.1.

Plan for actions to:

- **Avoid negative impacts on biodiversity**
- **Minimize negative impacts on biodiversity that were not avoided**
- **Restore negative impacts on biodiversity**
- **Offset impacts on biodiversity (as a last resort)**
- **Transform performance to achieve net positive outcomes**

Avoid: Proactive measures taken to prevent impacts from occurring in the first place. For example, this can include selecting an alternative location for development, adjusting the timing of construction activities to avoid interfering with key breeding or migration periods, or opting not to pursue activities that would cause irreversible biodiversity loss (e.g., canceling plans to expand a project site). These measures include nature-based solutions (NbS) that preserve existing ecosystems, as well as management activities that proactively protect against adverse impacts.

Minimize: Proactive measures taken to reduce the duration, intensity, extent and/or likelihood of impacts that cannot be completely avoided. These include NbS that implement soft measures to deliver ecosystem-based and ecosystem-sensitive solutions, as well as various abatement and control measures that address pressures on biodiversity (not necessarily based on NbS approaches).

Restore: Measures taken to rehabilitate and improve degraded ecosystems after exposure to negative impacts that cannot be entirely avoided or effectively minimized, including both permanent and temporary impacts (such as those occurring during construction).

Offset (offsite compensation): Measures taken to compensate for significant residual adverse impacts after the avoidance, mitigation, and restoration steps have been fully implemented. Biodiversity offsets are conservation actions designed to compensate for unavoidable impacts on biodiversity caused by projects.[71] They are a form of offsite compensation in which a disturbed habitat is recreated elsewhere. Offsets are intended "to achieve no net loss, and preferably a net gain of biodiversity."[72]

"Biodiversity offsets should conserve the same biodiversity values—species, habitats, ecosystems, or ecological functions—as those lost due to the original project, following the principle known as 'like-for-like or better'."[73] Examples of offsets could include:

- "Intentionally protecting an ecosystem, habitat, or species population (outside the original project's footprint) that is already in good condition or otherwise of high biodiversity value, but that lacks sufficient legal or on-the-ground protection."[74] The offset in this case provides conservation support to prevent this area from diminishing, degrading, or being lost
- "Deliberate actions to restore an ecosystem, habitat, or species population"[75]

Biodiversity offsets must be "measurable or at least verifiable, long-term, and additional to any other ongoing or planned conservation measures."[76]

The "offset" step does not apply to all types of biodiversity pressures. Impacts such as the exploitation of primary resources, as well as various forms of pollution (air, waste, noise, and light) and the introduction of invasive species, cannot be meaningfully compensated for through measures in another location.

Transform: Measures taken to achieve net positive outcomes for biodiversity. NbS drive transformative change,[77,78] particularly those that aim for biodiversity gain and the creation of new ecosystem assets, thereby contributing to the supply of new ecosystem services. To align with the concept of nature-positive infrastructure development, these actions also integrate climate considerations. Therefore, transformative biodiversity actions are inherently integrated climate-biodiversity actions.

In addition, similar to climate change mitigation, transformative action encompasses the ways in which projects and organizations can contribute to broader systemic change—both within and beyond their supply chains.

This subset of criteria classifies project strategies into two broad categories:

- Solutions that use the inherent processes of natural assets to mitigate project-driven pressures (impacts) on biodiversity
- Technical and/or technological prevention, abatement, and control measures that mitigate pressures without relying on nature

The first category refers to strategies that fall under the term nature-based solutions (NbS), while the second encompasses other types of abatement measures that respond to all pressures on biodiversity which cannot be addressed through NbS, within a given context.

Climate change and its related effects have been identified as key pressures on biodiversity and major drivers of change—from decline to loss. However, the complexity of how biodiversity change will manifest makes it hard to design targeted actions for each potential impact. Because climate change mitigation is a long-term effort, reducing other, non-climatic pressures represents one of the most practical near-term options. The criteria recognize two overarching responses to address the impacts of climate change on biodiversity:

- Implementation of climate change mitigation strategies, namely the greenhouse gas (GHG) reduction strategies of a project as presented in the climate change mitigation response criteria
- Strategies to maximize biodiversity's capacity to adapt and withstand climate change, including the application of NbS and measures that address other non-climatic stressors

BIO3.2.

Implement Nature-based Solutions (NbS)

Nature-based solutions (NbS) are defined as "actions to protect, sustainably use, manage, and restore natural or modified ecosystems that address societal challenges effectively and adaptively, while providing human well-being and biodiversity benefits."[79] According to the IUCN's compliance criteria NbS, they "must result in a net gain to biodiversity and ecosystem integrity."[80] NbS can be implemented independently—replacing traditional grey infrastructure—or integrated with other approaches, such as technological and engineering solutions. They are a necessary component of no net loss targets and the only type of solutions capable of achieving biodiversity net gain targets. The following list presents types of NbS, as included in the WWF/AECOM Playbook for Nature-Positive Infrastructure Development:[81]

- Ecosystem conservation
- Ecosystem restoration
- Soil infiltration systems for wastewater treatment
- Eco-engineering, vertipools, habitat tiles
- River restoration and catchment management
- Restoration of water bodies (lakes, ponds)
- Riparian buffers
- Sludge treatment reed beds
- Wetlands (natural and constructed)
- Sustainable Urban Drainage Systems (SuDS), bioretention areas
- Solar gardens
- Green roofs
- Afforestation, reforestation, and open space creation
- Rewilding[82] grasslands and meadows
- Bioengineering for landslide and erosion protection
- Green bridges, wildlife bridges, ecological corridors, and culverts
- Urban and peri-urban agriculture, urban gardens
- Bio-building materials and "soft" measures

This criterion encourages projects to explore potential NbS based on available ecosystem services, ecosystem type, project context, and the project's and community's dependencies on biodiversity. Projects should also consider their specific commitments and assess the compatibility of potential solutions with the local ecosystem type and activities.

Similar to the process for defining dependencies, projects can use the list of ecosystem services as a checklist to determine which types of NbS provide or support the ecosystem services that align with project needs and targets.

DEPENDENCIES ADDRESSED:
Various depending on the nature of the NbS

PRESSURES ADDRESSED:
Various depending on the nature of the NbS

BIO3.3.

Use prevention, abatement, and control measures against pressures on biodiversity (other than NbS)

This criterion encompasses technical and technology-based project strategies that do not rely on nature as part of the solution. They represent measures to address non-climatic pressures ("doing everything else better") and maximize the opportunity for species and ecosystems to adapt to and survive climate change. Such measures must be integrated into climate change-focused actions.[83]

They include siting, micro-siting and physical measures, as well as prevention abatement, and control measures in line with best available practices, which should be prioritized based on the mitigation hierarchy.

The ISI Envision Rating System credits, which assess performance against listed pressures on biodiversity, were consulted, and examples of relevant strategies are presented below. To facilitate connections, the examples are organized according to the type of biodiversity pressure they address:

Strategies against land use/freshwater/sea change:

- Reclaim brownfields[84] to preserve undeveloped land
- Proper siting of projects to intentionally avoid areas with special designations, such as conservation areas, national wildlife refuges, national parks, other ecologically critical or sensitive areas, wetlands, water bodies, shorelines, etc.
- "Limiting natural vegetation clearance to the minimum necessary during construction works"[85]

Project site design and micro-siting measures can also reduce onsite risks of fauna fatalities through:

- Changes in project layout to avoid barrier effects and collisions (e.g., re-routing, marking, or burying power lines)
- Arranging project elements in "clusters with corridors between them to provide passage through the site"[86]

Strategies against resource exploitation:

These strategies coincide with strategies described as part of the climate change mitigation and adaptation criteria, and more specifically:

- CCM3.7. Implement resource circularity strategies
- CCA3.3. Implement strategies to maximize resourcefulness
- CCA3.4. Implement strategies to maximize durability
- CCA3.5. Implement strategies to maximize adaptability
- CCA3.7. Implement strategies to maximize integration

Strategies against water pollution through engineering solutions

- Wastewater treatment measures
- Effluent control, including surface water and/or groundwater quality monitoring or contaminant source monitoring of receiving waters (e.g., pollutant loading, biological impact, water temperature, and impact on receiving water flow)
- Runoff control and prevention of sewage system overflows
- Provision of runoff interceptors and drainage channels for pollutants in stormwater runoff, ice melt, potential spills, and leakage
- Strategies to maximize water circularity (similar to criterion CCA.3.3. Implement strategies to maximize resourcefulness)
- Implementation of erosion, sediment, and pollutant controls during construction

- Minimization of potential impacts on surface water and groundwater quality by performing selected works offsite (e.g., applying alternative methods for the removal of existing structures, such as saw-cutting into segments and transferring the cut sections to an offsite location for demolition)
- Minimization of in-water works during construction through design optimization
- Installation of sufficient drainage systems

Air pollution prevention, abatement, and control measures:
- Compliance with all applicable air quality standards and regulations
- Limited use or controlled exposure to volatile organic compounds through the use of low-VOC and/or biodegradable coatings and treatments
- Direct removal of existing air pollutant sources or capture and sequestration of air pollutants

Soil pollution prevention, abatement, and control measures:
- Minimization of soil compaction to preserve natural infiltration capacity
- Restoration of disturbed soils
- Controlled application of fertilizers or pesticides during construction
- Reduced pesticide and fertilizer application rates, or elimination of their use through selection of soil tolerant, pest resistant, or native plant species

Solid waste reduction and diversion measures:
These strategies coincide with strategies described in the climate change mitigation and adaptation criteria:
- CCM3.7. Implement resource circularity strategies
- CCA3.3. Implement strategies to maximize resourcefulness
- CCA3.4. Implement strategies to maximize durability
- CCA3.5. Implement strategies to maximize adaptability
- CCA3.7. Implement strategies to maximize integration

Noise pollution prevention, abatement, and control measures:
- Relocating noise-generation sources away from sensitive receptors
- Noise abatement at source through the use of low-noise emitting technologies, equipment, and surface materials
- Proper equipment sizing, avoidance of idling, and other operational controls
- Noise transmission reduction strategies such as screening, enclosures, or silencing of sources
- Receptor abatement, e.g., use of noise barriers or buffers

During construction, additional measures include:
- Performing selected works offsite
- Timing works to consider "seasonal and diurnal activity patterns of sensitive species" (feeding, breeding, migration, movement patterns, etc.)[87]

Light pollution prevention, abatement, and control measures:
- Identifying areas where lighting is unnecessary
- Determining the minimum lighting requirements to meet safety and performance needs
- Restricting light spillage to sensitive areas or directing light only where needed.
- Reducing backlight, uplight, and glare (BUG), including performing BUG rating for exterior luminaires and using dark-sky compliant (or equivalent) luminaire types
- Compensating for lighting in one location by removing lighting elsewhere onsite

Invasive species prevention, abatement, and control measures:
- Balancing earthworks onsite to avoid introducing new soil
- Using native species known to be non-invasive
- Implementing eradication, monitoring, and management plans
- Long-term follow-up to control infestations

It is worth noting that when these strategies are implemented in conjunction with GHG reduction measures, they have an increased potential to maximize positive outcomes for biodiversity and climate.

HOW BIODIVERSITY STRATEGIES ADDRESS PRESSURES ON BIODIVERSITY

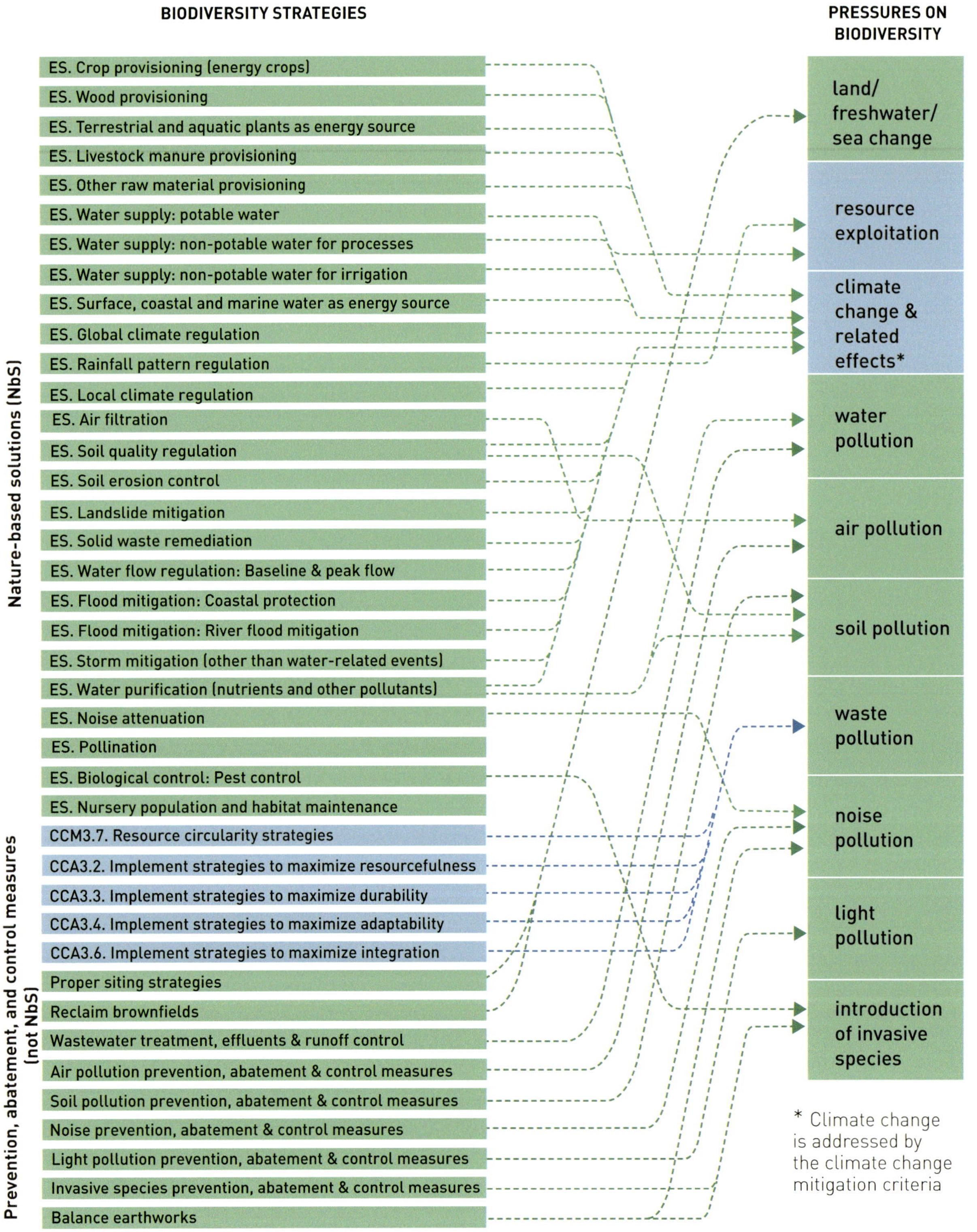

The potential contributions of NbS strategies are represented as ecosystem services (ES)

Opportunities:

- The inherent capacity of nature to deliver more than one benefit
- The joint benefit of GHG reduction and core resilience principles for biodiversity

BIO4.1

Monitor pressures and responses and track progress against targets set to verify selection of strategies or adapt them as needed

According to the IUCN, "given that state level indicators generally change slowly and companies may not be able to demonstrate improvements in species, habitats and ecosystem services, pressure and response indicators can demonstrate change and progress and can help companies verify their selection of strategies or adapt them as needed."[88]

Therefore, feasible and tractable performance indicators for biodiversity action are those related to existing and project-driven pressures on biodiversity, as well as how these pressures are managed using the steps of the biodiversity mitigation hierarchy steps as a guiding framework.

NOTES

1. Secretariat of the Convention on Biological Diversity, United Nations Environment Programme, *Convention on Biological Diversity (CBD). Text and Annexes* (Secretariat of the Convention on Biological Diversity, 2011), 4, https://www.cbd.int/doc/legal/cbd-en.pdf.
2. Cambridge Conservation Initiative and Natural Capital Coalition, *Biodiversity and Natural Capital* (flyer, May 2017), https://www.cambridgeconservation.org/wp-content/uploads/2017/05/Flyer_May2017_draft_v3_LD.pdf.
3. Taskforce on Nature-related Financial Disclosures, *The TNFD Nature-related Risk & Opportunity Management and Disclosure Framework Beta v0.1 release* (TNFD, 2023), 30, https://www.learningfornature.org/wp-content/uploads/2023/05/23-23882-TNFD_v0.4_Short_Summary_v5.pdf.
4. Secretariat of the Convention on Biological Diversity, *Biodiversity and the 2030 Agenda for Sustainable Development: Technical Note* (Secretariat of the Convention on Biological Diversity, 2016), 16, https://www.cbd.int/development/doc/biodiversity-2030-agenda-technical-note-en.pdf.
5. Ibid.
6. Convention on Biological Diversity (CBD) and United Nations Environment Programme (UNEP), *Updated Glossary for the Draft Post-2020 Global Biodiversity Framework* (Montreal: Open-Ended Working Group on the Post-2020 Global Biodiversity Framework, Fifth Meeting, December 3–5, 2022), 2, https://www.cbd.int/doc/c/c3ab/388d/950ddc02586468a814120acf/wg2020-05-04-en.pdf.
7. TEEB for National and International Policy Makers, *The Economics of Ecosystems and Biodiversity* (TEEB, 2009), Chapter 3: 2, https://www.cbd.int/financial/values/g-accounting-teeb.pdf.
8. Climate Disclosure Standards Board (CDSB), *CDSB Framework: Application Guidance for Biodiversity-Related Disclosures* (November 2021), 15, https://www.cdsb.net/biodiversity.
9. European Environment Agency, *EEA Technical Report No 12/2015: Exploring Nature-based Solutions: The role of green infrastructure in mitigating the impacts of weather- and climate change-related natural hazards* (EEA, 2015), 16, https://www.eea.europa.eu/publications/exploring-nature-based-solutions-2014.
10. Joseph Alcamo et al., *Ecosystems and Human Well-being: A framework for assessment* (Island Press, 2003), 63, http://pdf.wri.org/ecosystems_human_wellbeing.pdf.
11. Millennium Ecosystems Assessment, *Overview of the Millennium Ecosystem Assessment*, accessed November 24, 2025, https://www.millenniumassessment.org/en/About.html#1.
12. Convention on Biological Diversity, *Decision Adopted by the Conference of the Parties to the Convention on Biological Diversity: CBD/COP/DEC/15/4* (December 19, 2022), 10, https://www.cbd.int/doc/decisions/cop-15/cop-15-dec-04-en.pdf.
13. Science for Environment Policy, *Ecosystem Services and the Environment, In-depth Report 11 produced for the European Commission, DG Environment by the Science Communication Unit*, (UWE, Bristol, 2015), 3, https://uwe-repository.worktribe.com/output/834982.
14. United Nations et al., *System of Environmental Economic Accounting—Ecosystem Accounting (SEEA EA)* (White cover publication, pre-edited text subject to official editing, 2021), 131-134.
15. Taskforce on Nature-related Financial Disclosures (TNFD), *Recommendations of the Taskforce on Nature-related Financial Disclosures* (September 2023), https://tnfd.global/publication/recommendations-of-the-taskforce-on-nature-related-financial-disclosures/.
16. United Nations et al., *System of Environmental Economic Accounting—Ecosystem Accounting*, 357.
17. Ibid.
18. Ibid., 353.
19. Ibid., 355.
20. Ibid.

21. Ibid., 129.
22. Ibid., 131-135.
23. IUCN, *Defining Nature-based Solutions*, Resolution of the World Conservation Congress at its session in Hawaii, United States of America, September 1-10, 2016, 1, https://portals.iucn.org/library/sites/library/files/resrecfiles/WCC_2016_RES_069_EN.pdf.
24. Ibid.
25. Ibid., 1-2.
26. IUCN, *Global Standard for Nature-based Solutions: A User-friendly Framework for the Verification, Design and Scaling Up of NbS, First edition* (Gland, Switzerland: IUCN, 2020), https://portals.iucn.org/library/sites/library/files/documents/2020-020-En.pdf.
27. Natural England, *Biodiversity Net Gain: An introduction to the benefits*, accessed November 24, 2025, 4, https://naturalengland.blog.gov.uk/wp-content/uploads/sites/183/2022/04/BNG-Brochure_Final_Compressed-002.pdf.
28. H.O. Pörtner et al., *Scientific outcome of the IPBES-IPCC co-sponsored workshop on biodiversity and climate change*; IPBES secretariat (Zenodo, 2021), 59, DOI:10.5281/zenodo.4659158.
29. "What Is Ecosystem Restoration?," UN Decade on Restoration, accessed November 24, 2025, https://www.decadeonrestoration.org/what-ecosystem-restoration.
30. "What Is Ecological Restoration?," Society for Ecological Restoration, accessed November 24, 2025, https://www.ser-rrc.org/what-is-ecological-restoration/.
31. Pörtner et al., *Scientific outcome of the IPBES-IPCC co-sponsored workshop on biodiversity and climate change*, 17.
32. Ibid., 20.
33. Ibid., 47.
34. Ibid.
35. Based on UN Global Compact, *Water Action Hub's NbS Benefits Explorer*, accessed November 26, 2025, https://nbsbenefitsexplorer.net/tool#.
36. Peter.J. Stephenson and Giulia Carbone, *Guidelines for Planning and Monitoring Corporate Biodiversity Performance* (IUCN, 2021), 7, https://doi.org/10.2305/iucn.ch.2021.05.en.
37. United Nations Environment Programme (UNEP), *Expert Input to the Post-2020 Global Biodiversity Framework: Transformative Actions on All Drivers of Biodiversity Loss Are Urgently Required to Achieve the Global Goals by 2050*, CBD/WG2020/3/INF/11, Note by the Executive Secretary (14 January 2022), 19, https://www.cbd.int/doc/c/16b6/e126/9d46160048cfcf74cadcf46d/wg2020-03-inf-11-en.pdf.
38. Pörtner et al., *Scientific outcome of the IPBES-IPCC co-sponsored workshop on biodiversity and climate change*, 42.
39. Sophus O.S.E. Zu Ermgassen et al., "Are Corporate Biodiversity Commitments Consistent with Delivering 'Nature-positive' Outcomes? A Review of 'Nature-positive' Definitions, Company Progress and Challenges," *Journal of Cleaner Production* 379 (October 21, 2022): 134798, https://doi.org/10.1016/j.jclepro.2022.134798.
40. WWF, FIDIC, and AECOM (Technical Partner), *A Playbook for Nature-Positive Infrastructure Development: Part of the Sustainable Infrastructure Series* (WWF, FIDIC, and AECOM, October 2023), viii, accessed November 1, 2025, https://issuu.com/fidic/docs/wwf_playbook_-_single_pages?fr=xKAE9_zU1NQ.
41. Secretariat of the Convention on Biological Diversity, United Nations Environment Programme, *Convention on Biological Diversity (CBD). Text and Annexes*, 4.
42. Taskforce on Nature-related Financial Disclosures (TNFD), *Glossary Version 1.0* (September 2023), 16, https://tnfd.global/wp-content/uploads/2023/09/Glossary_of_key_terms_v1.pdf?v=1702506695.
43. Global Sustainability Standards Board (GSSB), *Consolidated Set of the GRI Standards* (Amsterdam, The Netherlands: GSSB, 2024), 519.
44. TNFD, *Glossary Version 1.0*, 16.
45. Endangered Wildlife Trust, *The Biological Diversity Protocol (BD Protocol)*, National Biodiversity and Business Network – South Africa (2020), 25, https://nbbnbdp.org/wp-content/uploads/2022/05/bdp_final_080321.pdf.
46. Lars Hein et al., *SEEA Experimental Ecosystem Accounting: Towards a definition and classification of ecosystem services for SEEA* (discussion paper, Final Report, 2018), 14, https://seea.un.org/sites/seea.un.org/files/lg_24_bg_ecosystem_services_classification.pdf.
47. Eduardo Brondizio et al., "Chapter 1: Assessing a Planet in Transformation: Rationale and Approach of the IPBES Global Assessment on Biodiversity and Ecosystem Services," in *IPBES Global Assessment Report on Biodiversity and Ecosystem Services* (IPBES secretariat, 2019), 25-26, https://doi.org/10.5281/zenodo.3831852.
48. Stephenson and Carbone, *Guidelines for Planning and Monitoring Corporate Biodiversity Performance* (IUCN, 2021), 15.
49. Pörtner et al., *Scientific outcome of the IPBES-IPCC co-sponsored workshop on biodiversity and climate change* 109.
50. Ibid.
51. Ibid., 38.
52. Ibid.
53. Ibid., 42
54. "Target 7: Reduce Pollution to Levels That Are Not Harmful to Biodiversity," Convention on Biological Diversity, accessed November 2, 2025, https://www.cbd.int/gbf/targets/7.
55. Ibid.
56. Institute for Sustainable Infrastructure (ISI), *Envision: Sustainable Infrastructure Framework Version 3* (ISI, 2018), 166, https://sustainableinfrastructure.org/wp-content/uploads/EnvisionV3.9.7.2018.pdf.
57. "Target 6: Reduce the Introduction of Invasive Alien Species by 50% and Minimize Their Impact," Convention on Biological Diversity, accessed November 2, 2025, https://www.cbd.int/gbf/targets/6.
58. Convention on Biological Diversity, *Decision Adopted by the Conference of the Parties to the Convention on Biological Diversity*, 10.
59. IPBES, *Summary for Policymakers of the Thematic Assessment Report on Invasive Alien Species and Their Control of the Intergovernmental Science-Policy Platform on Biodiversity and Ecosystem Services*, ed. Helen E. Roy et al. (Bonn, Germany: IPBES Secretariat, 2023), https://doi.org/10.5281/zenodo.7430692.

60. Pörtner et al., *Scientific outcome of the IPBES-IPCC co-sponsored workshop on biodiversity and climate change*, 41.
61. Adapted from Stephenson and Carbone, *Guidelines for Planning and Monitoring Corporate Biodiversity Performance*, x, https://doi.org/10.2305/iucn.ch.2021.05.en.
62. Neville Ash et al., "Conceptual Frameworks for Ecosystem Assessment: Their Development, Ownership, and Use Chapter 3," in *Ecosystems and Well-being: A manual for assessment practitioners* (Island Press, 2010), https://portals.iucn.org/library/efiles/documents/2010-029.pdf.
63. Convention on Biological Diversity, *Decision Adopted by the Conference of the Parties to the Convention on Biological Diversity*, 8.
64. Ibid., 5.
65. Ibid., 8-9.
66. Adapted from "Nature Positive Initiative launches to promote the integrity and implementation of the Global Goal for Nature", World Business Council for Sustainable Development (WBCSD), September 6, 2023, https://www.wbcsd.org/news/nature-positive-initiative/.
67. Natural England, *Biodiversity Net Gain*.
68. Convention on Biological Diversity, *Decision Adopted by the Conference of the Parties to the Convention on Biological Diversity*, 13.
69. United Nations Global Compact (UNGC) and International Union for Conservation of Nature (IUCN), *A framework for corporate action on biodiversity and ecosystem services* (UN Global Compact and IUCN, 2012), 14, https://portals.iucn.org/library/sites/library/files/documents/2012-036.pdf.
70. Science Based Targets Network, *Step 4: Act* (SBTN, 2023), https://sciencebasedtargetsnetwork.org/companies/take-action/act/.
71. UNGC and IUCN, *A framework for corporate action on biodiversity and ecosystem services*, UN Global Compact and IUCN, 2012, 14.
72. World Bank Group, *Biodiversity Offsets: A User Guide* (World Bank Group, October 2016), 3, https://documents1.worldbank.org/curated/en/344901481176051661/pdf/110820-WP-BiodiversityOffsetsUserGuideFinalWebRevised-PUBLIC.pdf.
73. Ibid., 11-12.
74. Ibid., 4.
75. Ibid.
76. Ibid.
77. Boipelo Tshwene Mauchaza et al., *What Is the State of Evidence on How Nature-Based Solutions Promote Transformative Change for the Sustainable Use and Management of Biodiversity in Socio-Ecological Systems?* (Biodiversa+, European Biodiversity Partnership, and UNEP-WCMC, March 2024), https://www.biodiversa.eu/wp-content/uploads/2024/07/Biodiversa_D4.8_Desk-Study-NBS-Transformative-change.pdf.
78. Ignacio Palomo et al., "Assessing Nature-based Solutions for Transformative Change," *One Earth* 4, no. 5 (May 1, 2021): 730–41, https://doi.org/10.1016/j.oneear.2021.04.013.
79. IUCN, *Defining Nature-based Solutions*, 1.
80. IUCN (2020), *Global Standard for Nature-based Solutions*, 10.
81. WWF, FIDIC, and AECOM, *Playbook for Nature-Positive Infrastructure Development*, 56-70.
82. Ibid., 68. Rewilding refers to the practice of restoring an area back to its uncultivated state. It could either take an active form (e.g., through species reintroduction, or active landscape management to restore native biodiversity) or passive (e.g. through natural regeneration).
83. Pörtner et al., *Scientific outcome of the IPBES-IPCC co-sponsored workshop on biodiversity and climate change*, 42.
84. A brownfield refers to a site previously developed for industrial, or commercial purposes that has been abandoned or underutilized, regardless if it carries pollution or risk of pollution from its previous use. Polluted or not the site requires further development before reuse.
85. L. Bennun et al., *Mitigating Biodiversity Impacts Associated with Solar and Wind Energy Development: Guidelines for Project Developers* (IUCN, 2021), 52, https://doi.org/10.2305/iucn.ch.2021.04.en.
86. Ibid., 69.
87. Ibid., 72.
88. Stephenson and Carbone, *Guidelines for Planning and Monitoring Corporate Biodiversity Performance* (IUCN, 2021), 7.

FRAMING INTEGRATED CLIMATE AND BIODIVERSITY ACTION IN INFRASTRUCTURE PROJECTS

Integrated

Climate-Biodiversity Action Criteria

Chapter 6

Integrated Climate-Biodiversity Action Criteria
comprise one merged set of criteria for framing integrated climate and biodiversity action in infrastructure projects that addresses all impact drivers and maximizes co-benefits.

MERGING THE CLIMATE AND BIODIVERSITY CRITERIA

INTEGRATED CRITERIA

Biodiversity and climate change mutually reinforce each other. Climate change and its related effects represent one of the main pressures on biodiversity and its contributions to people. Conversely, the decline and loss of biodiversity impact climate systems through their effects on nitrogen, carbon, and water cycles, posing significant risks to efforts aimed at mitigating and adapting to climate change.

The 2021 IPBES-IPCC Report—the key reference supporting the argument for the need for integrated action—discusses the interconnectedness of the two crises and presents findings. Its findings support the concept of integration and frame it as a necessary approach that benefits both climate and biodiversity by identifying synergies, avoiding conflicts, and linking policy measures. However, it does not provide a single, specific definition of "integrated climate-biodiversity action."

As part of the criteria, integrated climate and biodiversity project action is defined as follows:

- Action that recognizes the necessity of addressing the climate change and biodiversity crises in tandem, enabling projects to avoid unintended trade-offs and maximize co-benefits across climate change mitigation, adaptation, and biodiversity efforts
- Action that aims to tackle all relevant impact drivers of both climate change and biodiversity loss, as well as all climate-related risks pertinent to a given project and location
- Action that sets higher ambitions, striving for a measurably improved state for both climate and biodiversity, aiming for no loss of resilience and biodiversity, and fostering projects to pursue net gains
- Action that is sustainable and equitable

Integrated action is structured around three key components: climate change mitigation, adaptation, and biodiversity action. Initially, the three types of action were defined separately as standalone guidance for projects on how to address specific impact drivers and risks of climate change, as well as pressures on biodiversity. Together, they form a comprehensive, high-level framework for climate and biodiversity action within which considerable overlaps and linkages—both direct and indirect—highlight their interconnectedness. To avoid duplication of effort when applying the three sets of criteria, they were merged into a single, integrated set. This integrated set of criteria aims to represent a mix of the optimal actions for each type of action (mitigation, adaptation, and biodiversity) and their sum, thereby addressing the need for integrated action to prevent unintended trade-offs.

The co-benefits of integrated action are underscored by the linkages between climate and biodiversity strategies. Previous sections have highlighted the overlaps and linkages between climate change mitigation and adaptation criteria, referred to as "opportunities." This section seeks to illustrate the linkages and overlaps between climate action and biodiversity action. To this end, climate change mitigation and adaptation strategies, as included in the response criteria, are examined in relation to pressures on biodiversity (drivers of biodiversity loss), while biodiversity strategies are reviewed against the drivers of climate change and climate-related risks. This analysis identifies opportunities for joint benefits, highlighting strategies that have greater potential to deliver multiple contributions.

It is also important to note that overlaps exist between the impact drivers of climate change and biodiversity loss in the first place. Sharing common impact drivers, climate change and biodiversity loss can benefit from the same responses. This is a first overlap that exists between the climate change mitigation, adaptation, and biodiversity response criteria. More specifically, climate change and its related effects is among the identified pressures on biodiversity, along with resource exploitation, which also underlies various climate change mitigation and adaptation criteria. These two pressures on biodiversity are considered to be addressed mainly through the two sets of the climate action criteria.

The following sections aim to present the various overlaps identified between the three sets of criteria. The different criteria are now merged and presented based on the four action steps, assess, commit, respond, and track. The merged set of criteria takes into consideration the identified overlaps and highlights cross-cutting strategies that should be part of all types of actions, namely transformative action, inclusivity, local skills development, and reflective capacity.

OVERLAPS BETWEEN THE DRIVERS OF CLIMATE CHANGE AND RELATED RISKS AND THE PRESSURES ON BIODIVERSITY

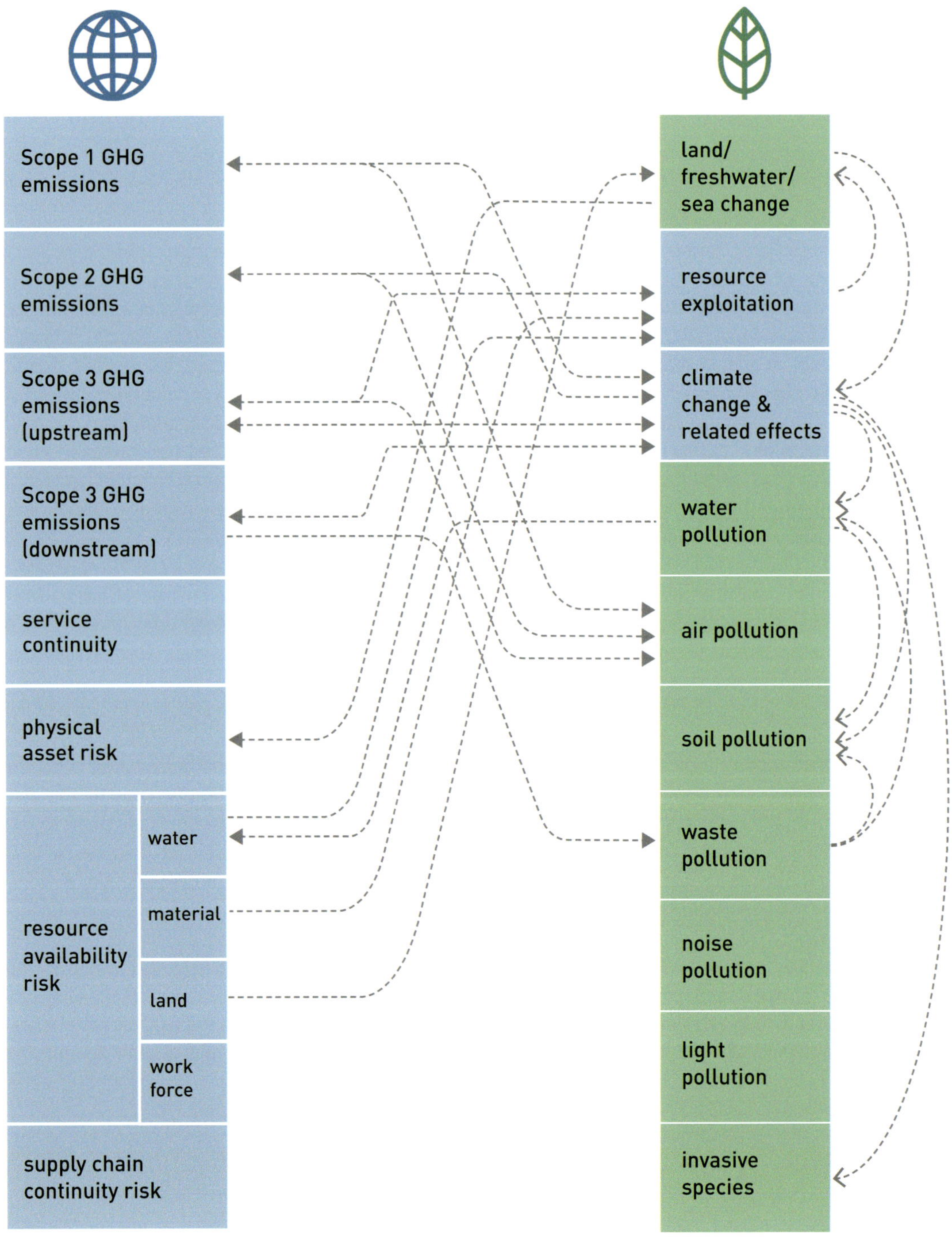

1. ASSESS

Assess the Project's Estimated GHG Inventory	CCM1.1. Assess the estimated Scope 1 GHG emissions CCM1.2. Assess the estimated Scope 2 GHG emissions CCM1.3. Assess the estimated Scope 3 GHG emissions (upstream and downstream) CCM1.4. Assess the estimated avoided emissions as a result of the project
Assess Exposure to Climate-Related Hazards/Threats	CCA1.1. Assess exposure to all location-relevant climate-related hazards/threats
Assess Vulnerability to Climate-Related Risks	CCA1.2. Assess climate-related service continuity risk CCA1.3. Assess climate-related physical asset risk CCA1.4. Assess climate-related resource availability risk (water, materials, land, workforce) CCA1.5. Assess climate-related supply chain continuity risk
Assess the State of Biodiversity	BIO1.1. Assess the state of ecosystems BIO1.2. Assess the state of species
Assess Project-Driven Pressures on Biodiversity	BIO1.3. Assess land/freshwater/sea change BIO1.4. Assess resource exploitation BIO1.5. Assess climate change and its related impacts BIO1.6. Assess pollution (water, air, soil, waste, noise, light) BIO1.7. Assess introduction of invasive species
Assess Dependencies on Biodiversity	BIO1.8. Assess project's dependency on ecosystem services BIO1.9. Assess community's dependency on ecosystem services

Where the assessment criteria overlap:

- Climate change represents a pressure on biodiversity (BIO1.5). Assessment of climate change impact drivers is part of the CCM assessment criteria (CCM 1.1-1.4)
- Resource exploitation (materials) (BIO1.4) overlaps with Scope 3 upstream emissions (CCM1.3)
- Resource exploitation (water) (BIO1.4) overlaps with water availability risk (CCA1.4)
- Water pollution (BIO1.6) contributes to water availability risk (CCA1.4)
- Waste pollution (BIO1.6) overlaps with Scope 3 downstream emissions (CCM1.3)
- Land/freshwater/sea change (BIO1.3) contributes to increase of GHG emissions by changing the carbon storage capacity of land, water, and marine carbon sinks
- Land/freshwater/sea change (BIO1.3) is related to physical asset risk, as alteration of the surrounding natural environment can exacerbate the impacts of hazards and threats
- Biodiversity loss is a chronic climate-related threat for projects and communities, posing risks to dependencies on biodiversity for both (BIO1.8, BIO1.9)

2. COMMIT

Commit to and Set Ambitious GHG Reduction Targets	CCM2.1. Commit to and set targets aligned with the Paris Agreement goals
Commit to and Set Ambitious Resilience Targets	CCA2.1. Commit to and set targets for no resilience loss and/or net resilience gain
Commit to and Set Ambitious Biodiversity Targets	BIO2.1. Commit to and set ambitious targets aligned with global policy goals to both halt and reverse biodiversity loss

Where the commitment criteria overlap:

- The commitment criteria overlap and reinforce one another. Article 7 of the Paris Agreement calls for actions to enhance resilience, recognizing adaptation as a key component of the long-term global climate response. It also highlights that greater mitigation efforts reduce the level of adaptation needed
- The CCA criterion expands on this goal by referring to no resilience loss and net resilience gain
- At the same time, the Paris Agreement acknowledges the importance of conserving and enhancing the carbon-storage capacity of ecosystems and protecting biodiversity as part of the climate solution. The Kunming-Montreal Global Biodiversity Framework (GBF) further specifies the required level of biodiversity conservation, restoration, and enhancement through quantitative targets for 2030
- The UAE Framework for Global Climate Resilience and its 11 new global adaptation targets for 2030 include a thematic target focused on reducing climate impacts on ecosystems and biodiversity through the accelerated use of nature-based solutions
- Target 8 of the GBF for 2030 highlights the need to increase biodiversity's resilience to climate change through mitigation, adaptation, and disaster reduction actions, including nature-based solutions. Target 11 links the restoration, conservation, and enhancement of ecosystems and their services to climate regulation and protection from natural hazards and disasters
- Nature-positive outcomes, by definition, integrate climate change mitigation and adaptation objectives

Prioritize Responses Based on the GHG Management Hierarchy	CCM3.1. Plan for actions to: • Avoid/eliminate GHG emissions • Reduce GHG emissions that were not avoided • Substitute/replace GHG emissions • Compensate/offset residual GHG emissions • Transform performance to achieve net negative emissions
Prioritize Responses Based on the Resilience Hierarchy	CCA3.1. Plan for actions to: • Avoid/eliminate threat • Reduce exposure/vulnerability to threat • Absorb/minimize impact of failure • Restore/recover after shocks or stresses • Transform performance
Prioritize Management Responses Based on the Mitigation Hierarchy	BIO3.1. Plan for actions to: • Avoid impacts on biodiversity • Minimize impacts on biodiversity • Restore impacts on biodiversity • Offset impacts on biodiversity • Transform performance

Where the response criteria (adherence to the mitigation hierarchy) overlap:

- In addition to the shared principle of following the mitigation hierarchy's prioritized steps, the criteria overlap and should coordinate actions for transformative action. In the integrated set of criteria, this step of the mitigation hierarchy is not labeled under climate change mitigation, adaptation, or biodiversity action, but is instead highlighted as a common integrated action that considers both climate and biodiversity
- The definition of transformative action itself bridges the different types of actions and refers to cross-cutting solutions, such as knowledge sharing, collaboration, research and development, and initiatives that extend beyond project boundaries to drive change across the supply chain, the market, and the industry practice. It involves addressing the indirect drivers of climate change and biodiversity loss that are common to both crises. In the context of biodiversity action, transformative action aligns with nature-positive outcomes, which by definition incorporate climate considerations. In the context of resilience, it aligns with reflective capacity, and may also involve high climate ambition, such as transforming infrastructure to meet net-zero targets. Reflective capacity is a quality that should be embedded in all types of action to address the uncertainty and complexity of both climate change and biodiversity loss, as well as the feedback loops between them

CLIMATE CHANGE MITIGATION AND ADAPTATION STRATEGIES AS CONTRIBUTIONS TO BIODIVERSITY

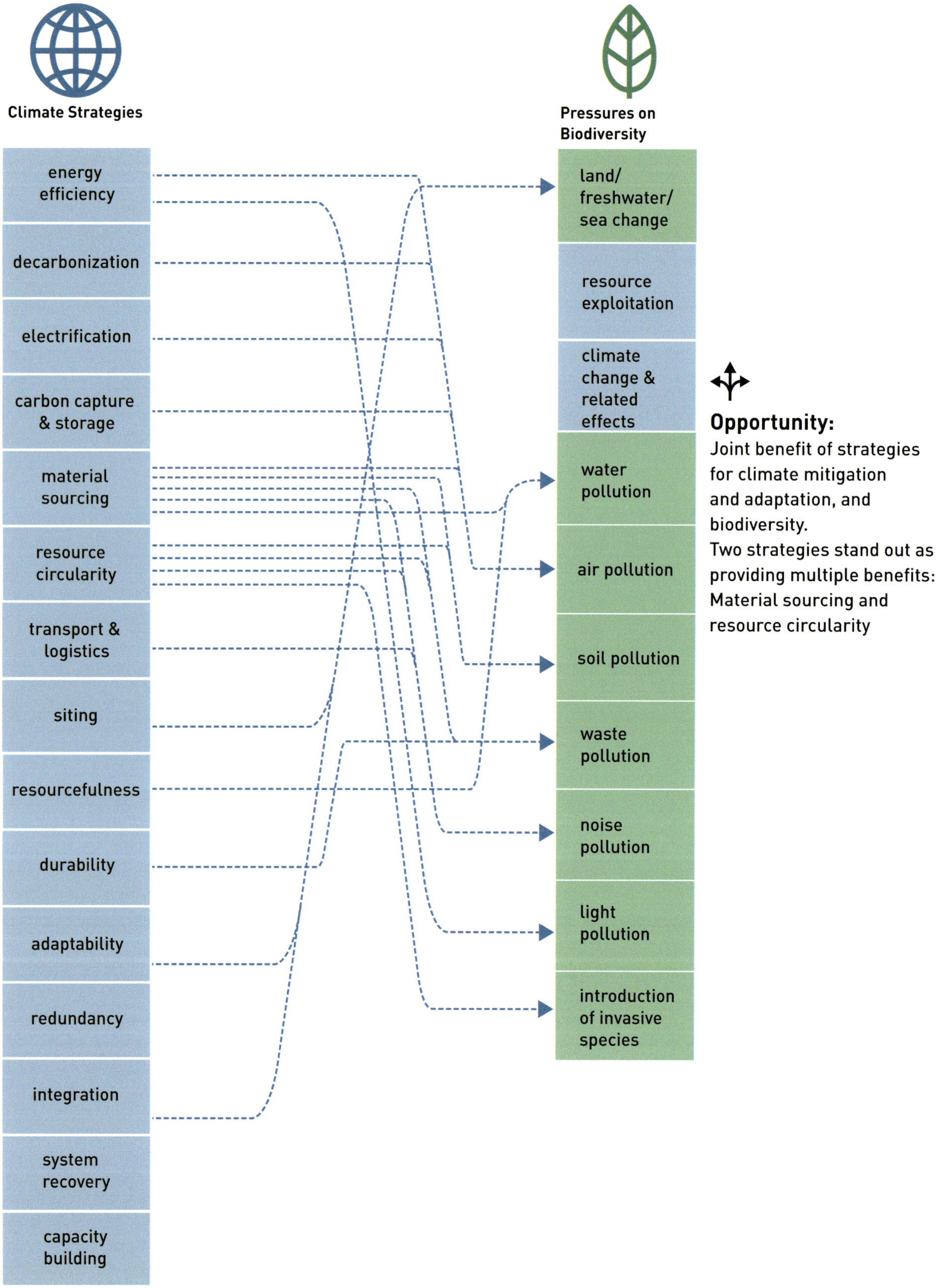

The diagram focuses on the additional contributions of climate strategies in addressing pressures on biodiversity, except for resource exploitation and climate change and its related effects. Addressing these pressures was the subject of the climate action criteria.

ECOSYSTEM SERVICES AS CONTRIBUTIONS TO CLIMATE CHANGE MITIGATION AND ADAPTATION

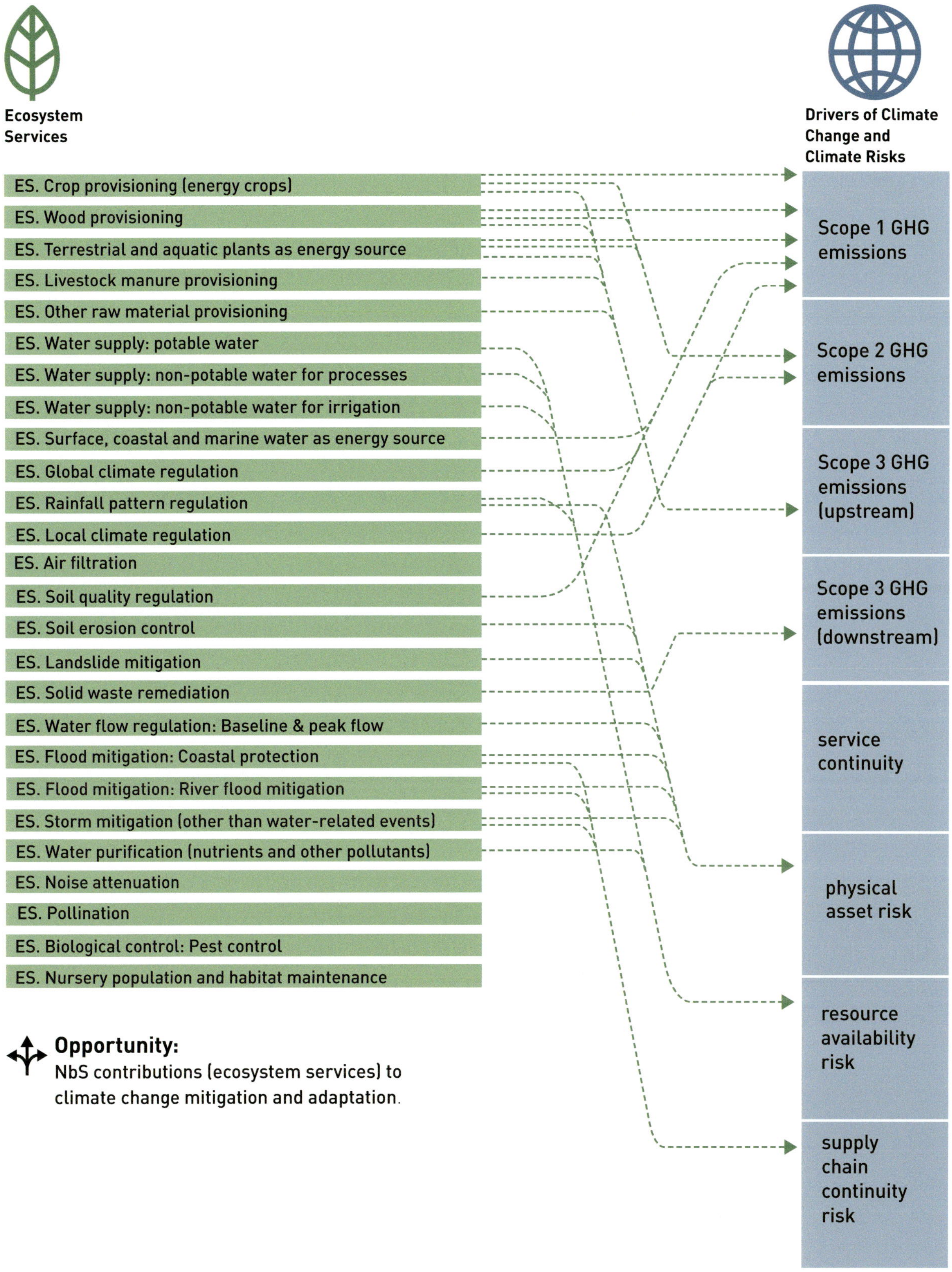

Opportunity:
NbS contributions (ecosystem services) to climate change mitigation and adaptation.

3. RESPOND

Implement GHG Reduction Strategies	CCM3.2. Implement energy efficiency strategies CCM3.3. Implement decarbonization strategies CCM3.4. Implement electrification strategies CCM3.5. Implement carbon capture & storage strategies CCM3.6. Optimize material sourcing practices CCM3.7. Implement resource circularity strategies CCM3.8. Optimize project-related transport and logistics
Implement Risk Reduction Strategies	CCA3.2. Optimize project siting and site design to avoid threats CCA3.3. Implement strategies to maximize resourcefulness CCA3.4. Implement strategies to maximize durability CCA3.5. Implement strategies to maximize adaptability CCA3.6. Implement strategies to maximize redundancy CCA3.7. Implement strategies to maximize integration CCA3.8. Implement strategies to optimize system recovery CCA3.9. Develop local skills and capabilities CCA3.10. Promote inclusivity
Implement Biodiversity Strategies	BIO3.2. Implement nature-based solutions (NbS) BIO3.3. Use prevention, abatement, and control measures against pressures on biodiversity (other than NbS)

Where the response criteria (strategies) overlap:

- Carbon capture and storage strategies (CCM3.5) overlap with nature-based solutions that enhance the carbon storage capacity of land, freshwater, and marine carbon sinks
- Resource circularity (CCM3.7) overlaps with resourcefulness (CCA3.3), durability (CCA 3.4), adaptability (CCA3.5)
- Resource circularity (CCM3.7), durability (CCA3.4), and adaptability overlap with prevention, abatement, and control measures against waste pollution (BIO3.3)
- Resource circularity (CCM3.7), resourcefulness (CCA3.3), durability (CCA3.4), and integration overlap with strategies addressing resource exploitation (materials, water)
- Resourcefulness (CCA3.3), when referring to water, overlaps with strategies against water pollution (BIO3.3)
- Optimized material sourcing (CCM3.6) can contribute to prevention, abatement, and control measures for various pressures on biodiversity during both the product stage and the operational phase
- Optimized project siting and site design (CCA3.2) overlap with nature-based solutions that preserve existing ecosystems and create new ecosystem assets (BIO3.3)

4. TRACK

Climate change mitigation	**Track Progress against Targets Set**	CCM4.1. Monitor and quantify actual GHG reductions to track progress against targets set
Climate change adaptation	**Track Progress against Targets Set**	CCA4.1. Implement strategies to promote reflective capacity
Biodiversity	**Track Progress against Targets Set**	BIO4.1. Monitor pressures and responses and track progress against targets set to verify selection of strategies or adapt them as needed

Where the track criteria overlap:

- The climate change adaptation criteria introduced the concept of "reflective capacity," which encapsulates what tracking progress against targets should represent for all types of actions. Implementing strategies to promote reflective capacity (CCA4.1) is therefore a cross-cutting criterion applicable to both climate and biodiversity action. As such, it provides an additional opportunity to track progress on GHG reductions, climate-related risk reduction, and biodiversity pressure management, coordinating and readjusting strategies for integrated responses

INTEGRATED SET OF CRITERIA FOR CLIMATE-BIODIVERSITY ACTION

ACTION STEPS	SUBSET OF CRITERIA	CRITERIA	RELEVANT CRITERIA
1.ASSESS	**Assess the Project's Estimated GHG Inventory**	CCM1.1. Assess the estimated Scope 1 GHG emissions	BIO1.5
		CCM1.2. Assess the estimated Scope 2 GHG emissions	BIO1.5
		CCM1.3. Assess the estimated Scope 3 GHG emissions (upstream and downstream)	BIO1.3-1.6
		CCM1.4. Assess the estimated avoided emissions as a result of the project	BIO1.5
	Assess Exposure to Climate-Related Hazards/Threats	CCA1.1. Assess exposure to all location-relevant climate-related hazards/threats	BIO1.1-1.2, BIO1.5
	Assess Vulnerability to Climate-Related Risks	CCA1.2. Assess climate-related service continuity risk CCA1.3. Assess climate-related physical asset risk CCA1.4. Assess climate-related resource availability risk (water, materials, land, workforce) CCA1.5. Assess climate-related supply chain continuity risk	BIO1.4, BIO1.6
	Assess the State of Biodiversity	BIO1.1. Assess the state of ecosystems BIO1.2. Assess the state of species	CCA1.1
	Assess Project-Driven Pressures on Biodiversity	BIO1.3. Assess land/freshwater/sea change BIO1.4. Assess resource exploitation BIO1.5. Assess climate change and its related impacts	CCA1.1
		BIO1.6. Assess pollution (water, air, soil, waste, noise, light) BIO1.7. Assess introduction of invasive species	CCM1.1-1.4
	Assess Dependencies on Biodiversity	BIO1.8. Assess project's dependency on ecosystem services	CCM3.5
		BIO1.9. Assess community's dependency on ecosystem services	CCA1.3-1.4
2.COMMIT	**Commit to and Set Ambitious, Integrated Climate and Biodiversity Targets**	CCM2.1. Commit to and set targets aligned with the Paris Agreement goals	CCA2.1, BIO2.1
		CCA2.1. Commit to and set targets for no resilience loss and/or net resilience gain	CCM2.1, BIO2.1
		BIO2.1. Commit to and set ambitious targets aligned with global policy goals to both halt and reverse biodiversity loss	CCM2.1, CCA2.1

3.RESPOND	**Prioritize Management Responses Based on the Mitigation Hierarchy**	CCM3.1. Plan for actions to: • Avoid/eliminate GHG emissions • Reduce GHG emissions that were not avoided • Substitute/replace GHG emissions • Compensate/offset residual GHG emissions	BIO3.1
		CCA3.1. Plan for actions to: • Avoid/eliminate threat • Reduce exposure/vulnerability to threat • Absorb/minimize impact of failure • Restore/recover after shocks or stresses	BIO3.1
		BIO3.1. Plan for actions to: • Avoid impacts on biodiversity • Minimize impacts on biodiversity • Restore impacts on biodiversity • Offset impacts on biodiversity	CCM3.1, CCA3.1
		CCM-CCA-BIO3.1. Plan for transformative actions	
	Implement GHG Reduction Strategies	CCM3.2. Implement energy efficiency strategies	BIO3.2
		CCM3.3. Implement decarbonization strategies	BIO3.2
		CCM3.4. Implement electrification strategies	
		CCM3.5. Implement carbon capture & storage strategies	BIO3.2
		CCM3.6. Optimize material sourcing practices	BIO3.3
		CCM3.7. Implement resource circularity strategies	CCA3.3-3.5, CCA3.7, BIO3.3
		CCM3.8. Optimize project-related transport and logistics	
	Implement Risk Reduction Strategies	CCA3.2. Optimize project siting and site design to avoid threats	BIO3.2
		CCA3.3. Implement strategies to maximize resourcefulness	CCM3.7, BIO3.3
		CCA3.4. Implement strategies to maximize durability	CCM3.7
		CCA3.5. Implement strategies to maximize adaptability	CCM3.7, BIO3.2, BIO3.3
		CCA3.6. Implement strategies to maximize redundancy	CCM3.6-3.7, BIO3.2
		CCA3.7. Implement strategies to maximize integration	CCM3.7
		CCA3.8. Implement strategies to optimize system recovery	
	Implement Biodiversity Strategies	BIO3.2. Implement nature-based solutions (NbS)	CCM3.5, CCM3.7, CCA3.2, CCA3.5
		BIO3.3. Use prevention, abatement, and control measures against pressures on biodiversity (other than NbS)	CCM3.2-3.8
	Implement Enabling Strategies	CCA3.9. Develop local skills and capabilities during all aspects of the project	
		CCA3.10. Promote inclusivity during assessment, commitment, and selection of strategies for climate and biodiversity	
4. TRACK	**Track Progress against Targets Set**	CCA4.1. Implement strategies to promote reflective capacity	CCM4.1, BIO4.1

KEY CHANGES IN THE INTEGRATED SET OF CRITERIA

The previous table merges the standalone sets of criteria into one integrated set—the Integrated Climate-Biodiversity Criteria. Despite the identified overlaps among the criteria, the decision was taken not to eliminate any climate change mitigation, adaptation, or biodiversity action criteria. Overlaps are highlighted by listing the relevant criteria next to each one. The key changes compared to the separate sets of criteria are as follows:

- The different subsets of the commitment criteria were merged into a single group to highlight the need for mutually reinforcing targets across climate change mitigation, adaptation, and biodiversity action
- The first subset of the response criteria, referring to the prioritization of responses based on the mitigation hierarchy, was also merged for all types of action to encourage accounting across all response mechanisms. In addition, the "Transform" step of the hierarchy is emphasized as a common step for all types of action, as transformative action transcends climate change mitigation, adaptation, or biodiversity action alone
- A new subset was introduced under the criteria grouping "enabling strategies," which includes the development of local skills and capabilities across all aspects of a project, as well as the promotion of stakeholder engagement (inclusivity) during all action steps for climate change mitigation, adaptation, and biodiversity. Capacity building and inclusivity—two qualities initially identified as part of the climate change adaptation criteria—are therefore expanded to all types of action. These strategies correspond to two of the cross-cutting principles that apply to all criteria
- The different subsets of the tracking criteria were merged into one common group for all types of action to highlight the need for integrated performance indicators across climate change mitigation, adaptation, and biodiversity action, ensuring that performance is tracked and evaluated in conjunction

The criteria maintain the same color coding, naming conventions (e.g., CCM for climate change mitigation, CCA for adaptation, and BIO for biodiversity), as well as the numbering used in the standalone sets, to indicate the origin of each criterion. Strategies that were identified as cross cutting—those not falling solely under a single type of action, such as transformative action, enabling strategies, and reflective capacity—were rephrased, as needed, to reflect their integrated nature.

THE OPPORTUNITY MATRIX

A key component of the analysis performed for each individual set of criteria was the identification of links between the proposed response criteria (strategies) and the impact drivers and risks they address. This analysis was conducted both to ensure that the proposed list of response criteria is comprehensive in addressing all identified impact drivers and risks (assessment criteria) and to confirm that the criteria can support higher ambition by implementing all steps of the mitigation hierarchy. The analysis revealed several strategies that address more than one impact at a time. A similar process was followed for the cross-analysis of climate action strategies (mitigation and adaptation) and biodiversity strategies. In this case, the mitigation and adaptation strategies were examined in relation to pressures on biodiversity, while biodiversity strategies were reviewed against the drivers of climate change and climate-related risks. This step of the analysis showed that several climate strategies can simultaneously address pressures on biodiversity, and that biodiversity strategies, in turn, can address drivers of climate change and climate risks—particularly those strategies that fall under the category of nature-based solutions. These overlapping strategies are considered opportunities within the present framework.

The matrix table presented in the following pages—Opportunity Matrix—compiles all strategies and all impact drivers and risks included in the separate sets of criteria to highlight the opportunities for joint, integrated climate change mitigation, adaptation, and biodiversity action.

To guide interpretation of the Opportunity Matrix, the image on the right illustrates how areas of opportunities are identified.

The grey-shaded areas in the table represent the expected primary outcomes of strategies—for example, GHG reduction strategies (CCM response criteria) addressing the drivers of climate change (GHG emissions), or climate risk reduction strategies (CCA response criteria) addressing climate-related risks, or biodiversity strategies (BIO response criteria) addressing pressures on biodiversity. When strategies address impacts outside these primary outcome areas, they represent opportunities.

HOW TO READ THE OPPORTUNITY MATRIX:

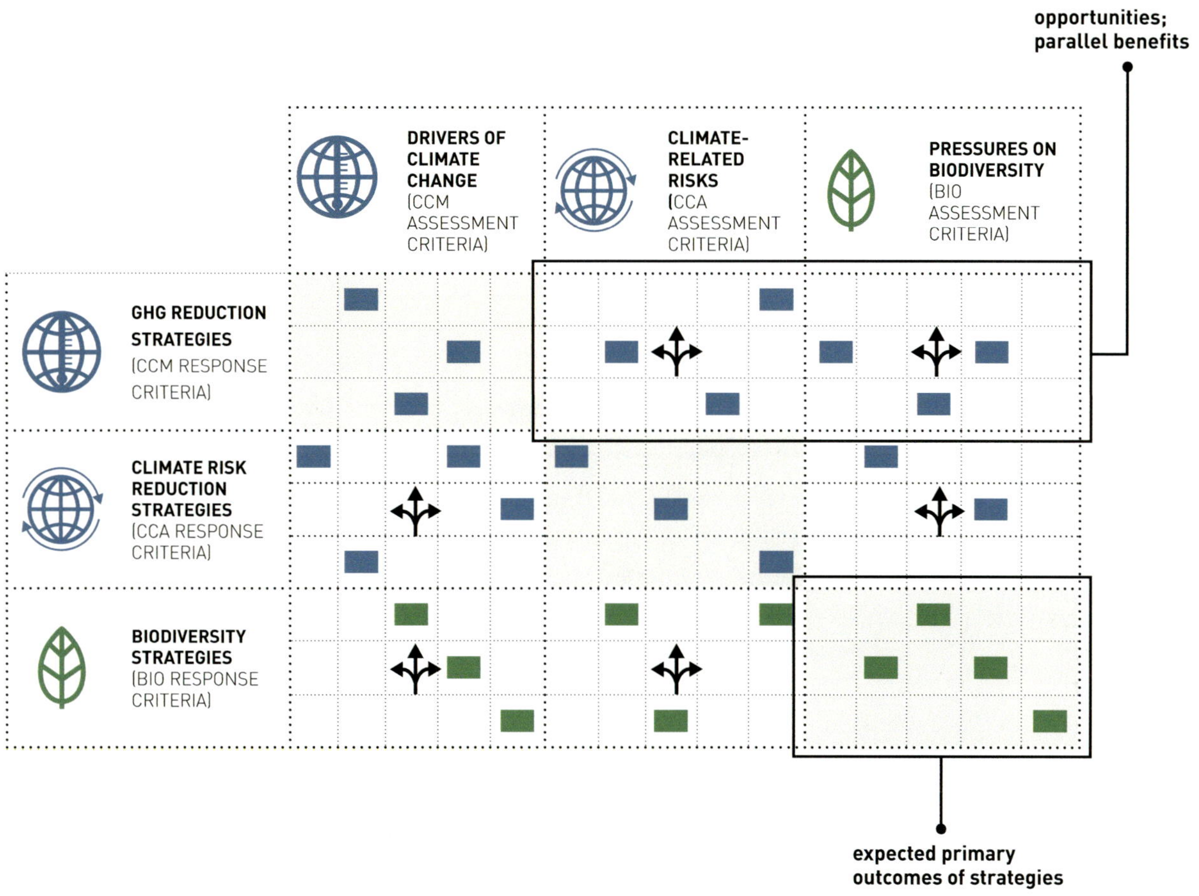

OPPORTUNITY MATRIX

DRIVERS OF CLIMATE CHANGE (CCM ASSESSMENT CRITERIA)

Strategy group	CLIMATE CHANGE MITIGATION, ADAPTATION, AND BIODIVERSITY STRATEGIES:	Scope 1 emissions	Scope 2 emissions	Scope 3 emissions	
				Upstream	Downstream
GHG REDUCTION STRATEGIES (CCM RESPONSE CRITERIA)	CCM3.2. Implement energy efficiency strategies	■	■		
	CCM3.3. Implement decarbonization strategies	■	■		
	CCM3.4. Implement electrification strategies	■			
	CCM3.5. Implement carbon capture & storage strategies	■	■	■	■
	CCM3.6. Optimize material sourcing practices			■	
	CCM3.7. Implement resource circularity strategies	■	■	■	■
	CCM3.8. Optimize project-related transport and logistics			■	■
CLIMATE RISK REDUCTION STRATEGIES (CCA RESPONSE CRITERIA)	CCA3.2. Optimize project siting and site design				
	CCA3.3. Implement strategies to maximize resourcefulness			■	■
	CCA3.4. Implement strategies to maximize durability	■	■	■	■
	CCA3.5. Implement strategies to maximize adaptability	■	■	■	
	CCA3.6. Implement strategies to maximize redundancy				
	CCA3.7. Implement strategies to maximize integration			■	
	CCA3.8. Implement strategies to optimize system recovery				
	CCA3.9. Develop local skills and capabilities				
BIODIVERSITY STRATEGIES (BIO RESPONSE CRITERIA)	BIO3.2. Implement nature-based solutions (NbS)				
	ES. Crop provisioning (energy crops)	■	■		
	ES. Wood provisioning			■	
	ES. Terrestrial and aquatic plants as energy source	■	■	■	
	ES. Raw material provisioning			■	
	ES. Water supply: potable water				
	ES. Water supply: non-potable water for processes				
	ES. Water supply: non-potable water for irrigation				
	ES. Surface water/coastal/marine water as energy source	■	■		
	ES. Global climate regulation	■			
	ES. Rainfall pattern regulation				
	ES. Local climate regulation	■	■		
	ES. Air filtration				
	ES. Soil quality regulation	■			
	ES. Soil erosion control				
	ES. Landslide mitigation				
	ES. Solid waste remediation				■
	ES. Water flow regulation: Baseline & peak flow				■

CLIMATE-RELATED RISKS
(CCA ASSESSMENT CRITERIA)

PRESSURES ON BIODIVERSITY
(BIO ASSESSMENT CRITERIA)

Service continuity	Physical asset	Resource availability				Supply chain	Land/freshwater/sea change	Resource exploitation	Pollution						Climate change	Introduction of invasive species
		Water	Material	Land	Workforce				Water	Air	Soil	Waste	Noise	Light		
														■	■	
															■	
															■	
	■														■	
						■	■	■	■	■	■	■			■	
	■		■			■	■	■	■			■			■	■
						■										
	■														■	
		■	■	■		■	■	■							■	
	■		■		■			■							■	
■			■	■								■			■	
■															■	
■			■	■												
■																
■					■	■										
															■	
		■	■					■							■	
															■	
			■			■		■							■	
■		■				■		■							■	
■		■				■		■							■	
■		■				■		■							■	
															■	
															■	
	■	■													■	
■					■										■	
										■						
											■					
	■										■				■	
	■														■	
												■			■	
■	■	■													■	

OPPORTUNITY MATRIX (CONTINUED)

CLIMATE CHANGE MITIGATION, ADAPTATION, AND BIODIVERSITY STRATEGIES:		DRIVERS OF CLIMATE CHANGE (CCM ASSESSMENT CRITERIA) Scope 1 emissions	Scope 2 emissions	Scope 3 emissions Upstream	Downstream
BIODIVERSITY STRATEGIES (BIO RESPONSE CRITERIA)	ES. Flood mitigation: Coastal protection				
	ES. Storm mitigation (other than water-related events)				
	ES. Flood mitigation: River flood mitigation				
	ES. Water purification: (nutrients and other pollutants)				
	ES. Noise attenuation				
	ES. Biological control: Pest control				
	BIO3.3. Use prevention, abatement, and control measures against pressures on biodiversity (other than NbS)				
	Reclaim brownfields				
	Proper siting strategies	■			
	Wastewater treatment, effluents, runoff control measures				
	Air pollution prevention, abatement, and control measures				
	Soil pollution prevention, abatement, and control measures				
	Noise prevention, abatement, and control measures				
	Light pollution prevention, abatement, and control measures				
	Invasive species prevention, abatement, and control measures				
	Balance earthworks	■	■	■	■

CLIMATE-RELATED RISKS (CCA ASSESSMENT CRITERIA)							PRESSURES ON BIODIVERSITY (BIO ASSESSMENT CRITERIA)									
Service continuity	Physical asset	Resource availability				Supply chain	Land/freshwater/ sea change	Resource exploitation	Pollution						Climate change	Introduction of invasive species
		Water	Material	Land	Work force				Water	Air	Soil	Waste	Noise	Light		
■	■					■									■	
■	■					■									■	
■	■					■									■	
		■							■		■				■	
					■								■			
																■
				■			■				■					
							■				■				■	
		■							■							
										■						
											■					
													■			
														■		
																■
			■					■				■	■			■

KEY STRATEGIES FOR INTEGRATED ACTION

The previous sections have already brought to the surface several strategies that do not fit strictly within climate change mitigation, climate change adaptation, or biodiversity action alone. Diagrams and tables have highlighted various strategies that address more than one impact driver and contribute to all three types of action, summarized in the Opportunity Matrix. Although all strategies listed under the climate change mitigation, adaptation, and biodiversity action response criteria are indispensable components of action, this section highlights four strategies as key opportunities for integrated action:

- Resource circularity
- Optimized material sourcing
- Resourcefulness
- Nature-based solutions

Although referred to as "strategies," these are, in fact, umbrella terms that encompass a wide range of actions. Although referred to as "strategies," these are, in fact, umbrella terms that encompass a wide range of actions. Moreover, they represent concrete examples of how transformative action, the higher level of climate and biodiversity ambition, can be operationalized.

Resource circularity and optimized material sourcing were identified as contributing primarily to climate change mitigation. Resourcefulness was one of the strategies included among the climate change adaptation response criteria, representing solutions aimed at addressing climate change-related risks; more specifically, it is one of the core resilience principles that form the basis of

THE POTENTIAL OF RESOURCE CIRCULARITY FOR CONTRIBUTIONS TO CLIMATE CHANGE MITIGATION, ADAPTATION, AND BIODIVERSITY ACTION

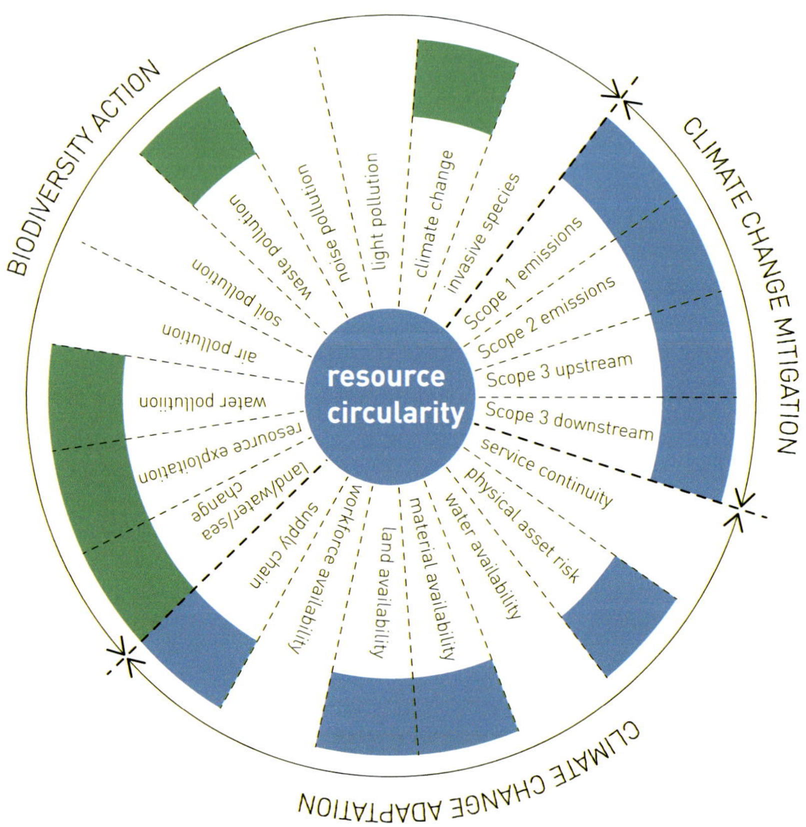

resilience action. Nature-based solutions form the core of the biodiversity action response criteria; however, by definition, they extend beyond the mere boundaries of biodiversity action, representing widely celebrated examples of integrated action.

The following graphs summarize the drivers of climate change, climate change-related risks, and pressures on biodiversity that each strategy can potentially address. In doing so, they illustrate why these strategies represent key components of integrated action.

Resource circularity aims to maintain a circular flow of resources by recovering, retaining, and adding to their value, while minimizing the need for primary material inputs and reducing waste outputs. It is widely recognized as a transformative action required to achieve a net-zero, nature-positive world. It is an inclusive concept that encompasses resource efficiency, zero-waste practices, adaptability, reparability, and durability. As such, it addresses multiple drivers of both climate change and biodiversity loss, while also minimizing climate-related risks. Resource circularity is particularly relevant to infrastructure projects, as it can minimize a project's overall environmental footprint. Only through its potential to extend a project's useful life does it help avoid multiple long-term impacts and risks.

Optimized material sourcing refers to securing sourcing from suppliers that implement environmentally and/or socially sustainable strategies. It helps control potential impacts occurring along the project's supply chain through selection criteria for suppliers and manufacturers of materials, products, and equipment used in the project. When selecting materials and products, projects should evaluate different options based on third-party verified Environmental Product Declarations (EPDs), Product Environmental Footprint (PEF) studies, and other sustainability certifications for products or companies. These established tools provide clearer insight into market practices. EPDs offer manufacturers' information on various environmental properties of building materials—such as carbon intensity, water intensity, and resource depletion—according to relevant ISO standards and Product Category Rules. Certification schemes for suppliers ensure that practices meet specific sustainable performance standards or requirements, such as those addressing site-specific impacts during raw material extraction. These certifications provide transparency regarding a company's products and can drive

THE POTENTIAL OF OPTIMIZED MATERIAL SOURCING FOR CONTRIBUTIONS TO CLIMATE CHANGE MITIGATION, ADAPTATION, AND BIODIVERSITY ACTION

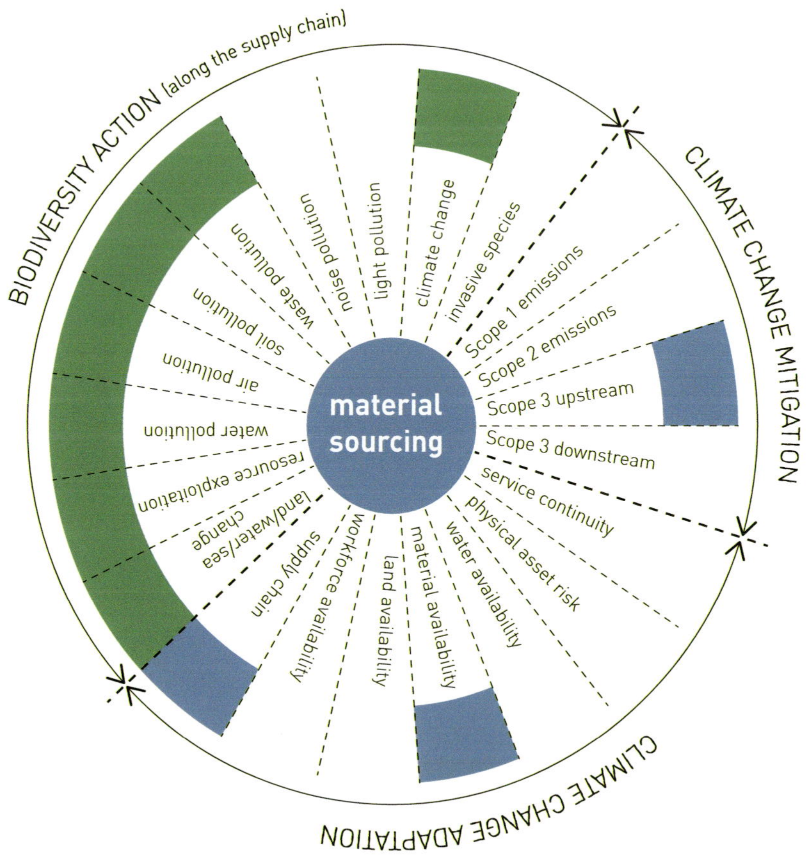

continuous improvement across the industry, enabling transformation.

Resourcefulness is the ability to deliver greater value with fewer inputs, thereby reducing pressure on limited natural resources such as water and primary materials. It encompasses alternative practices that treat the byproducts of processes as valuable resources. Resourcefulness is a broader and more inclusive concept than resource efficiency. It involves experimentation and innovation—testing alternative ideas and resourceful solutions to achieve project goals or meet project needs, such as diversifying sources of materials and water or minimizing the project's footprint.

Resourcefulness shares many common principles with resource circularity, including material efficiency, durability, and adaptability, while also addressing water circularity. Circularity principles offer an opportunity to recognize and capture the full value of water—as a service, an input to processes, a source of energy, and a carrier of nutrients and other materials. Resourcefulness is the strategy that best reflects the need for innovation to maximize project outcomes and can set new examples of best practice—an essential opportunity in a context where higher ambition is urgently required.

Nature-based solutions are widely recognized as a key component of integrated climate and biodiversity action. They are a principal driver of the transformative change needed to achieve global biodiversity goals and nature-positive goals, as they are the only strategies capable of delivering net biodiversity gain.

The need for their accelerated adoption is explicitly highlighted in the targets of key frameworks guiding adaptation and biodiversity action for 2030, including the UAE Framework for Global Climate Resilience and the Kunming-Montreal Global Biodiversity Framework. Although the Paris Agreement does not explicitly refer to NbS, it calls for action to conserve, enhance, and sustainably manage carbon sinks and reservoirs, such as forests and oceans. More recent climate negotiations have increasingly incorporated references to NbS as part of their calls for integrated climate-biodiversity action, recognizing that biodiversity loss can undermine climate mitigation targets.

NbS, if properly designed and managed—for example according to the IUCN criteria that define their intended qualities—can simultaneously support the cross-cutting principles that should underpin climate and biodiversity action, such as sustainability, inclusivity, equity, reflective

THE POTENTIAL OF RESOURCEFULNESS FOR CONTRIBUTIONS TO CLIMATE CHANGE MITIGATION, ADAPTATION, AND BIODIVERSITY ACTION

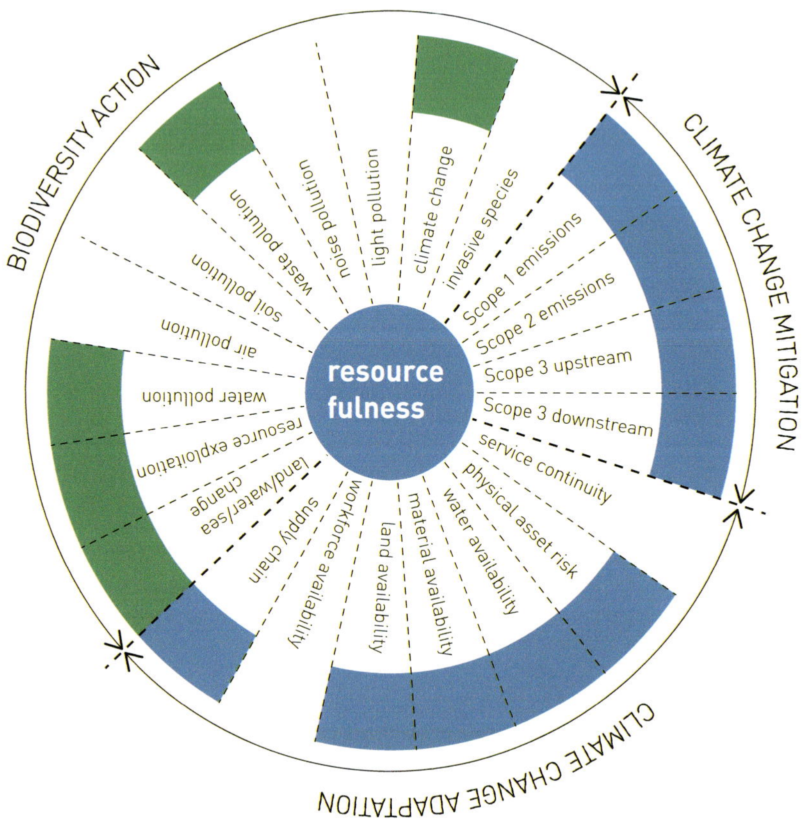

capacity, capacity building, and adherence to social safeguards.

In the present framework, NbS are represented through their contributions, expressed as the full list of ecosystem services. Given the wide variety of NbS, which differ according to site-specific and ecosystem type-specific conditions, the graph below illustrates the potential contributions they can provide. However, due to their inherent multi-functionality and the possibility of combining multiple solutions, NbS hold strong potential to deliver the co-benefits presented. To further support these co-benefits, the framework recommends that project teams assess both the project's and the community's dependencies on biodiversity and design NbS tailored to these needs.

Finally, the four highlighted strategies hold even more opportunities due to their interconnection. For example, optimized material sourcing is a key enabler of resource circularity. NbS due to the inherent capacity of ecosystems for adaptability, can support both resource circularity and resourcefulness.

The present framework calls on project teams to identify and leverage opportunities available to a specific project or location, as these can address multiple impacts and risks at once and maximize positive outcomes, driving higher ambition.

THE POTENTIAL OF NATURE-BASED SOLUTIONS (NbS) FOR CONTRIBUTIONS TO CLIMATE CHANGE MITIGATION, ADAPTATION, AND BIODIVERSITY ACTION

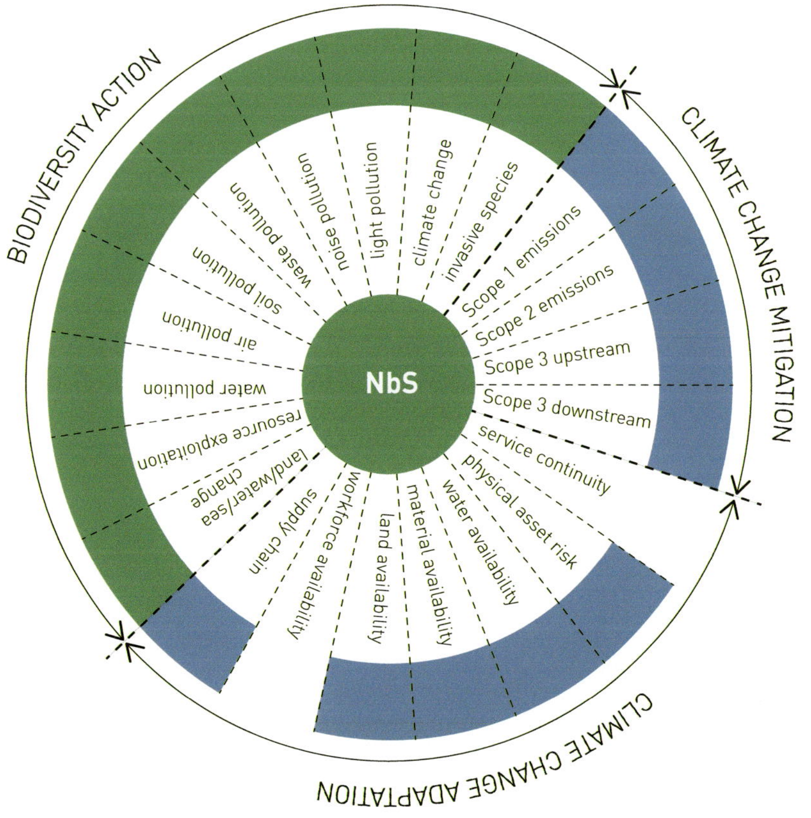

C

CONNECTING THE CRITERIA WITH SUSTAINABLE PROJECT STRATEGIES

Part C connects the integrated criteria to examples of sustainable infrastructure project strategies. Chapter 7 introduces the methodology used to analyze two real-world projects, presented in Chapters 8 and 9.

CONNECTING THE CRITERIA WITH SUSTAINABLE PROJECT STRATEGIES

Methodology

for Project Analysis

Chapter 7

METHODOLOGY FOR PROJECT ANALYSIS

The purpose of this section is to highlight how climate- and biodiversity-related actions are already embedded within sustainable infrastructure projects. The inherent interconnection of sustainability, climate, and biodiversity demonstrates that project teams do not need to start from scratch—many existing strategies already deliver co-benefits for climate change mitigation, adaptation, and biodiversity protection. Yet, achieving a net-zero and nature-positive world requires a higher level of ambition and integration going forward.

Two sustainable infrastructure projects are presented through an alternative analysis based on the climate and biodiversity criteria: one renewable energy generation and one public transit project:

- The Berryessa Transit Center in Santa Clara Valley, California (completed in 2020), part of the Phase I of VTA's Silicon Valley BART extension project
- The Upland Prairie wind farm in Poweshiek County, Iowa (completed in 2018)

At their core, both projects are climate change mitigation initiatives designed to contribute to significant avoided GHG emissions. Yet, they also incorporated adaptation strategies and sought to go beyond simply mitigating negative environmental impacts—actively creating positive biodiversity outcomes. Each leveraged the opportunities provided by natural systems and introduced new biodiversity assets through the use of nature-based solutions.

The Berryessa Transit Center enhances regional connectivity and alleviates congestion by providing green, high-frequency transit service and multiple sustainable mobility alternatives for local communities. Importantly, it successfully integrated a creek restoration project into the station design—serving both as a natural flood defense and a catalyst for restoring a severely degraded habitat.

The Upland Prairie project, meanwhile, increased renewable energy generation capacity at the state level while demonstrating compatibility with ongoing agricultural activities. This successful coexistence was enabled by careful siting and design, and a rigorous local stakeholder engagement process.

Both projects received Envision Platinum Awards from the Institute for Sustainable Infrastructure (ISI), recognizing their outstanding sustainable performance.

The aim here is not to evaluate the projects' direct climate or biodiversity outcomes, but to illustrate how integrated climate and biodiversity strategies are inherently embedded within sustainable project development. The analysis explores the extent to which integrated criteria are already addressed in projects guided primarily by sustainability commitments.

Project descriptions and analyses draw on documentation submitted during ISI's Envision third-party verification process, with information generously provided by the Santa Clara Valley Transportation Authority and Alliant Energy, the respective project owners and leads.

Each project analysis follows four action steps—Assess, Commit, Respond, and Track—corresponding to the structure of the integrated climate-biodiversity

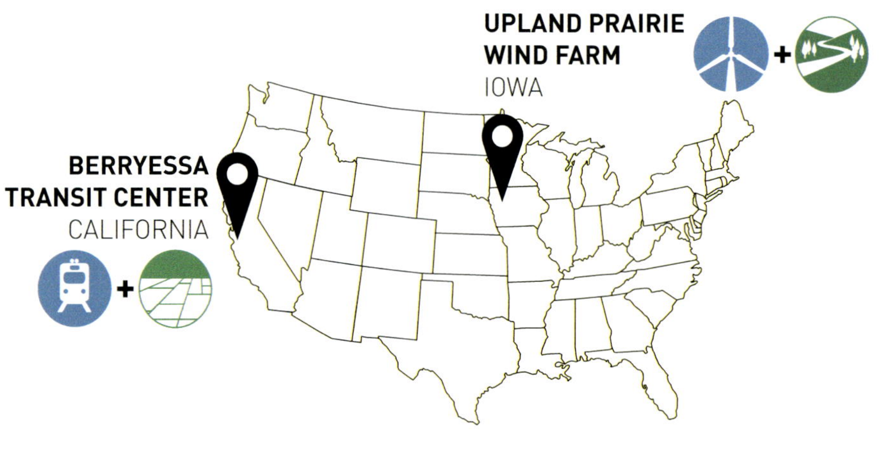

criteria and key stages of sustainable project delivery. The depth of each analysis depends on the availability of project documentation, and the limitations of this publication. Since the integrated criteria were not yet established when these projects were developed, a criterion-by-criterion assessment is not feasible. Instead, the analysis focuses on how specific project strategies address the main drivers of climate impacts, climate-related risks, and pressures on biodiversity identified within the mitigation, adaptation, and biodiversity criteria.

Following a brief description of key impact drivers and risks ("Assess"), each project's main sustainability commitments are outlined in the "Commit" section. In the "Respond" step, the primary sustainable strategies are presented and categorized under broader sustainability goals. These strategies are then systematically compared with the response strategies defined in the integrated criteria framework. This comparison highlights where project actions already align with integrated responses.

Finally, project strategies are mapped against the climate and biodiversity impacts and risks identified in the integrated criteria. This mapping offers a new perspective on the added value of each sustainable strategy when viewed through the lens of integrated action. The value of each strategy is determined by the number of criteria it addresses, revealing opportunities for multiple co-benefits. The analysis shows that several existing sustainability strategies contribute simultaneously to climate mitigation, adaptation, and biodiversity enhancement—emphasizing that while strong foundations exist, achieving a truly net-zero and nature-positive future will require even greater ambition and integration in sustainable infrastructure development.

OVERVIEW OF PROJECT ANALYSIS BASED ON THE INTEGRATED CRITERIA

1. ASSESS

The "Assess" section provides an overview of existing & project-driven impacts, as well as climate change- and biodiversity-related risks identified by the project team requiring mitigation.

2. COMMIT

The "Commit" section outlines the broader context of sustainability, climate, and biodiversity commitments—at state, regional, city, organizational, and project levels—that shaped the project's targeted performance goals.

3. RESPOND

The "Respond" section lists the key sustainable project strategies addressing climate change mitigation, adaptation, and biodiversity enhancement.
A matrix links these strategies to the main drivers of climate change (GHG emissions), biodiversity loss (pressures on biodiversity), and climate-related risks they mitigate, highlighting opportunities for co-benefits across multiple objectives.
In addition, nature-based solutions are mapped to the ecosystem services they support, demonstrating how project strategies respond to both project- and community-level dependencies on biodiversity.

4. TRACK

The "Track" section provides an overview of monitoring mechanisms established by the project to ensure the long-term effectiveness of implemented strategies.

KEY TAKEAWAYS

The "Key Takeaways" illustrate how sustainable project strategies align with the integrated action response criteria (generic strategies), demonstrating that climate and biodiversity action are already embedded in sustainable project design and delivery.

CONNECTING THE CRITERIA WITH SUSTAINABLE PROJECT STRATEGIES

Climate-Biodiversity Action in a

Transit Project

Chapter 8

PROJECT NAME: BERRYESSA TRANSIT STATION
LOCATION: SAN JOSE, CALIFORNIA
AREA: 118,980 SQ M (29.4 ACRES)
PROJECT OWNER: SANTA CLARA VALLEY TRANSPORTATION AUTHORITY (VTA)
PROJECT OPERATOR: BAY AREA RAPID TRANSIT DISTRICT (BART)
BART EXTENSION DESIGN-BUILD TEAM: JOINT VENTURE OF SKANSKA, SHIMMICK, HERZOG
CREEK RESTORATION CONSULTANTS: KIMLEY HORN AND ASSOCIATES; KEISH ENVIRONMENTAL

THE BERRYESSA TRANSIT CENTER

Integration of a creek restoration project into a transit center

PROJECT OVERVIEW

The Berryessa Transit Center, also known as the Berryessa/North San José Station, is an intermodal transit center located in the Berryessa district, northeast of downtown San José, California. It serves as a multimodal regional transportation hub, connecting two Bay Area Rapid Transit (BART) heavy rail lines and several Santa Clara Valley Transportation Authority (VTA's) bus lines.

BART is a rapid transit system serving the San Francisco Bay Area in California through five lines. The station is currently the southern terminus of the Orange and Green BART lines.

The project was completed as part of VTA's BART Silicon Valley Extension Program (BSV).[1] Before the project, the BART network was a nearly 110-mile system serving the counties of Alameda, Contra Costa, San Francisco, and San Mateo, reaching just north of the Santa Clara County limit. The BSV program fills a significant gap in the regional rail system. It will ultimately extend BART service by 16 miles into Santa Clara County, along the former Union Pacific Railroad right-of-way, bringing frequent, convenient, and reliable regional rail service to more than 1.7 million county residents. The project is being built in two phases:

- The completed Phase I extended service for 10 miles from Alameda County and the Warm Springs Station, into Santa Clara County, along the former Union Pacific Railroad (UPRR) right-of-way, with two stations at Milpitas and Berryessa. The $2.4 billion project began in 2012 and opened to passenger service in 2020
- Phase II is currently under implementation, with construction activities having started in 2023 and is anticipated to be completed in 2036. It is a six-mile, four-station extension of the BART system from Berryessa Transit Center through downtown San José to the City of Santa Clara[2]

The Berryessa BART station is one of two stations that are part of the 10-mile BART extension south from Warm Springs to Berryessa.[3] The station is situated in a transition zone between suburban residential development to the north and east, and industrial development to the south and west. The station campus occupies a 29.4-acre site between Berryessa and Mabury Roads, adjacent to the San José Flea Market in East San José.

Project Purpose

The Phase I BART extension was the first step in addressing the increased and pressing demand for public transit connections in the San Francisco Bay Area Region. Overall, the two-phase BART extension will fulfill

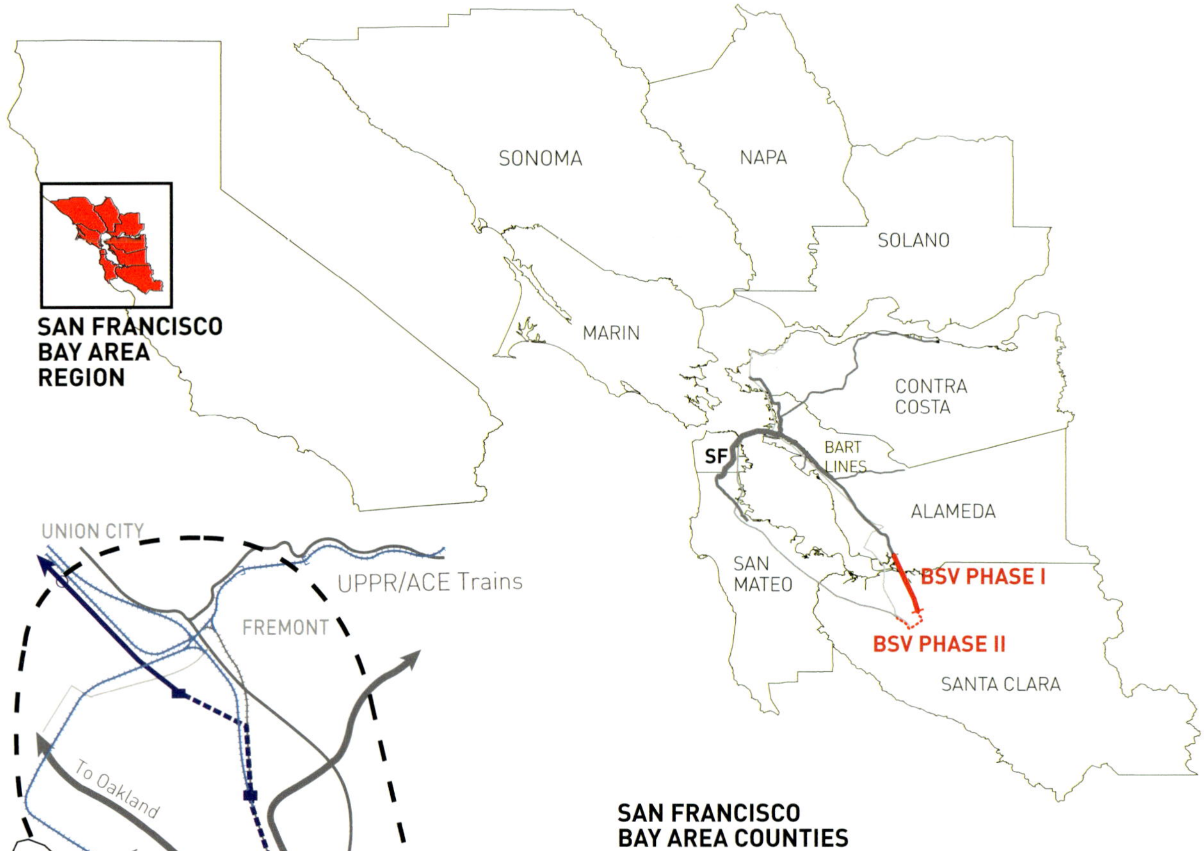

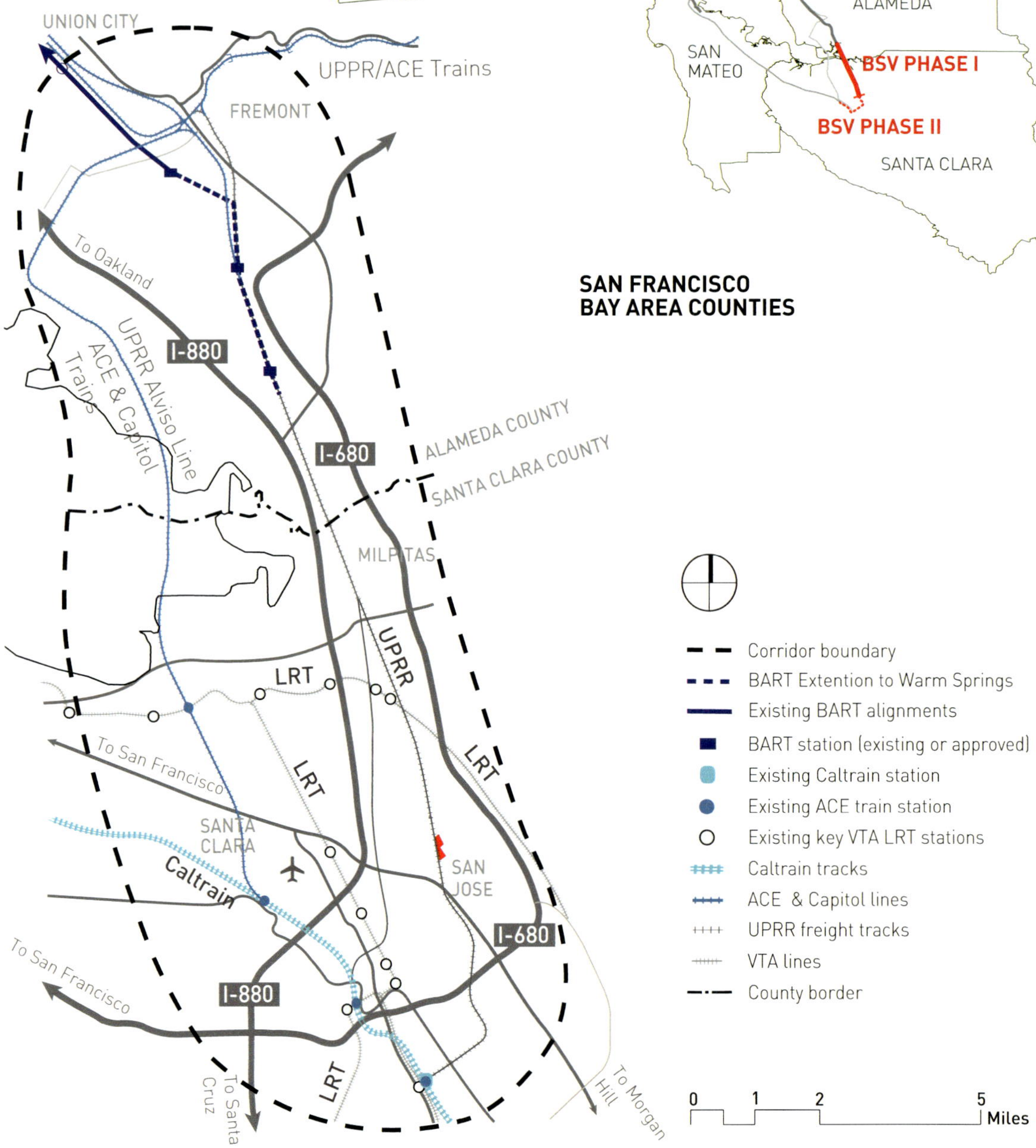

Diagram showing the Silicon Valley Corridor

the vision of "ringing the bay" with high-quality, green transit.[4] Transit improvements are intended to expand mobility options for Santa Clara County and Bay Area residents and help address serious transportation needs that will only become more critical as the region continues to grow in both population and employment. In the City of San José alone, population is expected to increase by 45% by 2030, with 37% more jobs created. With these population increases and new job opportunities, more people will be commuting. The BART extension and new transit centers will provide a new link in the regional rail network, offering direct access to the central business district of the region's largest city, San José, and the core employment areas of Silicon Valley, thereby supporting the region's economic viability and development.

By enabling intermodal connections to Caltrain, intercity light rail, and regional bus service, the project will connect 1.7 million transit-dependent riders annually, including several historically underserved communities in the Bay Area. It represents a clean, green alternative to severely congested commute corridors between the Alameda and Santa Clara County, such as the I-880 and I-680 freeways, and will shorten transit travel times by an average of 30 minutes for a 50-mile trip. Studies show that the project's reduction of 56.4 million vehicle miles traveled (VMT) will result in the avoidance of 50 tons of CO_2 and 19,500 tons of GHG emissions.

It is a forward-looking project designed to absorb future travel demand and an anticipated 52% growth in travel along the Silicon Valley corridor by 2030. The proposed BART extension will improve mobility options to employment, education, medical, retail, and entertainment centers for corridor residents, particularly low-income, youth, the elderly, people with disabilities, and historically underserved ethnic minority populations that have been historically underserved.

The Berryessa Transit Center is designed to accommodate a projected BART daily ridership of up to 25,000 passengers by 2030. This is equivalent to 1,000,000 passenger miles per day, or 365,000,000 passenger miles per year. Moreover, the station enhances the area's environmental resiliency, long-term sustainability, and mobility options for local communities, while improving quality of life. As a multimodal transit facility it strengthens connectivity through the Bay by integrating various modes of public and active transportation.

The future vision for the Berryessa Station is to serve as the center of a transit-oriented, higher-density, mixed-use development, the Berryessa BART Urban Village. The planned 272.5-acre area is part of San José City's General Plan, which aims to guide the city's growth through 2040 toward more efficient and sustainable development patterns. The plan directs business and residential investments in transit-oriented development. It proposes transit urban villages as strategically located growth areas with direct access to major transit facilities. This strategy aims to support the intensification of station areas as regional employment destinations and new residential neighborhoods. The need for housing is heightened in the area due to the ongoing job expansion in Downtown San José. The Berryessa BART Urban Village will create the capacity for over 5,000 new homes, 25% of which are targeted as affordable housing.[5]

The station as a transit node supports compact urban design within walking distance and encourages increased use of the public transportation system. In doing so, it contributes to a lower impact on overcrowded freeways and, by extension, to avoided GHG emissions.

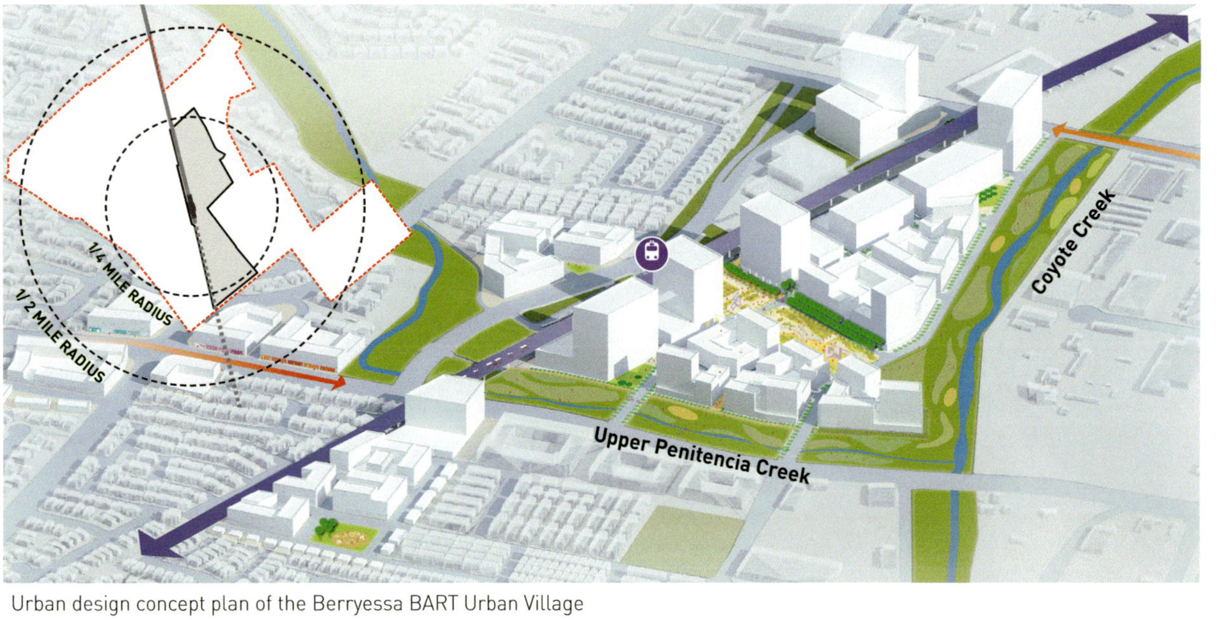

Urban design concept plan of the Berryessa BART Urban Village

1. BART station
2. BART aerial guideway
3. Upper Penitencia Creek restoration project
4. New Station way
5. Bridge over creek
6. Bus Transit Center
7. Outdoor parking lot
8. Parking structure
9. New Police station
10. Bicycle storage facility
11. Traction Power Station
12. Future Parking expansion
13. Contemplative Garden
14. Upper Penitencia Creek trail

KEY PROJECT COMPONENTS

The Berryessa Transit Center project is a combination of technical, technological, and nature-based solutions. To meet its intended function, it incorporates a range of integrated components, including:

- The aerial BART guideway, the heavy rail track system that runs along the former Union Pacific Railroad alignment. Grade-separation of the tracks provides safe crossings beneath the train tracks for pedestrians, cyclists, bus commuters, and drivers
- A BART commuter rail station with connections to VTA's light rail system. The station consists of a ground-level concourse with an aerial boarding platform 35 feet above the ground. Trains arrive every 7.5 minutes, because the station serves two BART lines that operate every 15 minutes
- A bus transit center connecting to regular and express bus lines
- A new road, the approximately 2,700-foot-long Berryessa Station Way located between Berryessa Road and Mabury Road, which connects cars, transit, bicycles and pedestrians. It includes a 10-foot-wide sidewalk and a 12-foot-wide bike path. An approximately 212-foot-long bridge spans the Upper Penitencia Creek
- Bicycle storage area and bicycle parking spaces
- Shared-use trails for bicyclists and pedestrians with connections to the regional bicycle and pedestrian trail called the Bay Area Ridge Trail.[6] Though some portions are yet to be developed, the long-term vision for the Bay Area Ridge Trail is a 550-mile trail that rings San Francisco Bay on the ridgeline, connecting green spaces, and offering recreational opportunities and scenic views
- A drop-off and pick-up area for passengers and employer shuttles
- Park-and-ride facilities, including a 330-space surface parking lot and a seven-level parking structure with 1,164 parking spaces, equipped with 25 EV charging stations. The design of the station includes provision for future expansion of the parking structure that would increase the parking capacity to 2,000 parking spaces
- Public spaces featuring locally created public art
- A Native American contemplative garden designed to honor the Ohlone Indian tribe and educate the public about how they once lived and worked. The garden includes a stone monument, interpretive panel, educational signage, and native plants that were traditionally used by the Ohlone. The project team worked extensively with local community members of the Muwekma Ohlone Indian Tribe of the San Francisco Bay Area to develop the language of the interpretive panel, the plant identification signs, and the monument
- A new BART police station
- The Upper Penitencia Creek Improvement Project
- A VTA ancillary building and station plaza
- Extension of the local recycled water line to the site
- 20 bioretention basins, as well as tree wells to promote groundwater recharge and proper drainage during large storm events

The station is accessible from a public plaza on the eastern side of the site, as well as from the San Jose flea market on the northern and southern sides. The flea market is an important aspect of the area's local character. The project pays tribute to this landmark through a featured landscaping art piece.

View of the station's aerial boarding platform and the ground-level concourse

THE UPPER PENITENCIA CREEK RESTORATION PROJECT

The station is positioned approximately 1,400 feet upstream from the confluence of Upper Penitencia Creek with Coyote Creek in eastern San Jose, before discharging to the San Francisco Bay. A section of Upper Penitencia Creek runs approximately 920 linear feet over the project site. In addition to encouraging a modal shift from single-occupancy vehicles to transit, the project also restored and enhanced the creek.

The Upper Penitencia Creek Improvement project is a compensation measure to offset permanent and temporary impacts on Federal and State waters and wetlands due to the construction of the BART Silicon Valley Berryessa extension (SVBX) project, which occurred outside the Berryessa station project site. A total of 63 linear feet of creek and 0.5 acres of ditch wetlands have been affected by bridge replacement, culvert replacement, and loss of earthen drainage ditches along the 10-mile railroad alignment. VTA committed to compensating for the permanent loss of wetlands at a minimum 2:1 ratio (replacement area: loss area) and the temporary loss of wetlands at a minimum 1:1 ratio.

Before the project, the creek was a narrow (approximately 30 to 50 feet wide) and shallow confined man-made channel with minimal floodplain, characterized by failing stream banks. The channel that extends to Coyote Creek was confined by existing infrastructure and property lines. The previous site use was extensively developed, with commercial/industrial buildings and parking lots covering 27.4 acres of the 29.4-acre total site.

The goals of this restoration project were to:

- Create a more natural channel with enhanced floodplain and restored habitat
- Enhance the hydrologic and geomorphic functions of the creek by creating a primary channel sized to convey the approximate annual flood, a high-flow channel to convey sediment and water during floods greater than the annual flood, and a floodplain sized to convey the 2-year recurrence interval flood
- Improve fish passage and refuge for the federally and state-listed Central California Coast steelhead

The need for creek improvements was identified as part of a watershed-wide plan, the Resilient Landscape Vision for Upper Penitencia Creek, prepared by the

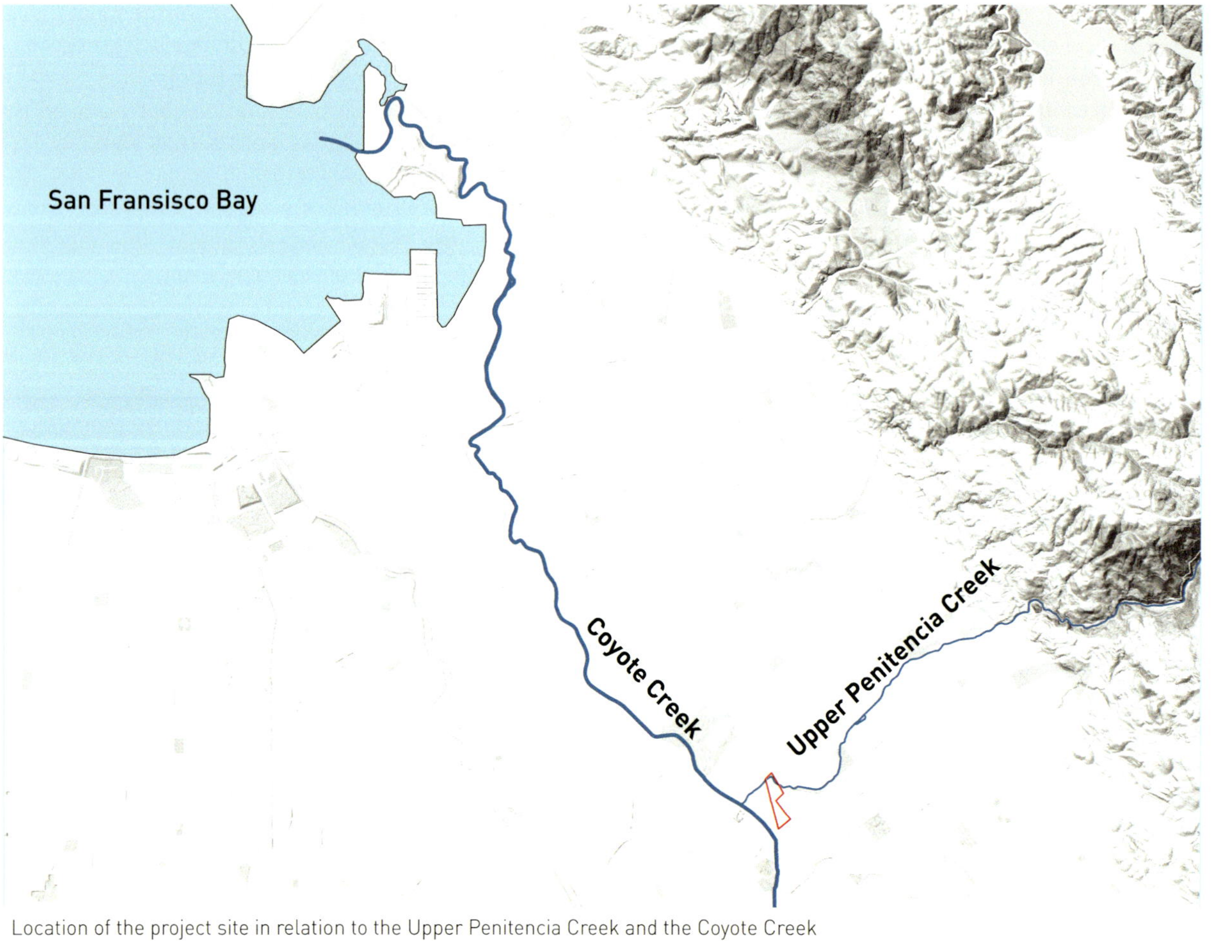

Location of the project site in relation to the Upper Penitencia Creek and the Coyote Creek

San Francisco Estuary Institute. This conceptual plan suggested some improvement actions that were fully integrated by the project.

The Upper Penitencia Creek Improvement project, covering an area of 3.74 acres includes:

- The realignment, restoration, and enhancement of a portion of the creek (920 linear feet)
- The creation of approximately 1.06 acres of seasonal floodplain wetlands
- The restoration of approximately one acre of riparian habitat
- Creation of appropriate geomorphology (riffles and pools) for steelhead migration, refuge, and resting
- The repair of the highly degraded condition of the eastern and northern banks of the creek
- Installation of bioengineered bank treatment structures to minimize significant channel alterations under flood conditions (such as the erosion of a section of bank that causes a shift in the stream location) and encourage hydraulic and habitat complexity
- Reconstruction of existing stormwater outfalls to connect to the new channel alignment and enhancement using bioengineering structures to minimize erosion or scour
- New irrigation lines for plant establishment

The construction of the creek improvement started in 2012, preceding the construction of the station campus.

The project removed constraints and restored and expanded the creek and its floodplain. Thus, from a confined man-made channel with little, if any, suggestion of functioning stream corridor, Upper Penitencia was converted into a free spanning creek with a 200-foot-wide floodplain through regrading into a more natural configuration.

The project removed 4.5 acres of concrete and industrial buildings. It replaced this hardscape with a 200-foot-wide creek and adjacent areas habitat restoration that included realigning the creek channel, daylighting approximately 0.02 acres of creek that was previously enclosed in a double box culvert, replanting riparian and upland habitat with native plantings, and creating over one acre of new wetland habitat.

After the revegetation of the creek in 2013 and through 2019, annual monitoring and adaptive management practices were performed to evaluate the condition of the restoration site and document the success of plant establishment. Moreover, it is worth highlighting that the project has already been tested during a 10-year storm event in 2017, the wettest year on record in the Bay Area, which resulted in severe flooding events. The project performed as it was supposed to; the low- and high-flood channels were activated, and flooding was successfully contained within the project's boundary.[7]

The creek restoration project that was integrated into the Berryessa BART station design is one of the main reasons this project was selected to be presented as part of this publication. During the Envision third-party verification process, the strategies implemented as part of the creek restoration addressed multiple Envision criteria, exemplifying the multi-functionality and parallel co-benefits of nature-based solutions, as well as what a meaningful compensation measure should represent.

Flood event during 2017, while construction of the station was still underway

Flood event after construction completion

The Berryessa station site before (left) and after project completion (right)

ANALYSIS BASED ON THE INTEGRATED CRITERIA

ASSESS

The assessment of the project's impact on GHG emissions, the main driver of climate change, cannot be considered in isolation from the overall BART extension project. A comprehensive analysis would require a regional-scale assessment encompassing the entire Bay Area, which is beyond the scope of this publication. Therefore, this section provides only a concise overview of existing conditions related to GHG emissions, air quality, congestion, or equitable access to public transport—factors that defined the project's purpose and need. These conditions also established the baseline against which the required performance of the BART extension project was determined.

A more detailed assessment of the baseline conditions at the Berryessa Transit Center site focuses on exposure to climate change-related threats and risks, and the state of biodiversity prior to project implementation.

Assessment of the Project's Contribution to the Bay Area's GHG Inventory

The Silicon Valley Rapid Transit Corridor is considered a geographically constrained and severely congested freeway corridor due to current and anticipated future growth in employment and population. The studies that were conducted as part of the project indicated:

- Limited options for low-income and transit-dependent populations: The socioeconomic assessment in the area served by the corridor highlighted the concentration of minority and low-income populations. 11% of the households did not have a private vehicle, and 10% had incomes below the poverty level. Both percentages indicated strong transit dependency. Also, there were high percentages of minority groups, predominantly of Hispanic ancestry
- Need to support the 2030 projected population growth in the cities and counties along the alignment and their residential and employment growth as envisioned by local general plans: For the City of San José, which is served by the Berryessa Station, a 43.2% growth in population, 45.1% in housing, and 37% in jobs was projected as compared to 2000
- Worsening regional air quality and excessive GHG emissions: The Bay Area did not comply with federal and state air quality standards for ozone, particulate matter, PM10 and PM2.5, which can cause respiratory problems. Automobile emissions were identified as the primary source of these and other air pollutants in the area. With steadily increasing travel, it would become more difficult to substantially improve air quality. Congestion and slowing travel speeds tend to worsen the problem because there is a direct relationship between vehicle miles traveled (VMT), travel speed, and air pollution

The BART extension project was selected after the completion of a Major Investment Study in 2000. The study evaluated "eleven alternatives representing various modes of travel (express bus, bus rapid transit, commuter rail, diesel and electric light rail, and BART) and various alignments and stations located in the cities of Fremont, Milpitas, San José, and Santa Clara."[8] All alternatives were assessed for 2030 conditions and service levels to determine their long-term consequences. After an extensive screening and multi-criteria evaluation process, the BART extension was adopted as the preferred strategy by the VTA Board of Directors in 2001.[9] Based on projections for the year 2030, roadways and other transit modes, such as express bus and light rail, are not viable alternatives for meeting future travel demand. When compared to the other alternatives, the BART on the former UPRR alignment alternative offered the fastest travel times to passenger destinations; the greatest congestion relief; improved air quality; best regional connectivity; lowest traffic safety impacts due to the fully grade-separated guideway; and the most consistency with local land use plans and policies.

As a federally and state-funded project, the Silicon Valley extension project was subject to both the US National Environmental Protection Act (NEPA) and the state-level California Environmental Quality Act (CEQA) review process. They are regulatory environmental review processes during which a federal or California state agency assesses the potential effects of a proposed major action before approving the action. In accordance with NEPA and CEQA requirements, a combined Draft Environmental Impact Statement/Environmental Impact Report (EIS/EIR) was prepared and released for public review and comment. A Supplemental EIR, an update to the prior EIR to address project design refinements, was also prepared.

As part of the environmental review process, an assessment of the existing location-specific conditions and the expected project-driven impacts was conducted. Among the impacts considered were various drivers of climate change and biodiversity loss, as well as climate change-related risks. The assessment included, among others, the potential consequences on transportation and transit, air quality, vegetation communities, wetlands and other US waters, special status species, community services, environmental justice, land use, noise and vibration, water resources (groundwater and surface), water quality and floodplains, and GHG emissions. Short-term impacts, including construction impacts, were also considered.

The BART project was concluded to have a positive impact on GHG emissions as compared to the "no project" conditions and other alternative transport mode options. The project, by definition, contributes to avoided GHG emissions by providing a low-emission, green alternative to private vehicles, alleviating congestion, and providing connections to active transportation. The Berryessa Station, which is the focus of the present analysis, is an integral part of the project that contributes to and supports the overall project benefits.

Climate Change-related Threats and Risks Assessment

The project is located in an area where the impacts of climate change are already present. California's Fourth Climate Change Assessment: San Francisco Bay Area Region Report of 2018 highlights the area's exposure to major climate-related threats based on observed and projected climate factors:[10]

- "Overall, the Bay Area's average annual maximum temperature increased by 0.95 °C from 1950 to 2005"
- "The coastal fog along the California coast, so critical to the Bay Area climate, is less frequent than before"
- "Sea level in the Bay Area has risen over 20 cm in the last 100 years"
- "The powerful 2015-16 El Niño, one of the three largest in the historical record, resulted in winter wave energy that was over 50% larger than the typical winter in the Bay Area, driving unprecedented outer coast beach erosion"
- "The 2012-2016 California drought led to the most severe moisture deficits in the last 1,200 years and a 1-in-500-year low in Sierra snowpack "

"Even with substantial global efforts to reduce GHG emissions, the Bay Area will likely experience a significant temperature increase by mid-century."[11] These increases are expected to result in longer and deeper droughts, posing a major risk to water supplies and natural ecosystems and increasing the frequency of extreme wildfires. Precipitation in the Bay Area is expected to "continue to exhibit high year-to-year variability, with very wet and dry years. The Bay Area's largest winter storms will likely become more intense and potentially more damaging."[12]

Moreover, the California coast is exposed to "median sea level rise between 0.74 m (RCP4.5) and 1.37 m (RCP8.5) by 2100."[13]

Both BART and VTA's transit networks are exposed to locally known vulnerabilities and hazards associated with flooding, extreme heat, wildfire, and drought. The agencies have developed plans to respond and adapt to the unavoidable consequences of climate change.

CLIMATE CHANGE-RELATED THREATS AND RISKS

CLIMATE CHANGE-RELATED ACUTE SHOCKS AND CHRONIC STRESSORS	CLIMATE CHANGE-RELATED RISKS
Increased severity and duration of flooding	Risk of exacerbating existing and projected onsite and adjacent communities' flooding
	Increased risk of spills and stormwater contamination
	Safety risk for ridership and service continuity during rain events
Extreme heat	Risk of exacerbating existing and projected heat island effect
	Safety risk for workers and public
Increased frequency and duration droughts	Risk of insufficient capacity to address project water needs for landscaping
	Risk of exacerbating water shortages in a water-stressed region

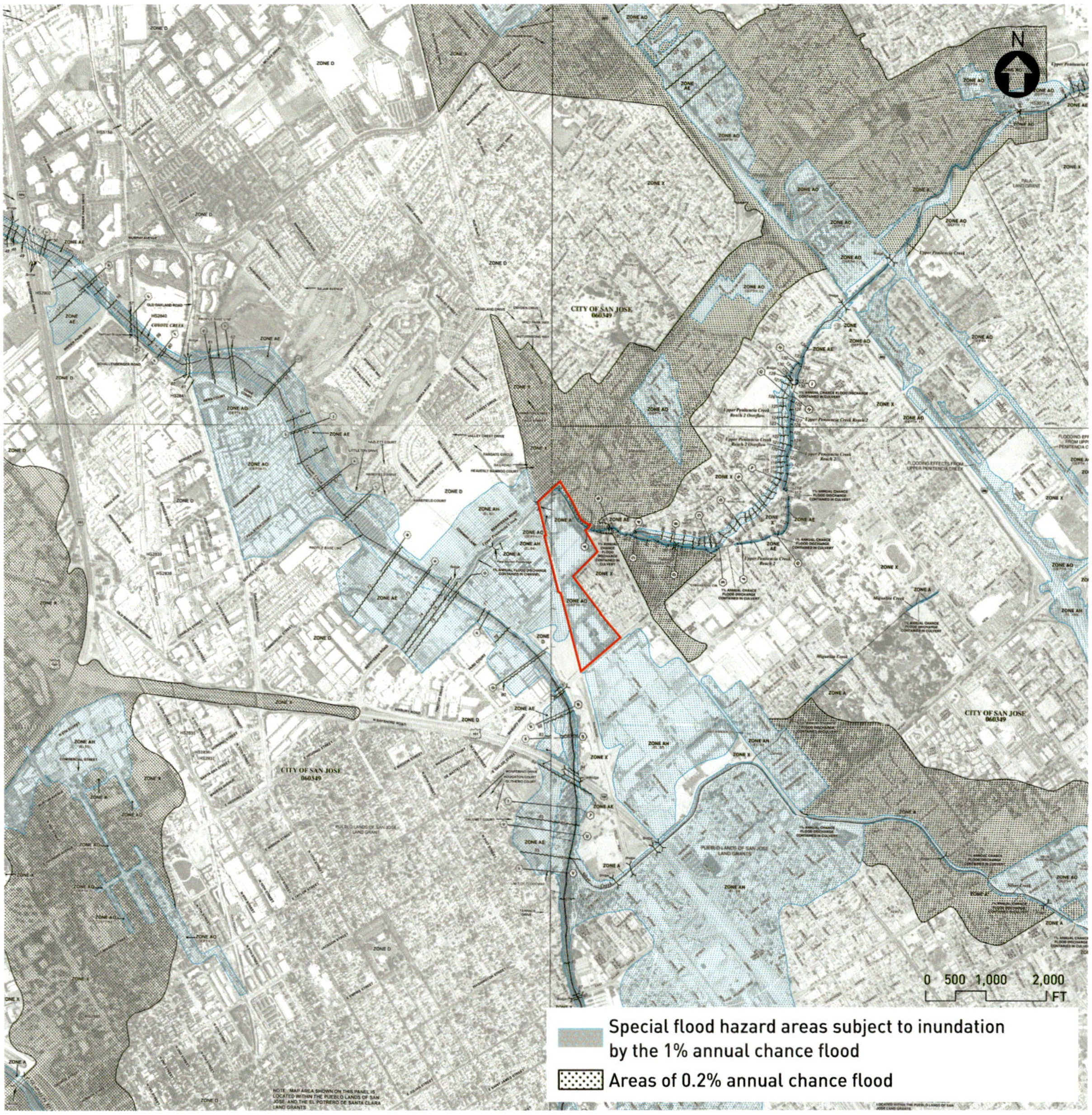

Flood hazard areas based on FEMA FIRMs (2009)

The site-specific climate threats that were identified for the Berryessa station included extreme heat, temporary inland flooding, and increased frequency and duration of droughts. These threats are not only project-specific but also pose a risk to the connected infrastructure and adjacent communities.

Assessment of Flooding Risk

To assess existing drainage conditions and potential flooding at the Berryessa station campus, the 2009 Federal Emergency Management Agency (FEMA) Flood Insurance Rate Maps (FIRMs) and Flood Insurance Study (FIS) were reviewed. According to the FEMA FIRMs, the station site lies within Zone AO—an area subject to 1-percent-annual-chance shallow flooding—with average flood depths ranging from one to three feet. The project site specifically experienced an approximate flood depth of one foot.

A detailed hydrologic and hydraulic analysis was also conducted, which included a review of available data from reports and studies, reconnaissance surveys, and determination of 100-year and 500-year design flows.[14] The Hydrologic Engineering Center's River Analysis System (HEC-RAS) computer model was utilized to estimate design water surface elevations for both existing conditions and future scenarios incorporating the proposed project.

VTA partnered with the Santa Clara Valley Water

OVERVIEW OF THE STATE OF BIODIVERSITY AT THE PROJECT SITE

	BART station campus site	**Upper Penitencia creek restoration site**
Site of high ecological value	No	No
Proximity to ecosystem	Adjacent to freshwater body (creek)	Containing freshwater body (creek)
Pre-existing site condition	Business park and warehouses	Degraded creek channel and adjacent floodplain
STATE OF BIODIVERSITY (EXISTING)		
Species		Sensitive status species: The federally and state-listed Central California Coast steelhead (*Oncorhynchus mykiss*)
Ecosystem	Ornamental non-native plants	Riparian habitat highly altered and degraded by urbanization and instream debris from adjacent construction works

The creek upper bank with informal structures to protect the adjacent properties foundations

(left and middle) The existing Business Park access double box culvert; (right) the existing UPRR bridge

District (SCVWD) and the US Army Corps of Engineers (USACE), which were already planning upstream and downstream improvements to the Upper Penitencia Creek. The reach of the creek extending through the urbanized area of San José—from Dorel Road to Coyote Creek, approximately 3.3 miles—was identified as under capacity to convey the 100-year flood flow. Progressive urbanization had converted the historical alluvial fan and floodplain surfaces into a confined creek channel with minimal floodplain.

Similar to the upstream and downstream conditions, the portion of Upper Penitencia Creek channel that runs through the site also lacked the capacity to convey the 100-year design flow of 4,750 cubic feet per second (cfs), based on a hydrological model for the Upper Penitencia Creek developed by the USACE Hydrologic Engineering Office in 2001. Two bridge crossings—a Union Pacific Railroad (UPRR) wooden trestle with two bents and a roadway double concrete box culvert—further narrowed the channel's section.

With a limited conveyance capacity of 760 cfs, the channel was prone to overtopping, and adjacent properties historically experienced flooding during high-flow events. Such flooding could affect the safety of ridership at the station and disrupt revenue service during storm events.

Assessment of the State of Biodiversity

The Upper Penitencia Creek restoration project was included in the project's scope to restore a creek with sensitive-status species' habitats in degraded condition. The Upper Penitencia Creek area was surveyed for areas of high ecological value and areas within 5 miles were inventoried for special status species. It was identified as a perennial stream that had been highly altered. The industrial park that existed prior to the project included multiple buildings, a bridge to Berryessa Road, a parking lot, and ornamental plantings. The previous site was impervious, except for a triangular rail laydown area near Mabury Road, and a narrow corridor within Upper Penitencia Creek, which was a linear, earthen channel.

The native riparian habitat, limited to the immediate southwest bank of the stream, was altered and degraded by urbanization and instream debris, the construction of the business park and parking lot, vegetation maintenance activities, and the introduction of non-native species. Canary ivy was growing in much of the riparian habitat, strangling many of the native riparian trees. No riparian habitat remained along the northeast corner of the site because most of the creek's upper bank was lined with riprap and floodwalls to protect the foundations of adjacent properties and adjacent parking areas. Moreover, there were signs of failed bank protection and a deteriorating condition. There was no floodplain vegetation for wildlife dispersal through the area to the habitats located east of the project along the creek.

At the same time, studies by the San Francisco identified the Penitencia Creek as a regionally important corridor for habitat connectivity from the Diablo Range to Santa Clara Valley. Historically, the Upper Penitencia Creek riparian corridor has provided pathways for a

SUMMARY OF EXISTING AND PROJECT-DRIVEN PRESSURES ON BIODIVERSITY

PRESSURES ON BIODIVERSITY		EXISTING	POTENTIALLY PROJECT DRIVEN
Land change, freshwater, sea change		Freshwater change • Former historically alluvial fan and floodplain surfaces converted into a narrow creek channel with minimal floodplain • Riparian habitat highly altered and degraded by urbanization. Land change • Ornamental non-native plants	
Pollution	**Water**	In-stream debris from adjacent construction works	Runoff
	Air		Temporary impacts during construction
	Soil		Application of fertilizers and pesticides
	Waste	In-stream debris from adjacent construction works	During construction, maintenance, and end-of-life
	Noise		Operational noise and vibration
	Light		Exterior lighting during operations
Introduction of invasive species		Canary ivy had invaded much of the riparian habitat	Potential introduction during construction works through imported soils

DEPENDENCIES ON ECOSYSTEM SERVICES

ECOSYSTEM SERVICES			DEPENDENCIES	
			project	community
PROVISIONING SERVICES	Biomass provisioning	Food crops		
		Grazed biomass provisioning		
		Livestock provisioning		
		Aquaculture provisioning		
		Wood provisioning		
	Terrestrial and aquatic plants as energy source			
	Other raw material provisioning			
	Genetic material			
	Water supply	Potable water		■
		Non-potable water for processes		
		Non-potable water for irrigation	■	■
REGULATING AND MAINTENANCE SERVICES	Global climate regulation		■	■
	Rainfall pattern regulation (at sub-continental scale)			
	Local (micro and meso) climate regulation		■	■
	Air filtration		■	■
	Soil quality regulation		■	
	Soil and sediment retention	Soil erosion control	■	■
		Landslide mitigation		
	Solid waste remediation			
	Water purification (water quality amelioration)	Retention and breakdown of nutrients	■	■
		Retention and breakdown of other pollutants	■	■
	Water flow regulation	Baseline flow maintenance	■	
		Peak flow mitigation	■	
	Flood mitigation	Coastal protection		
		River flood mitigation	■	
	Storm mitigation			
	Noise attenuation		■	■
	Pollination			■
	Biological control	Pest control	■	■
		Disease control		
	Nursery population and habitat maintenance			■
CULTURAL SERVICES	Recreation-related			■
	Visual amenity			■
	Education, scientific and research			■
	Spiritual, artistic and symbolic			■
	Ecosystem and species appreciation			

variety of mammals and birds, such as bobcats and oak titmice, to disperse from the Diablo Range westward into the urbanized Santa Clara Valley.

The Upper Penitencia Creek site was considered potentially suitable habitat for three federally listed threatened species: the California red-legged frog, the California tiger salamander, and the Central Coast steelhead. The California red-legged frog and the California tiger salamander are federally listed as threatened species by the US Fish and Wildlife Service (USFWS) and the California Department of Fish and Game (CDFG). The Central Coast steelhead is a medium to large-sized fish that spends part of its lifecycle in freshwater and part of it in the ocean. It is federally listed as a threatened species by the National Marine Fisheries Service (NFMS). It is known to occur in Upper Penitencia and Coyote creeks.

Therefore, the project presented an opportunity to support and maximize habitat connectivity and wildlife movement and provide onsite habitat for threatened species.

Assessment of Other Project-driven Impacts that Required Mitigation Measures

As mentioned, during the regulatory environmental review and compliance process, an assessment of the expected project-driven impacts was also conducted. The assessment identified the following potential impacts that needed to be mitigated through project strategies:

- Consumption of primary resources
- Temporary impacts on air quality due to construction works
- Operational and construction noise and vibration
- Light emissions due to outdoor lighting
- Impacts on water resources (groundwater and surface) due to project water needs
- Impacts on water quality and floodplains
- Impacts on soil quality
- Introduction of invasive species

With the exception of air quality, all other potential impacts could be permanent or temporary (during the construction period). To ensure compliance, efforts should be made to mitigate the identified impacts to the extent possible and on a best-effort basis, in line with the mitigation hierarchy. Moreover, environmental compliance requires evidence that mitigation measures have been implemented as part of a project, as well as monitoring of their actual performance and efficacy of measures.

All these impacts coincide with the pressures on biodiversity that are listed under the biodiversity action assessment criteria.

COMMIT

The Berryessa BART extension exemplifies a project implemented in line with state-, organization-, and project-level leadership and commitment to tackle climate change. California is a global leader in setting proactive climate change policy and has set its first major climate change mitigation goal in 2006 with the passing of the California Global Warming Solutions Act, or Assembly Bill 32 (AB 32). VTA and BART, the lead project agencies, developed a project consistent with state-level commitments. Project design and construction policies set the specific climate-related and sustainable performance requirements that the project had to comply with.

Commitment to Broader Climate-Related Goals and Target Setting

The Santa Clara Valley Transportation Authority (VTA), the lead agency responsible for the project, is an independent special district that provides sustainable, accessible, community-focused transportation options throughout Santa Clara County, including bus, light rail, and paratransit services. VTA is focused on reducing GHG emissions from both its internal operations and the countywide transportation system. VTA has a high degree of jurisdictional control over its own operations, and it also plays a significant role in shaping and influencing countywide transportation in the broader community and region through its role as a congestion management and countywide transportation agency.[15]

Given its role, VTA supports legislative efforts to avert climate change by federal, state, regional, and local entities and contribute to reduced greenhouse gas emissions in the transportation sector. VTA established a Sustainability Program in 2008 with commitment "to proactively reduce the consumption of natural resources, the creation of greenhouse gases, and the generation of pollution in the provision of public transportation services."[16] The Sustainability Plan, adopted in 2020, established specific objectives and targets to measure VTA's performance through 2040, including greenhouse gas (GHG) emissions, criteria air pollutants, energy use (buildings and fleet), water, and waste.[17]

In response to the increase in greenhouse gases, ocean warming and acidification, floods, droughts, wildfires, the global rise in temperatures, and public health consequences due to climate change, in 2020 VTA formally declared a climate change emergency,[18] becoming one of the first transit agencies in the country to make such a declaration and joining with local governments to demand immediate climate action. Additionally, the declaration directed VTA to take several specific actions, including developing a plan to guide its climate change emergency response. VTA committed to identifying specific metrics, including objectives and targets, to prioritize opportunities to reduce greenhouse gas emissions and to develop an action plan to guide its climate emergency response.

VTA's Climate Action and Adaptation Plan (CAAP) was released in February 2024 to reduce greenhouse gas emissions (GHG) associated with transportation use and operations and prepare for climate change impacts that are already occurring and projected to worsen over time.[19]

BART, the station operator, is one of the USA's cleanest transit systems, emitting fewer pounds of CO_2e per passenger mile than any other transit system in its class. BART trains are 100% electric, with over 97% of that power coming from zero- and low-carbon sources, including solar and hydro energy. BART is increasing its use of low-carbon, zero-carbon and renewable energy sources, with the goal of transitioning to 100% carbon-free by 2035 and 100% renewable energy by 2045.[20]

As part of the BART extension, three plans have been adopted for the purposes of reducing GHG emissions:

- The state-level AB 32 Climate Change Scoping Plan of 2008,[21] which outlined actions to reduce California's GHG emissions to 1990 levels by 2020 (equivalent to a 15% reduction below 2008 levels), in accordance with the California Global Warming Solutions Act of 2006 (Assembly Bill 32). The Act also included a long-term target for 2050, to reduce emissions 80% below 1990 levels
- The City of San Jose GHG Reduction Strategy and City of Santa Clara Climate Action Plan (CAP),[22] which identify city-level strategies to achieve the state-recommended GHG emissions reduction target for 2020

Additionally, the project meets the goals established in the long-range Santa Clara Valley Transportation Plan for 2030[23] for improved transit options, congestion relief, promotion of active transport; addresses issues identified in the Metropolitan Transportation Commission's Regional Transportation Plans for 2030, and is consistent with the Envision San Jose 2040 General Plan,[24] which sets the city's blueprint for sustainable growth through transit-oriented communities.

Commitment to Environmentally Sustainable Projects

Apart from the project's alignment with climate change mitigation goals and targets, the project's design and construction are consistent with specific state, city, and agency plans and policies.

The project follows the requirements of EPA's National Pollutant Discharge Elimination System (NPDES) Stormwater Program, which mandates that all transportation authorities manage stormwater discharges from their projects via regulated Municipal Separate Storm Sewer Systems (MS4s). The NPDES program also requires operators of regulated transportation systems to implement MS4s to reduce the discharge of pollutants to the maximum extent practicable. Operators must establish a comprehensive Stormwater Management Program (SWMP) under their NPDES permit.[25] The SWMP developed for the project addresses both long-term stormwater management and short-term runoff control during construction works.

VTA has internal policies that apply to all capital projects and set specific sustainability requirements for project teams to meet. Examples include:

- Policy to conserve water resources, reduce or prevent pollution, and improve wildlife habitat and biodiversity through the use of sustainable landscaping practices
- Environmentally Preferable Purchasing (EPP) Policy and associated checklist for procurement of goods and services. The checklist specifies criteria that must be met, including alternative energy sources, energy efficiency, local sourcing, low/no VOCs, pollution reduction, high recycled content materials, waste reduction, water efficiency, and more. Performance against criteria is tracked
- Green Building Policy outlining the agency's commitment to creating structures and using processes that are environmentally responsible and resource-efficient throughout a building's lifecycle, from siting to design, construction, operation, maintenance, renovation, and deconstruction
- Sustainable Landscape Policy, which calls for sustainable plant types, irrigation needs reduction, and minimized herbicide and fertilizer use
- Complete Streets Policy, which calls for designs to incorporate mobility of all forms into transportation projects while ensuring safe access for all users within the project area

In addition, BART has its own Sustainability Policy, which was used during the design of this project. The policy emphasized resource efficiency, service quality, environmental preservation, and cost effectiveness. The project included an overall sustainability specification calling for mandatory and voluntary sustainability measures. All measures identified as mandatory were prescriptive requirements of the contract. At the same time, sustainability requirements were also included in the BART Facility Standards design specifications. The specifications for the sustainability requirements included measurement and tracking through a Sustainability Measures Matrix and quarterly Sustainability Reports.

The Design-Build contractor created a Sustainability Matrix for the BART extension project to document and track sustainability requirements. This Matrix outlined a commitment to implement, assess, and prioritize mandatory and voluntary sustainable practices with quantified goals, including energy efficiency, water efficiency and conservation, material conservation and resource efficiency, and environmental quality.

VTA Environmental Oversight field inspectors verified compliance with all environmental and sustainability commitments.

Moreover, it is worth highlighting that the project committed to going above and beyond minimum requirements for the protection of surface waters and wetlands, and exceeded the City of San José's Riparian Corridor Policy, which delineates a riparian setback of at least 100 feet, by extending it to 200 feet.

RESPOND

The Berryessa Transit Center project is part of a larger transportation project. Only Phase I, under which the station was constructed, includes the construction of the BART 10-mile corridor (line, track, systems) and one additional station. Thus, it is hard to isolate the project strategies that are relevant to the design, construction, and operation of this station alone, especially regarding climate change mitigation.

The following section outlines selected key project sustainability strategies for the campus and the creek restoration project; however, they do not aim to capture all the sustainable strategies implemented in the project. Increased focus is given to the nature-based solutions implemented within the station campus and the Upper Penitencia Creek. The implemented strategies address multiple objectives, provide co-benefits, and could potentially be listed under a variety of other sustainability strategies, such as water quality enhancement, soil quality enhancement, and erosion control, among others.

To avoid duplication, the creek improvement-related strategies were grouped under three categories that reflect the project's main goals:

- Enhancement and protection of freshwater bodies and wetlands
- Increased floodwater storage capacity and stormwater control
- Riparian habitat enhancement

LIST OF SUSTAINABLE PROJECT STRATEGIES

GHG EMISSIONS REDUCTION

Provision of multi-modal transit options

The project contributes to reduction of GHG emissions locally through a modal shift from single-occupant vehicles to bike, foot and other public transportation modes with construction of:

- Pedestrian pathways
- Bike paths and bicycle storage rooms
- Bus-only lanes
- Bus transfer centers
- Private shuttle areas
- Direct connection to the Milpitas Light Rail Station

The Berryessa Transit Center project is in close proximity to several redevelopments where former industrial sites are being transformed into high-density, mixed-use neighborhoods, providing residents with expanded multimodal transit options. This transformation improves accessibility for both residents and employers within a half-mile radius of the Berryessa Transit Center. The anticipated modal shift from automobile use to regular BART ridership is expected to alleviate congestion, reduce regional bus and private vehicle miles traveled by approximately 12%, and decrease GHG emissions by an estimated 3,464 metric tons per year across the region.
[CCM1.4] [CCM3.3] [CCM3.5] [CCM3.6]

Onsite generation of renewable energy

Installation of onsite solar power generation (PVs) on top of the parking garage facility to supply power at the utility level.
[CCM1.1] [CCM1.2] [CCM3.3]

Use of clean energy procurement for electricity and fuels

The project electricity needs are addressed through wholesale low-carbon, zero-carbon, and renewable electricity purchases. Over two thirds of BART's power is comes from clean hydro and renewable sources.
[CCM1.1] [CCM1.2] [CCM3.3]

Electrification of BART fleet

Use of all-electric BART trains that offer up to 7% energy efficiency compared to current trains.
BART challenged Bombardier, the train manufacturer, to build trains that are at least 7% more efficient than the current trains. While the new trains are more efficient during acceleration and maintaining speed, they gain most of their efficiency by improving the regenerative braking system that the trains currently use. BART trains convert their kinetic energy of motion into electrical energy as they slow down. The energy regenerated during this process is returned to the power distribution system, where it is used by other trains.[26]
[CCM1.2] [CCM3.4]

Promotion of zero-emission vehicles

Promotion of zero-emission vehicles through the installation of 24 220-volt electric vehicle (EV) charging stations on the first level of the parking garage.
[CCM1.4] [CM3.4]

Promotion of zero-emission equipment for O&M works
Development of policy for the purchase of electric (battery) tools and replacement of gas-powered tools and fleet with electric, including:

- Prioritization of landscaping tool replacement
- Outfit hi-rail crew trucks with outlets and charging areas

[CCM1.1] [CCM3.4]

Construction Scope 1 and 2 emissions reduction
The Project Technical Specifications document included requirements for well-maintained, low-emission fleet vehicles to reduce emissions during construction.
[CCM1.1] [CCM1.2] [CCM3.2] [CCM3.3] [BIO3.3]

Carbon sequestration through revegetation
The project implemented revegetation practices resulting in increased number of trees and quality of vegetation enhancing the overall carbon storage potential:

- Revegetation of the Upper Penitencia creek site with 3,413 native trees and shrubs and 1,434 native herbaceous plantings, appropriate for wetlands and riparian habitats
- Revegetation of the station campus site with 474 trees, an increase of 349 trees from the 125 already in the site. The additional 349 trees at the station campus site alone have the potential to sequester 5 tons of CO_2 annually in a temperate climate

[CCM1.1] [CCM3.5] [BIO3.2]

ENERGY USE REDUCTION

Use of energy-efficient indoor and outdoor lighting systems for reduced operational energy use
Use of energy-efficient light emitting diodes (LED) for all indoor and outdoor lighting needs results in 75% less operational energy consumption than traditional incandescent lighting.
[CCM1.2] [CCM3.2] [BIO3.3]

Use of advanced lighting and mechanical systems' controls for avoiding energy wastage
The project incorporates preprogrammed demand-response strategies that can reduce the total lighting and electricity load by a minimum of 30% through:

- Dimming and bi-level switching in offices
- Energy-efficient intermittent escalators, each equipped with infrared motion sensors
- Light systems at the parking structure that have sensors, which when triggered by pedestrians, turn the lights to full power

[CCM1.2] [CCM3.2]

Use of passive design strategies for reduced lighting needs
Use of skylights and other light-permeable surfaces to increase natural light levels and take advantage of daylight for reduced lighting demand.
[CCM1.2] [CCM3.2] [BIO3.3]

Electric and mechanical systems' commissioning for operational efficiency
Initial commissioning of all electrical, CCTV, and Fire Life and Safety accounting was conducted for at least 50% of the total energy consumption/generation. All commissioning includes a detailed log of issues. The Commissioning Plan required detailed O&M instructions and copies of guaranties/warranties for each system prior to final inspection.
[CCM1.2] [CCM3.2]

Specifications for reduced construction energy use
Construction Technical Specifications were established for the contractor and subcontractors, including requirements for construction energy reduction through:

- Minimization of idling time by shutting equipment off when not in use or limiting idling to five minutes
- Use of low-energy consumption generators (Whisperwatt) and hydraulic tools instead of pneumatic and electric ones
- Use of well-maintained fleet vehicles and equipment through regular checks by a certified mechanic to determine if they were properly tuned in accordance with manufacturer's specifications before use

[CCM1.2] [CCM3.2] [BIO3.3]

USE OF LOCALLY AND REGIONALLY SOURCED MATERIALS

Use of locally and regionally sourced materials
Building materials or products for permanent installation have been harvested or manufactured in California or within a 500-mile radius from the project site.
[CCM1.3] [CCA1.5] [CCM3.8] [BIO3.3]

MATERIAL INPUT REDUCTION

Use of recycled-content materials

The concrete (accounting for 15% of the total material value) and cement used in the project were made with recycled products (recycled aggregates and fly ash). [CCM1.3] [CCA1.4] [BIO1.4] [CCM3.7] [CCA3.3] [BIO3.3]

Leverage of byproduct opportunities

VTA was awarded a grant from CalRecycle, the State's champion for waste reduction, recycling, and reuse, to install Tire-Derived Aggregate (TDA). TDA converts scrap vehicle tires into a vibration-reducing material. Approximately 3,008 tons of TDA were installed beneath tracks at four locations along the Berryessa Extension corridor to reduce vibration from passing BART trains, diverting the equivalent of 300,800 California tires from the waste stream. This method has a proven track record: in 2005, TDA was installed along 3,000 feet of a VTA light-rail extension project in the City of Campbell, making VTA the first public agency in the US to do so. The material has reduced vibrations from light-rail trains as intended. [CCM1.3] [CCA1.4] [BIO1.4] [CCM3.7] [CCA3.3] [BIO3.3]

Onsite reuse of demolition material

Since the project site was previously developed and various buildings and concrete surfaces had to be removed, it provided the opportunity to salvage and reuse part of the generated demolition materials. More specifically:

- Concrete was crushed onsite, stored and reused for onsite concrete production
- Salvaged fiberglass insulation of the existing buildings was reused as landscape cover
- A number of large trees that had to be removed as part of the restoration project, were reused as bank log structure to anchor the creek and prevent the creek migrating towards private property lines
- Leftover trees and vegetation were mulched onsite and reused as landscape cover. The excess mulch from vegetation clearings, not reused onsite, was hauled off site for use in other projects.

[CCM1.3] [CCA1.4] [BIO1.4] [CCM3.7] [CCA3.3] [BIO3.3]

END-OF-LIFE REPURPOSING

Use of recyclable or reusable materials and products

These include reinforced concrete, cast-in-place concrete, concrete flooring, painted Concrete Masonry Unit (CMU) walls at the office/restroom/utilities room cores, stainless steel restroom accessories, epoxy paint on restroom walls.
[CCM1.3] [CCA1.4] [BIO1.4] [CCM3.7] [CCA3.3] [BIO3.3]

SOLID WASTE DIVERSION FROM LANDFILLS

Provision for collection, sorting, and recycling of operational waste

The project team identified waste streams and byproducts from project operations, most of which expected to be generated by ridership. To divert operational waste from landfills, ten recycling bins were installed throughout the station. Based on the waste hauler's contract and past data, at least 25% waste will be recycled.

Republic Services, the contracted waste hauler, processes and sorts all waste material from the station offsite, at the Newby Island Resource Recovery Park, approximately nine miles from the site. They use a Wet/Dry system to process recycling and waste. The Berryessa Transit Center receives a "Wet" collection service for organics, such as food waste, and a "Dry" collection service for recyclables and everything else. This processing system allows sorting of wet and dry waste streams to maximize diversion. This innovative wet/dry system has nearly tripled the business recycling rate—from less than 25% to over 70%, since it started in 2012. [CCM1.3] [CCM3.7] [BIO1.4] [BIO3.3]

Setting minimum thresholds for construction waste diversion

BART Facility Standards required the contractor to provide a final summary of solid waste disposal and diversion against initial target goals. Specific minimum thresholds were specified as part of the requirements for the contractor and subcontractors for a total diversion rate of 75% or greater:

- Diversion rate of steel and concrete demolition waste: 100%
- Diversion rate of demolition waste (excluding concrete and steel): Greater than 50%

The contractor exceeded the initial 75% target goal by diverting over 90% of total demolition waste project-wide. [CCM1.3] [BIO1.4] [CCM3.7] [BIO3.3]

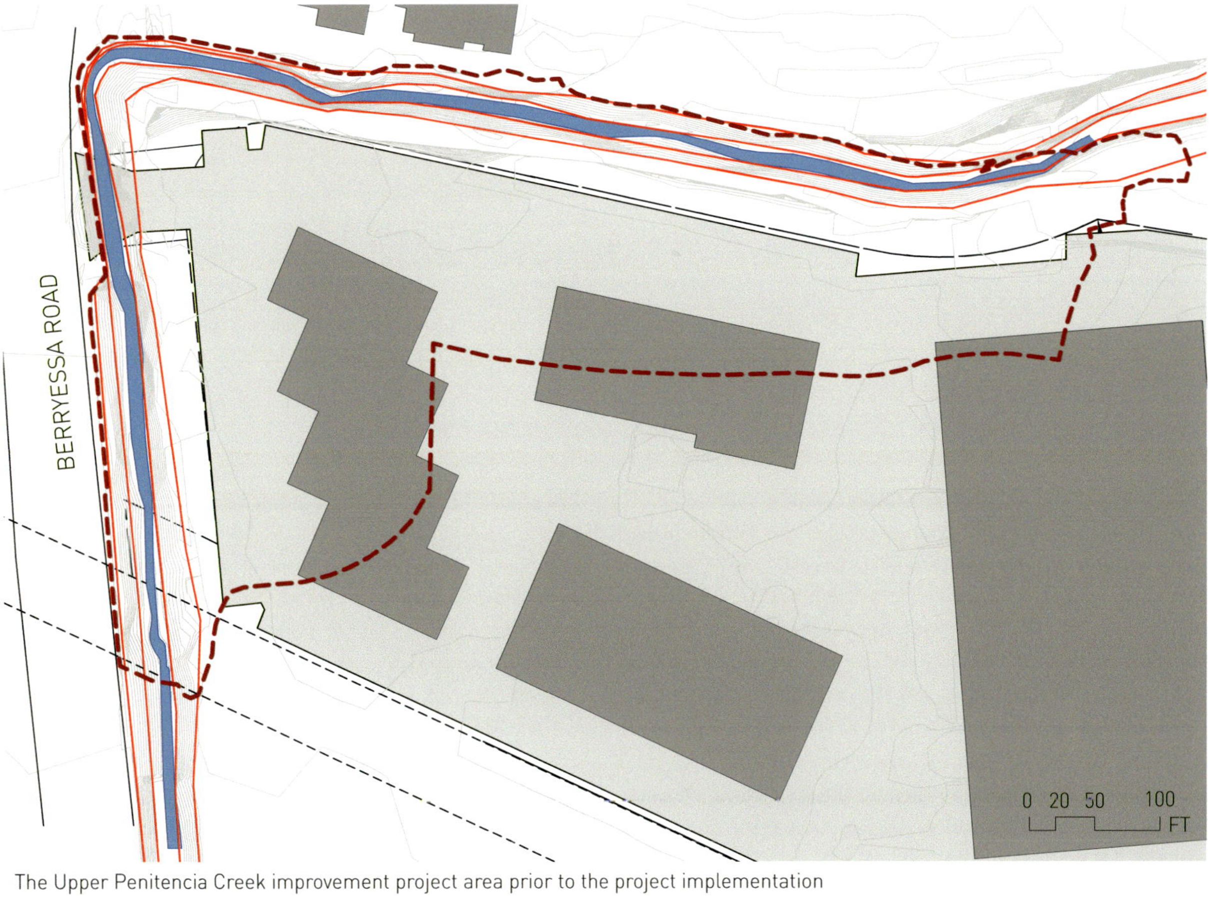

The Upper Penitencia Creek improvement project area prior to the project implementation

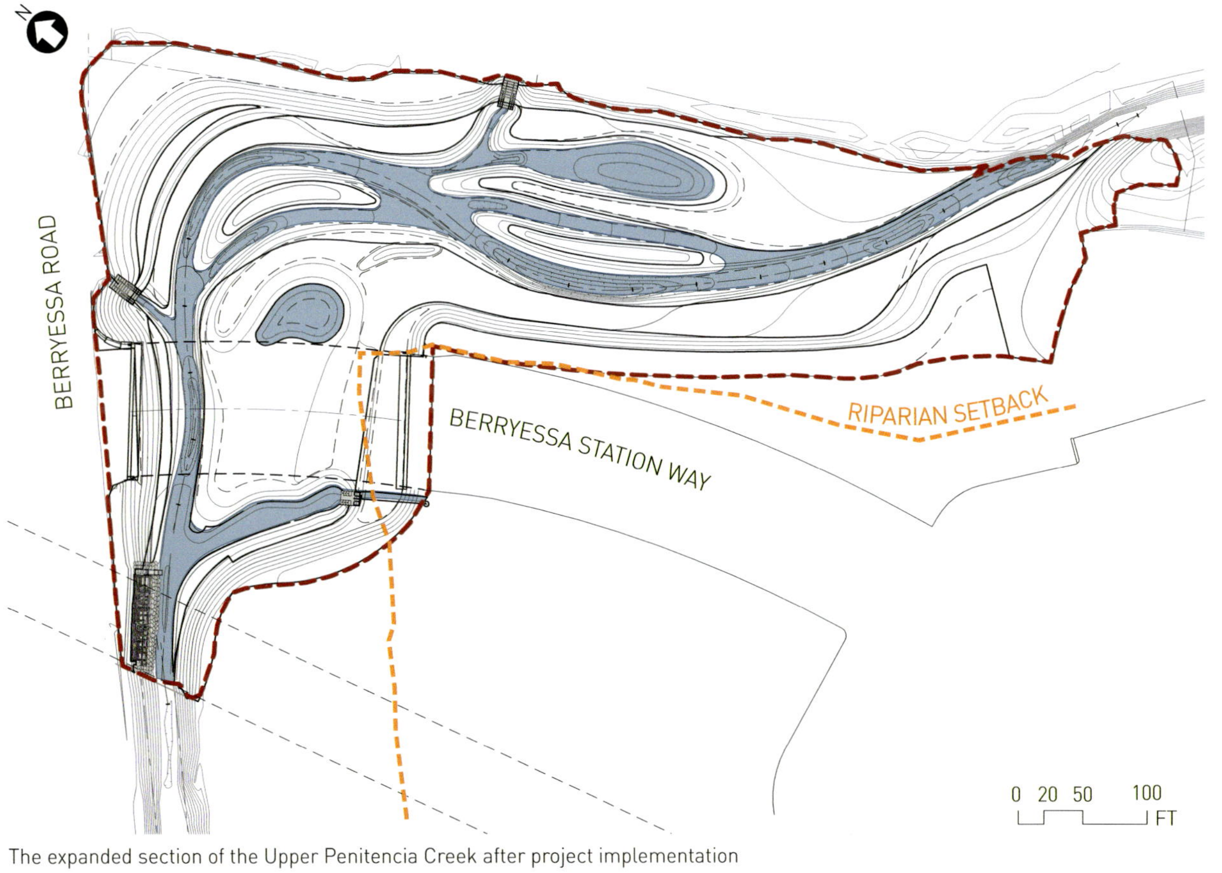

The expanded section of the Upper Penitencia Creek after project implementation

NOISE AND VIBRATION REDUCTION

Minimization of operational noise through design

- Design of tracks to be relatively straight so that trains running on them generate significantly less noise
- Installation of sound walls to minimize station noise

[CCM3.7] [CCA3.3] [BIO3.3]

Minimization of operational noise through innovative solutions

- Use of a new wheel profile in new BART trains and updating the profiles of existing fleet to reduce the noise produced by the rolling of steel wheels on rail
- Installation of 3,008 tons of Tire-Derived Aggregate beneath the tracks in four locations to help reduce vibration from trains

[CCM3.7] [CCA3.3] [BIO3.3]

INCREASED FLOODWATER STORAGE CAPACITY AND STORMWATER CONTROL

Widening of the existing confined creek's section

The Upper Penitencia Creek was widened and improved to increase its flood storage capacity. The channel's cross-section was significantly increased by 170 feet, from 45 to 215 feet. The creek now has sufficient capacity to convey the 100-year flood. As a result the project contributes to upstream and downstream efforts in the portion from Coyote Creek to Dorel Road.

Channel improvements included constructing a graded low-flow channel sized to contain the 1-year storm event and a high-flow channel to convey sediment and water during floods greater than the annual flood.

Moreover, approximately 920 linear feet of stream channel were enhanced, adding 62 ft to the overall stream length.

[CCA1.1] [CCA1.3] [CCA3.2] [CCA3.5] [BIO1.5] [BIO3.2]

Removal of existing constructions within the floodplain

Replacement of the two existing creek crossings—a Union Pacific Railroad (UPRR) trestle bridge and a 70 ft long roadway double box culvert—with a clear-span roadway bridge and BART aerial guideway to avoid wetlands and floodplains to the maximum extent practicable. The removal of the undersized bridge increased flood conveyance capacity.

Overall, permanent removal of 2.11 acres of impervious pavement from the 3.74-acre creek restoration site represents a 29% increase of permeable surfaces within historical alluvial fan and floodplain surfaces.

[CCA1.1] [CCA1.3] [CCA3.2] [CCA3.5] [BIO1.5] [BIO3.2]

Creation of additional floodplain wetland habitat

Creation of additional 1.06 acres of floodplain wetland habitat through the removal of approximately 4.5 acres of concrete and industrial buildings from former alluvial fan and floodplain surfaces. The reclaimed land was carefully regraded with appropriate soils and planted with diverse vegetation to support its restored function.

[CCA1.1] [CCA1.3] [CCA3.2] [CCA3.5] [BIO1.1] [BIO1.2] [BIO1.5] [BIO3.2]

Creation of bioretention basins on the station campus site

The project implemented bioretention basins to infiltrate, evapotranspirate, and reuse more than 100% of the 95th percentile local 24-hour rain event. A total of 20 stormwater bioretention basins were created across the station campus site that minimize and clean stormwater runoff from impervious surfaces, such as parking lots and roads. As a result, there is no net increase in stormwater runoff from existing to developed conditions. Moreover, 15 out of these 20 bioretention basins are sized larger than the minimum numeric sizing criteria of the Municipal Separate Storm Sewer System (MS4) Permit. These basins decrease the volume and velocity of stormwater flow to downstream receiving water bodies.

[CCA1.1] [CCA1.3] [CCA3.2] [CCA3.5] [BIO1.5] [BIO3.2]

Increase of permeable surfaces in the station campus

The station campus's compact site design contributes to reduced impervious areas by 10% as compared to existing conditions. Approximately 4.5 acres of concrete and industrial buildings were removed.

[CCA1.1] [CCA1.3] [CCA3.2] [CCA3.5] [BIO1.5] [BIO3.2]

Avoidance of the 100-year flood elevation for critical station components

Pad elevations of track and critical station facilities and equipment were set above the 100-year storm flood elevation. The BART aerial guideway, also entrance to the station, was set at an elevation with 1.1 feet of freeboard.

[CCA1.1] [CCA1.4] [CCA3.2] [CCA3.4] [CCA3.5]

Stormwater control during construction

During construction the use of retained cut sections, retained fill sections, station entrances, and access points that maintain 6 inches to 1 foot of freeboard above the base 100-year flood elevation, was required by VTA.

[CCA1.3] [CCM3.7] [BIO3.3]

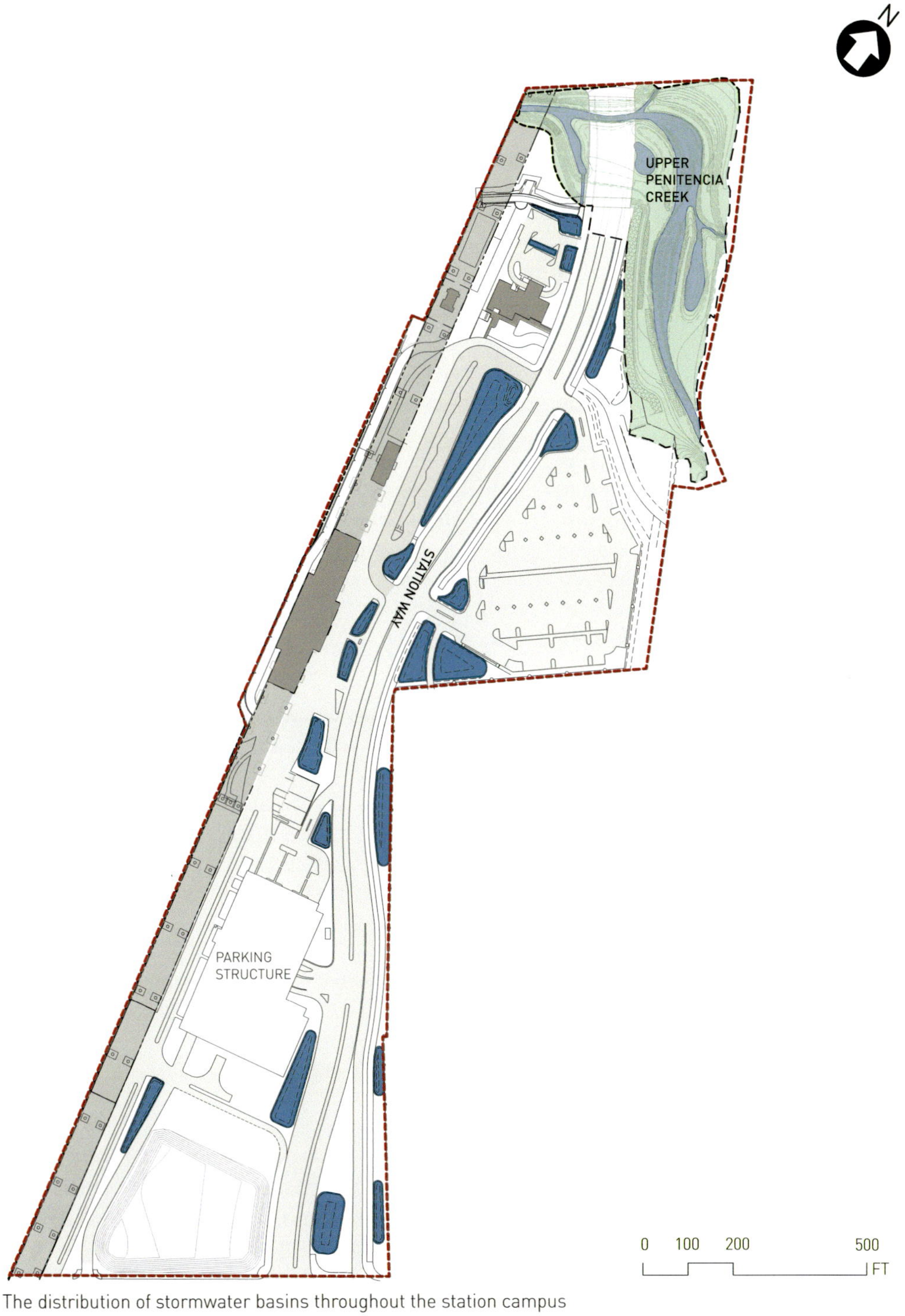

The distribution of stormwater basins throughout the station campus

View of stormwater basins at the station plaza. A panel is placed to educate on the functions of a stormwater basin

EROSION CONTROL

Realignment of the channel away from adjacent private properties

The alignment of the channel through the 90 degree bend has been modified with a larger radius to move the creek alignment further away from properties where existing erosion and undermining of the right bank had occurred. The creek was diverted into a multi-threaded channel and the risk of large-scale failure of artificially protected stream banks was eliminated.
[CCA1.3] [CCA3.2] [CCA3.3] [CCA3.4] [CCA3.5] [BIO3.2]

Stabilization of the creek's new configuration through bioengineering

Two large culverts draining nearby urban areas discharge to the enhanced Upper Penitencia Creek. As a result, the ultimate enhancement design had to incorporate appropriate elements to protect against local erosion at the outfall locations, as well as to protect against outfall destabilization. Bioengineered bank treatment structures, such as root wads, logs, and boulders, were installed for bank stabilization and to provide protection of the new creek configuration, as well as to provide aquatic habitat. Large wood structures incorporated into the design were mechanically anchored to prevent potential structure movement to downstream reaches, which could lead to large debris jams or rafting of debris, and thereby cause localized flooding or property damage.
[CCA1.3] [CCA3.2] [CCA3.4] [BIO3.2]

Stabilization of the creek's new configuration through revegetation

The project used diverse and multi-layered vegetation with more than 15 different tree and shrub species, 30 grass species, and four groundcovers appropriate for the different zones of the creek: streamside, floodplain, wetlands, bars, logs, boulder banks, wrapped lift, and upper slope. All planting zones were hydroseeded with native herbaceous species for erosion control. In total, the site was revegetated with 3,413 native trees and shrubs and 1,434 native herbaceous plantings, which were sourced from upstream.
[CCM3.5] [CCA1.3] [CCA3.2] [CCA3.4] [BIO3.2]

POTABLE WATER USE REDUCTION

Minimization of irrigation needs

The project incorporated various design strategies to minimize irrigation needs, including:

- Use of native drought-tolerant and low-water use plants for water conservation and improved watershed health. The Santa Clara Valley Water District's Approved Plant List was used as reference
- Grouping of plants with similar water needs into climate and water zones ("hydrozones") to maximize water efficiency
- Use of mulch in planted areas to reduce the amount of water evaporation from soil and thus reduce plant water needs
- Irrigation of the Upper Penitencia Creek only for three years during the plant establishment period
- Use of an irrigation system with an ETwater smart weather-based controller to adjust irrigation times and durations to ensure efficient water use, including rain shut-off devices, flow sensors, while providing proper coverage and even distribution

[CCA1.4] [BIO1.4] [CCA3.3] [BIO3.3]

Use of recycled water for irrigation

As part of the project, recycled water irrigation lines were installed, connecting and integrating the site with the existing recycled water pipeline network. The estimated total water use (potable water savings) for all project landscape areas is 4,411,846 gallons per year. All parking spaces drain to bioretention basins, ensuring that any leaks and drips from parked cars will be infiltrated, evapotranspirated, and reused.
[CCA1.4] [BIO1.4] [CCA3.3] [BIO3.3]

Use of water-efficient fixtures in the station facilities

Water conserving fixtures were installed in all staff and public areas in the station for minimizing potable water use. An estimated one million gallons of water will be saved annually.
The bathroom facilities operate at 0.5 and 1.28 gallons per flush (gpf), respectively, and meet the HET (High Efficiency Toilet), which defines water closets operating in the range of 1.1 gpf to 1.6 gpf.
[CCA1.4] [BIO1.4] [CCA3.3] [BIO3.3]

Reduction of water use during construction

The contractor prepared a Potable Water Reduction Plan, which outlined strategies to minimize the use of potable water during construction.
Reuse of groundwater collected during dewatering operations for dust control and soil compaction operations reduced potable water use during construction by 25% for dust control and 15% for soil compaction and demolition. Overall, dewatering reuse resulted in a 40% reduction in potable water consumption.
[CCA1.4] [BIO1.4] [CCA3.3] [BIO3.3]

Bioengineered outfall protection

Use of root wads for anchoring the creek

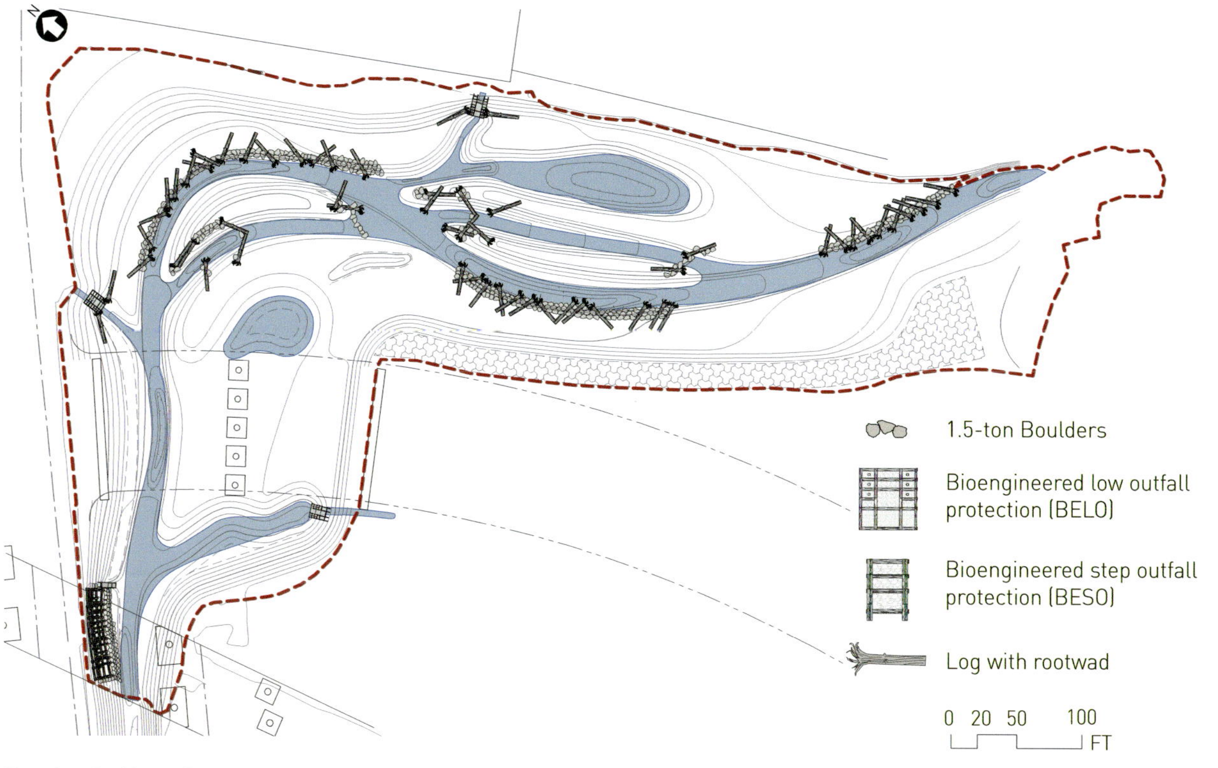

Site plan the Upper Penitencia Creek improvement project showing the location of bioengineered bank and outfall protection measures

Bioengineered bank protection

Construction of the Station Way bridge spanning the creek

PROPER PROJECT SITING AND SITE DESIGN

Maximizing habitat restoration potential through siting and site design

The project was sited to minimize impact on terrestrial riparian and upland habitats and maximize the restoration potential.
[CCA1.3] [CCA3.2] [CCA3.5] [BIO1.3] [BIO3.2]

ENHANCEMENT AND PROTECTION OF FRESHWATER BODIES AND WETLANDS

Provision of buffer zones for wetlands and surface waters

The project team determined the type and width of buffer zones necessary to protect wetlands and surface waters as part of the initial Design Basis and feasibility study. The Hydrologic Engineering Center River Analysis System (HEC-RAS) computer model was used to determine the appropriate channel cross-section and riparian setback area in terms of better hydrologic connectivity, flood capacity, and improved steelhead passage, as well as improved water quality.
The City of San José's Riparian Corridor Policy delineates a riparian setback of 100 ft. The recommended setback was exceeded for the project, and an approximately 200 ft wide creek corridor/riparian setback was incorporated into the design of the facility to avoid temporary and permanent impacts to the high ecological value stream to the maximum extent practicable. No project facilities were permitted within the creek riparian setback other than the required bridge supports to span the channel. The creek habitat/wetland restoration widened Upper Penitencia Creek channel to the full riparian setback limit.
[CCA1.3] [CCA3.2] [CCA3.5] [BIO1.3] [BIO3.2]

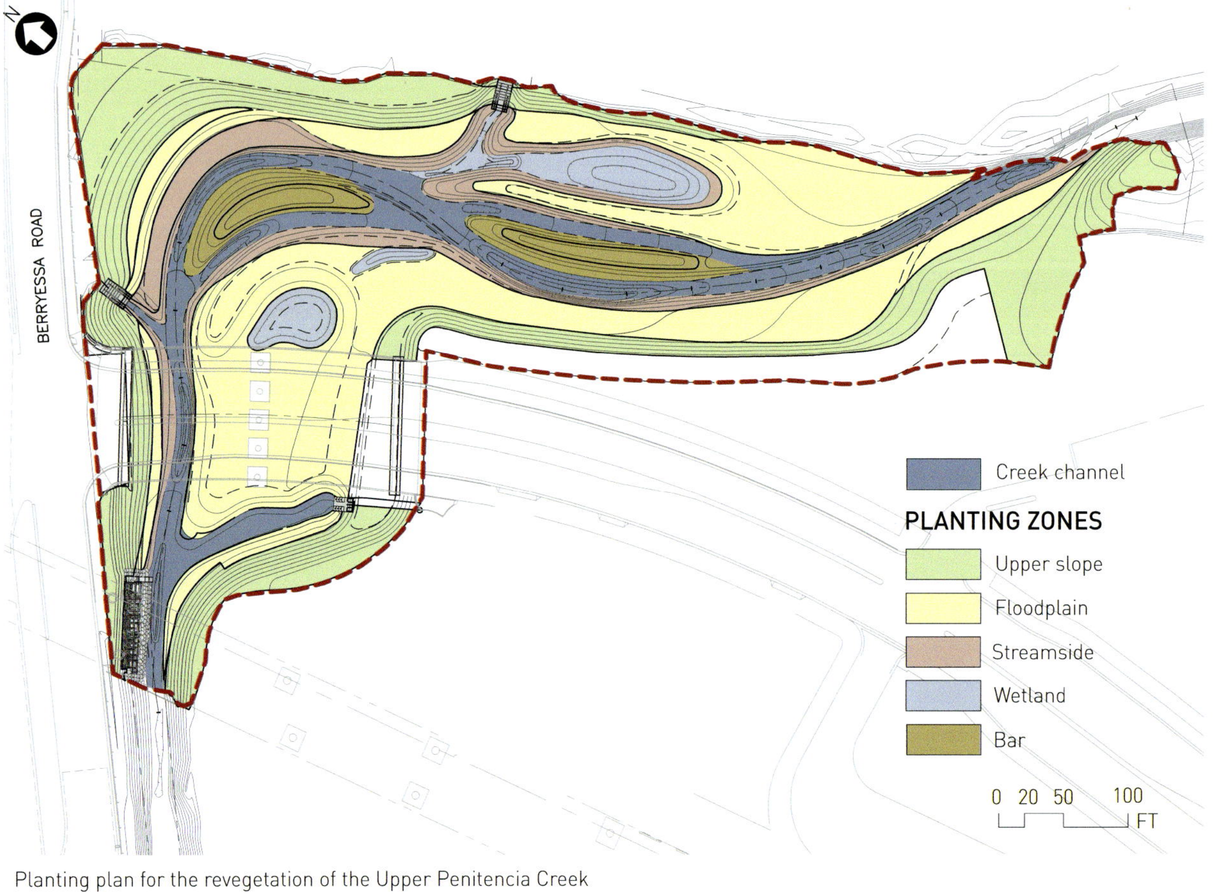

Planting plan for the revegetation of the Upper Penitencia Creek

RIPARIAN HABITAT ENHANCEMENT

Restoration and enhancement of a degraded creek

The project includes physical and ecological interventions to Upper Penitencia Creek to improve wetland and surface water quality. A total of 920 linear feet of the creek were realigned, restored, and enhanced. Approximately 1.06 acres of seasonal floodplain wetlands were created, and one acre of riparian habitat was restored. In addition, the project involved regrading 1,940 linear feet of earthen channels and daylighting approximately 0.02 acres of creek as a result of the double box culvert removal. The project helped reduce instream velocities in the creek, thus benefitting native fish populations. Secondary channels and floodplain wetlands were constructed to accommodate high flows. This constructed habitat improves habitat connectivity by creating various water and terrestrial features with different connections, such as ponds, intermittent flows, and other features, which provide specific habitat types. Riffles and pools were created for steelhead migration and resting.

[CCM3.5] [CCA3.2] [CCA3.5] [BIO1.1] [BIO1.2] [BIO3.2]

Proper plant selection

The site planting plan incorporates only native species occurring within the upper watershed of Upper Penitencia Creek and downstream reaches of Coyote Creek, promoting ecological continuity and visual cohesion throughout the creek corridor.

[CCM3.5] [BIO1.5] [BIO3.2]

Use of multi-layered diverse vegetation

Use of diverse and multi-layered vegetation with more than 15 different tree and shrub species, 30 grass species and four types of groundcovers appropriate for the different zones of the project: streamside, floodplain, wetlands, bars, logs, boulder banks, wrapped lift and upper slope.

[CCM3.5] [CCA3.2] [CCA3.5] [BIO1.1] [BIO1.2] [BIO3.2]

Installation of appropriate soil type fills

Extensive analysis of soils in the stream channel during design to provide specifications for restoration of the appropriate soil type, structure, and function to support healthy plant and tree growth. The areas whose pre-existing condition was parking and buildings, were stripped down 8-12 feet to install a designed creek bed consisting of controlled density fill, engineered fill, channel bed fill, and top dressings.

[CCA3.3] [BIO3.3]

MINIMIZATION OF TEMPORARY DISTURBANCE ON WILDLIFE DURING CONSTRUCTION

Application of construction period limits

Construction activities occurring within 200 feet of, or directly within, designated habitat for Chinook salmon or steelhead were restricted to the summer low-precipitation period (June 1 to October 15). This timing limitation minimized the likelihood of adverse effects on juvenile rearing steelhead and Chinook salmon, as well as on adult fish spawning and migration.

[BIO1.2] [BIO1.3] [BIO3.3]

Installation of wildlife exclusion fencing during construction

Given that the creek restoration project was implemented prior to the construction of the station, measures had to be taken to protect the restoration site from ongoing activities, including:

- The installation of orange fencing for protecting terrestrial habitat at the upstream and downstream ends of the project limits during construction of the Upper Penitencia Creek Improvement Project. This fencing served to protect terrestrial habitat
- Wildlife exclusion fencing was installed around the creek corridor to delineate the sensitive areas and provide a buffer from any construction or project activities

[BIO1.2] [BIO1.3] [BIO3.3]

CONTROL AGAINST THE INTRODUCTION OF INVASIVE SPECIES

Prevention, abatement and control measures for plant species listed as invasive by the California Invasive Plant Council's Invasive Plant Inventory were implemented before, during, and after construction:

Invasive species abatement before construction
The demolition plan for the pre-existing creek channel required clearing and grubbing all plant material in the creek down to subgrade. Removed plant material was not reused onsite.
[BIO1.7] [BIO3.3]

Control of imported soils during construction
The mitigation of the creek channel included extensive soil import to improve and widen the existing stream and build a floodplain that was specifically constructed to match the appropriate type for an alluvial channel. Care was taken during construction to prevent invasive plants from being introduced through imported soil. The top-dressing materials were brought in prior to the planning to provide sufficient time for checking and ensuring that there was no non-native vegetation left onsite.
[BIO1.7] [BIO3.3]

Post-construction invasive species controls
Following construction, Ecological Concerns, Inc. (ECI) conducted monthly invasive species management. Maintenance reports from 2015 through 2019 document that ECI biologists implemented invasive plant removal, trash and debris cleanup, and plant replacement. Management methods (e.g., hand weeding versus weed trimming) were evaluated for effectiveness by species.
The target for invasive species was to ensure that the percent cover remained below 5%, through removal and survey of the likely trajectory of plant community composition. Invasive plant maintenance test plots were established where removal was avoided to help discern the likely proportion of native versus alien species in the absence of invasive plant control.
[BIO1.7] [BIO3.3]

WATER POLLUTION PREVENTION

Runoff reduction and treatment onsite
By increasing permeable surfaces, the project reduced both the volume and rate of stormwater runoff compared to the previously impervious conditions. Surface water is now redirected to strategically distributed bioretention basins designed to detain, infiltrate, and treat stormwater before it enters nearby creeks and contributes to groundwater recharge within the watershed. All parking spaces are designed to drain into these bioretention basins, ensuring that any pollutants are naturally filtered out as the runoff filters through the system, and replenishes the groundwater supply.
[BIO1.6] [BIO3.3]

Stormwater Pollution Prevention Plan (SWPPP) for spill and leak prevention
During construction, potential water quality impacts could occur mainly due to:

- Dewatering for the installation of structural footings in the station campus area
- Sediment-laden discharges from drainage areas, or pollutant-laden discharges from storage and work areas. The drainage areas were areas of exposed, erodible soil, including stockpiles, within the construction limits and that result from construction activities

To mitigate these risks, the project implemented a Stormwater Pollution Prevention Plan (SWPPP) that identified best management practices (BMPs) for erosion, sedimentation, and pollutant control. The SWPPP included water pollution control drawings that mapped construction activities, erosion and sediment controls, potential pollutant sources, and stormwater flow directions. Runoff sampling was conducted to verify that pollutants did not leave the construction site. Chemicals were stored in watertight containers with appropriate secondary containment or within fully enclosed sheds to prevent spills or leaks. Across the project limits, the contractor also implemented spill and leak prevention procedures for all chemicals and hazardous materials encountered onsite.
Dewatering effluent was collected and either reused for dust control or discharged to the sanitary sewer system in accordance with water quality requirements. In some instances, discharge water meeting SWPPP criteria—pH between 6.5 and 8.5 and turbidity below 250 NTU—was infiltrated onsite.
[BIO1.6] [BIO3.3]

Avoidance of herbicide and pesticide use
Herbicide and pesticide use was not permitted in the mitigation area or the bioretention basins. Only minimal amounts were permitted for other landscaping features, when necessary. Moreover, the application had to be performed by a licensed pest control contractor trained in integrated pest management techniques.
[BIO1.6] [BIO3.3]

SUMMARY OF HOW PROJECT RESPONSES ALIGN WITH THE MITIGATION HIERARCHY (AS ADAPTED FOR CLIMATE CHANGE MITIGATION AND ADAPTATION WITHIN THE CRITERIA)

The following tables summarize how the implemented project strategies align with the steps of the mitigation hierarchy, as adapted for climate change mitigation, adaptation, and biodiversity loss within the criteria. Efforts were made to avoid, reduce impacts and exposure to threats and risks. The transit project, by definition, substitutes existing emissions, while the creek restoration project represents to a degree a transformative action.

			RESPONSES BASED ON THE GHG MANAGEMENT HIERARCHY STEPS					
			AVOID/ ELIMINATE	REDUCE	SUBSTITUTE/ REPLACE	COMPENSATE/ OFFSET: offsite	COMPENSATE/ OFFSET: onsite	TRANSFORM
GHG EMISSIONS	Scope 1 GHG emissions	Existing			■			
		Project-driven	■	■				
	Scope 2 GHG emissions	Existing						
		Project-driven	■	■				
	Scope 3 GHG emissions (upstream)	Project-driven	■	■				
	Scope 3 GHG emissions (downstream)	Project-driven	■	■				

	RISKS DRIVEN BY THE IDENTIFIED CLIMATE CHANGE-RELATED THREATS (FLOODING, EXTREME HEAT, AND DROUGHT)		RESPONSES BASED ON THE RESILIENCE HIERARCHY STEPS				
			AVOID THREAT	REDUCE EXPOSURE	ABSORB/ MINIMIZE IMPACT	RESTORE/ RECOVER	TRANSFORM
CLIMATE RISKS	Service continuity risk		■	■	■		
	Physical asset risk		■	■	■		■
	Resource availability risk	Water		■			
		Materials		■			
		Land		■			
		Workforce		■			
	Supply chain continuity risk			■			

Aerial view of the parking structure with the roof-mounted PVs

Interior view of the parking structure and the EV charging stations

SUMMARY OF HOW PROJECT RESPONSES ALIGN WITH THE MITIGATION HIERARCHY (AS ADAPTED FOR BIODIVERSITY WITHIN THE CRITERIA)

			RESPONSES BASED ON THE MITIGATION HIERARCHY STEPS					
			AVOID	MINIMIZE	RESTORE	OFFSET offsite	OFFSET onsite	TRANSFORM
PRESSURES ON BIODIVERSITY	Land change	Existing						
	Land change	Project-driven	■	■	■			
	Freshwater change	Existing			■			■
	Freshwater change	Project-driven	■	■				
	Resource exploitation							
	Pollution – Water	Existing			■	N/A		
	Pollution – Water	Project-driven		■		N/A		
	Pollution – Air	Existing				N/A		
	Pollution – Air	Project-driven		■		N/A		
	Pollution – Soil	Existing			■	N/A		
	Pollution – Soil	Project-driven		■		N/A		
	Pollution – Waste	Existing				N/A		
	Pollution – Waste	Project-driven		■		N/A		
	Pollution – Noise	Existing				N/A		
	Pollution – Noise	Project-driven		■		N/A		
	Pollution – Light	Existing				N/A		
	Pollution – Light	Project-driven		■		N/A		
	Climate change							
	Introduction of invasive species	Existing				N/A		
	Introduction of invasive species	Project-driven	■			N/A		

The gradual transformation of the creek from a confined, man-made channel to a widened, more natural configuration

The restored creek as documented in 2019

To summarize the Berryessa Transit Station's contribution to integrated climate-biodiversity action, the selected project strategies are presented in relation to the drivers of climate change, climate-related risks, and pressures on biodiversity that were identified within the integrated action criteria. The following matrix table is a project-specific version of the generic Opportunity Matrix presented as part of the Integrated Action Criteria Chapter.

Similarly to the generic matrix, this table illustrates the opportunities of most of the strategies to address more than one impact at a time and, by extension, contribute to the three types of action simultaneously.

The matrix particularly underscores the potential for integrated action of various groups of sustainable project strategies, such as:

- Material input reduction
- End-of-life repurposing
- Increased floodwater storage capacity and stormwater control
- Protection of freshwater bodies and wetlands through adequate buffer zones
- Proper project siting and site design
- Enhancement of riparian habitat

Many of these integrated potential strategies fall under the category of nature-based solutions (NbS). These strategies are then presented in a separate table in relation to their potential to support ecosystem services both for the project and the adjacent community. These ecosystem services mirror, to a high degree, nature's responses to the impact drivers and risks of climate change and biodiversity loss, and, within the present framework, represent project's and community's dependencies on biodiversity.

PROJECT STRATEGIES IN RELATION TO IMPACT DRIVERS AND RISKS OF CLIMATE CHANGE AND BIODIVERSITY LOSS

	DRIVERS OF CLIMATE CHANGE			
	Scope 1 emissions	Scope 2 emissions	Scope 3 emissions	
			Upstream	Downstream
GHG EMISSIONS REDUCTION				
Provision of multi-modal transit options	■			
Establishment of connections and integration to wider networks of active transport	■			
Onsite generation of renewable energy		■		
Electrification of BART fleet	■	■		
Promotion of zero-emission vehicles	■			
Promotion of zero-emission equipment and fleet for O&M works	■			
Carbon sequestration through revegetation	■			
Construction Scope 1 and 2 emissions reduction	■	■		
ENERGY USE REDUCTION				
Use of energy-efficient indoor and outdoor lighting systems for reduced operational energy use		■		
Use of advanced lighting and mechanical systems' controls for avoiding energy wastage		■		
Use of passive design strategies for reduced lighting needs		■		
Electric and mechanical systems' commissioning for operational efficiency		■		
Specifications for reduced construction energy consumption		■		
USE OF LOCALLY AND REGIONALLY SOURCED MATERIALS				
Use of locally and regionally sourced materials			■	
MATERIAL INPUT REDUCTION				
Use of recycled-content materials			■	
Leverage of byproduct opportunities			■	■
Onsite reuse of demolition material			■	■
END-OF-LIFE REPURPOSING				
Use of recyclable or reusable materials and products				■
SOLID WASTE DIVERSION FROM LANDFILLS				
Provision for collection, sorting, and recycling of operational waste				■
Setting minimum thresholds for construction waste diversion				■
NOISE AND VIBRATION REDUCTION				
Minimization of operational noise				
INCREASED FLOODWATER STORAGE CAPACITY AND STORMWATER CONTROL				
Widening of the existing confined creek's section for increased flood conveyance capacity				
Removal of existing constructions within the floodplain				
Creation of additional floodplain wetland habitat				
Creation of bioretention basins on the station campus site				
Increase of permeable surfaces in the station campus			■	

CLIMATE-RELATED RISKS							PRESSURES ON BIODIVERSITY									
Service continuity	Physical asset	Resource availability				Supply chain	Land/freshwater/ sea change	Resource exploitation	Pollution						Climate change	Introduction of invasive species
		Water	Material	Land	Workforce				Water	Air	Soil	Waste	Noise	Light		
										■					■	
										■					■	
■															■	
															■	
															■	
															■	
															■	
												■			■	
														■	■	
															■	
															■	
															■	
						■									■	
			■					■				■				
			■					■				■	■			
			■					■								
			■					■				■				
												■				
												■				
													■			
■	■						■		■						■	
■	■						■		■						■	
■	■						■		■						■	
■		■							■							
■	■	■						■	■							

PROJECT STRATEGIES IN RELATION TO IMPACT DRIVERS AND RISKS OF CLIMATE CHANGE AND BIODIVERSITY LOSS (CONTINUED)

	DRIVERS OF CLIMATE CHANGE			
	Scope 1 emissions	Scope 2 emissions	Scope 3 emissions	
			Upstream	Downstream
Avoidance of the 100-year flood elevation for critical station components				
Stormwater control during construction				
EROSION CONTROL				
Realignment of the channel away from adjacent private properties				
Stabilization of the creek's new configuration through bioengineering				
Stabilization of the creek's new configuration through revegetation	■			
POTABLE WATER USE REDUCTION				
Minimization of irrigation needs				
Use of recycled water for irrigation				
Use of water-efficient fixtures in the station facilities				
Reduction of water use during construction				
ENHANCEMENT AND PROTECTION OF FRESHWATER BODIES AND WETLANDS				
Provision of buffer zones for wetlands and surface waters				
PROPER PROJECT SITING AND SITE DESIGN				
Maximizing habitat restoration potential through siting and site design				
RIPARIAN HABITAT ENHANCEMENT				
Reconfiguration of a channel to active stream				
Proper plant selection				
Use of multi-layered diverse vegetation	■			
Installation of appropriate soil type fills	■			
MINIMIZATION OF TEMPORARY DISTURBANCE ON WILDLIFE DURING CONSTRUCTION				
Application of construction period limits				
Installation of wildlife exclusion fencing during construction				
CONTROL AGAINST THE INTRODUCTION OF INVASIVE SPECIES				
Invasive species abatement before construction				
Control of imported soils during construction				
Post-construction invasive species controls				
WATER POLLUTION PREVENTION				
Runoff reduction and treatment onsite				
Stormwater Pollution Prevention Plan (SWPPP) for spill and leak prevention during construction				
Avoidance of herbicide and pesticide use				

CLIMATE-RELATED RISKS							PRESSURES ON BIODIVERSITY									
		Resource availability							Pollution							
Service continuity	Physical asset	Water	Material	Land	Workforce	Supply chain	Land/freshwater/ sea change	Resource exploitation	Water	Air	Soil	Waste	Noise	Light	Climate change	Introduction of invasive species
■	■								■							
	■	■							■							
	■		■													
	■														■	
	■		■												■	
		■						■							■	
		■						■								
		■						■								
		■														
	■	■					■		■						■	
	■	■		■					■						■	
	■	■					■	■	■						■	
							■								■	■
	■						■								■	
	■						■								■	
							■						■			
							■									
															■	■
															■	■
															■	■
									■		■					
									■		■					
									■		■					

IMPLEMENTED NATURE-BASED SOLUTIONS IN RELATION TO THEIR POTENTIAL CONTRIBUTIONS (ECOSYSTEM SERVICES)

	ECOSYSTEM SERVICES (available to the project and/or community)										
	PROVISIONING SERVICES										
	Biomass provisioning					Terrestrial and aquatic plants as energy source	Other raw material provisioning	Genetic material	Water supply		
IMPLEMENTED NATURE-BASED SOLUTIONS	Food crops	Grazed biomass provisioning	Livestock provisioning	Aquaculture provisioning	Wood provisioning				Potable water	Non-potable water for processes	Non-potable water for irrigation
Creation of a 200-ft buffer/ riparian setback from Upper Penitencia Creek											
Permanent removal of 2.11 acres of permeable pavement from the 3.74-acre creek restoration site											
Enhancement of approximately 920 linear ft of stream channel, adding 62 ft to the overall stream length									■		
Regrading of 1,940 linear feet of earthen channels to provide the appropriate channel cross-section									■		
Construction of secondary channels and floodplain wetlands to accommodate high flows, as well as a widened floodplain with restored riparian habitat									■		
Daylighting of approximately 0.02 acres of creek through the removal of an undersized culvert											
Creation of riffles and pools for steelhead migration and resting.									■		
Installation of bioengineered bank treatment structures such as root wads, logs and boulders for bank stabilization											
Removal of non-native species and revegetation of the Upper Penitencia Creek site with of 3,413 native trees and shrubs and 1,434 native herbaceous plantings, appropriate for wetlands and riparian habitats											
All planting zones were hydroseeded with native herbaceous species											
Use of diverse and multi-layered vegetation appropriate for the different zones of the project: streamside, floodplain, wetlands, bars, logs, boulder banks, wrapped lift and upper slope.									■		
Reduction of impervious areas by 10% as compared to existing conditions in the BART station campus									■		
Creation of 20 stormwater bioretention basins to minimize and treat stormwater runoff from impervious surfaces and protect the creek habitats											
Creation of a Native American Contemplative Garden											
Revegetation with 474 trees, an increase of 349 trees from the 125 already in the campus site											

ECOSYSTEM SERVICES
(available to the project and/or community)

Global climate regulation	Rainfall pattern regulation	Local (micro and meso climate) regulation	Air filtration	Soil quality regulation	Soil and sediment retention: Soil erosion control	Soil and sediment retention: Landslide mitigation	Solid waste remediation	Water purification: Retention and breakdown of nutrients	Water purification: Retention and breakdown of other pollutants	Water flow regulation: Baseline flow maintenance	Water flow regulation: Peak flow mitigation	Flood mitigation: Coastal protection	Flood mitigation: River flood mitigation	Storm mitigation	Noise attenuation	Pollination	Biological control: Pest control	Biological control: Disease control	Nursery population and habitat maintenance	Recreation-related	Visual amenity	Education, scientific and research	Spiritual, artistic and symbolic	Ecosystem /species appreciation
REGULATING & MAINTENANCE SERVICES																				CULTURAL SERVICES				
		■								■	■		■						■		■			■
		■			■					■	■								■		■			■
		■			■				■	■	■								■		■			■
									■	■			■	■					■		■			■
		■								■	■								■		■			■
		■											■	■					■		■			■
								■		■	■								■		■			■
		■		■	■			■	■	■	■							■	■		■			■
■		■	■	■	■		■	■	■	■	■		■	■	■	■	■	■	■		■			■
		■	■	■	■			■	■	■	■							■	■		■			
■		■	■	■	■		■	■	■	■	■							■	■					
		■		■				■	■												■			
		■						■	■	■	■			■					■		■			■
		■																			■		■	■
■		■		■	■		■	■	■	■	■				■	■	■	■	■					■

TRACK

The project included internal monitoring systems to track performance across renewable energy generation, energy and water use, bioretention basin functionality, and the adaptive capacity of the creek restoration effort.

Energy systems: The roof-mounted solar panels at the parking structure are continuously tracked by the Also Energy monitoring platform, which provides real-time data visualization and system efficiency verification. Integrated energy management systems independently monitor all primary energy uses—including lighting, elevators, and escalators—capturing at least 50% of total energy consumption.

Water systems: The irrigation system is outfitted with an ETwater smart controller, which integrates real-time weather data, recent irrigation history, and dynamic adjustments for watering schedules, restrictions, or suspensions. This system prevents overwatering, halts irrigation during rainfall, and automatically shuts off when leaks are detected. VTA also employs EnergyCAP utility management software (UMS) to import and analyze water consumption data across irrigation (potable and non-potable) and domestic systems. While the UMS tracks water consumption rather than quality, it runs automated audits that alert VTA if costs increase more than 10% over the highest cost in the previous 12 months. When triggered, the alert initiates an internal response protocol, including staff notification, issue tracking, and post-resolution verification.

Bioretention monitoring: Semi-annual inspections are conducted at the station's bioretention basins to assess hydraulic and vegetative performance. Each inspection evaluates plant health, mulch and soil condition, infiltration capacity during the wet season, and sediment accumulation. Results are reported to the facilities manager for timely maintenance and corrective action.

Upper Penitencia Creek improvement project monitoring: The project engaged an independent ecological consultant to implement a 10-year monitoring and maintenance program for Upper Penitencia Creek. Another consultant was contracted to perform monthly maintenance for this period. A monitoring plan established specific success criteria aligned with the initial goals of the restoration project to ensure they were met:

- Restoration of hydrologic and geomorphic functions, including sediment transport/deposition
- Restoration of floodplain connectivity and flood storage
- Restoration of fish and wildlife habitats, particularly for the federally and state-listed Central California Coast steelhead (*Oncorhynchus mykiss*)
- Improvement of water quality

Annual reports documented performance related to vegetation establishment, geomorphic and hydrologic conditions, fish habitat and populations, and water quality, providing adaptive management recommendations as necessary (e.g., yearly replanting of woody plants to bring survival rates up to 90% as a minimum, or actions to ensure that the invasive species percent cover remained below 5%).

During the 10-year monitoring period, several key outcomes were documented:

- Restored abiotic conditions, such as increased floodplain connectivity to the low-flow channel, were suitable for natural recruitment of the target riparian plant species installed at the site
- Maintenance activities effectively suppressed invasive species, improving terrestrial habitat quality
- Increased cover of overhanging vegetation and tree height met and exceeded performance criteria, with riparian woodland and wetland habitats surpassing target percent cover
- Riparian woodland and wetland habitat exhibited rapid growth and were capable of becoming self-sustainable in the absence of irrigation, as vegetation did not show signs of drought stress and appeared healthy
- Streamflow events mobilized bedload sediments, driving natural channel aggradation, morphological evolution, and sediment transport across the channel and floodplain
- The creek developed multi-thread channel patterns, increased wood loading, and substrate coarsening, reflecting restored geomorphic dynamism

KEY TAKEAWAYS

This section seeks to provide additional examples of implementing the criteria in infrastructure projects. The following table presents the Berryessa Transit Center strategies in relation to the response criteria—the generic strategies that are proposed within the integrated action criteria. This process provides examples of real-world solutions as implemented in a transit project and links them with the response criteria they relate to.

Most of the project's sustainable strategies are relevant to more than one generic strategy. This alternative reading aligns with the analysis of project strategies in relation to the impact drivers and risks of climate change and biodiversity loss they address, as presented in the project-specific Opportunity Matrix in the previous section. Again, it is the same groups of sustainable strategies that stand out as more relevant to the criteria, many of which coincide with nature-based solutions:

- Material input reduction
- End-of-life repurposing
- Increased floodwater storage capacity and stormwater control
- Protection of freshwater bodies and wetlands through adequate buffer zones
- Proper project siting and site design
- Enhancement of riparian habitat

That is no coincidence since in the present framework, the criteria are interlinked and designed in such a way that the proposed response criteria (strategies) address specific impact drivers and risks.

Moreover, the strategies that the following table highlights as relevant to all types of action are related to the four key strategies for integrated action presented in the integrated action criteria:

- Resource circularity
- Optimized material sourcing
- Resourcefulness
- Nature-based solutions

Despite the key role of the above strategies and the opportunities they present for all types of action, all strategies listed under the climate change mitigation, adaptation, and biodiversity action response criteria, along with all project strategies presented, are indispensable components of climate and biodiversity action. Representative examples are strategies that involve prevention, abatement, and control measures that fall under the category of "doing everything else better." These are typical sustainable actions to minimize environmental impact. When properly implemented, following best practices, they can have a significant positive effect on biodiversity action, especially in the current context of changing climate conditions and the need to ensure that ecosystems maintain their capacity to adapt.

In general, the analysis performed for this transit project reaffirms that sustainable project strategies are integral to the criteria for climate and biodiversity action.

PROJECT STRATEGIES IN RELATION TO THE INTEGRATED CLIMATE-BIODIVERSITY RESPONSE CRITERIA

CLIMATE CHANGE MITIGATION

	CCM3.2 Implement energy efficiency strategies	CCM3.3 Implement decarbonization strategies	CCM3.4 Implement electrification strategies	CCM3.5 Implement carbon capture & storage strategies	CCM3.6 Optimize material sourcing practices	CCM3.7 Implement resource circularity strategies	CCM3.8 Optimize project-related transport and logistics
GHG EMISSIONS REDUCTION							
Provision of multi-modal transit options		■					
Establishment of connections and integration to wider networks of active transport		■					
Onsite generation of renewable energy		■					
Electrification of BART fleet			■				
Promotion of zero-emission vehicles		■					
Promotion of zero-emission equipment and fleet for O&M works		■					
Carbon sequestration through revegetation				■			
Construction Scope 1 and 2 emissions reduction	■	■					
ENERGY USE REDUCTION							
Use of energy-efficient lighting systems for reduced operational energy use	■						
Use of advanced lighting and mechanical systems' controls	■						
Use of passive design strategies for reduced lighting needs	■						
Electric and mechanical systems' commissioning for operational efficiency	■						
Specifications for reduced construction energy consumption	■						
MATERIAL INPUT REDUCTION							
Use of recycled-content materials						■	
Leverage of byproduct opportunities						■	
Onsite reuse of demolition material						■	
END-OF-LIFE REPURPOSING							
Use of recyclable or reusable materials and products						■	■
SOLID WASTE DIVERSION FROM LANDFILLS							
Provision for collection, sorting, and recycling of operational waste						■	
Setting minimum thresholds for construction waste diversion						■	
NOISE AND VIBRATION REDUCTION							
Minimization of operational noise through design							
Minimization of operational noise through innovative solutions						■	

CLIMATE CHANGE ADAPTATION							BIODIVERSITY ACTION											CROSS-CUTTING		
CCA3.2 Optimize project siting and site design to avoid threats	CCA3.3 Implement strategies to maximize resourcefulness (materials, water, land)	CCA3.4 Implement strategies to maximize durability	CCA3.5 Implement strategies to maximize adaptability	CCA3.6 Implement strategies to maximize redundancy	CCA3.7 Implement strategies to maximize integration	CCA3.8 Implement strategies to optimize system recovery	BIO3.2 Implement nature-based solutions (NbS)	BIO3.3. Use prevention, abatement, and control measures against pressures on biodiversity (other than NbS): Land/freshwater/sea change	Resource Exploitation	Water pollution	Air pollution	Soil pollution	Waste pollution	Noise pollution	Light pollution	Climate change and related impacts	Introduction of Invasive species	CCA3.9 Develop local skills and capabilities	CCA3.10 Promote inclusivity	CCA4.1 Implement strategies to promote reflective capacity
			■	■							■					■				
			■	■							■					■				
																■				
																■				
																■				
																■				
			■				■									■				
											■			■		■				
															■	■				
																■				
																■				
																■				
																■				
	■								■							■				
	■								■							■				
	■								■							■				
	■		■						■				■			■				
													■			■				
													■			■				
	■													■						
	■								■				■	■						

PROJECT STRATEGIES IN RELATION TO THE INTEGRATED CLIMATE-BIODIVERSITY RESPONSE CRITERIA (CONTINUED)

CLIMATE CHANGE MITIGATION

	CCM3.2 Implement energy efficiency strategies	CCM3.3 Implement decarbonization strategies	CCM3.4 Implement electrification strategies	CCM3.5 Implement carbon capture & storage strategies	CCM3.6 Optimize material sourcing practices	CCM3.7 Implement resource circularity strategies	CCM3.8 Optimize project-related transport and logistics
INCREASED FLOODWATER STORAGE CAPACITY AND STORMWATER CONTROL							
Widening of the existing creek's section for increased flood conveyance capacity							
Removal of existing constructions within the floodplain						■	
Creation of additional floodplain wetland habitat				■			
Creation of bioretention basins on the station campus site							
Increase of permeable surfaces in the station campus							
Avoidance of the 100-year flood elevation for critical station components						■	
Stormwater control during construction							
EROSION CONTROL							
Realignment of the channel away from adjacent private properties							
Stabilization of the creek's new configuration through bioengineering							
Stabilization of the creek's new configuration through revegetation				■			
POTABLE WATER USE REDUCTION							
Minimization of irrigation needs							
Use of recycled water for irrigation							
Use of water-efficient fixtures in the station facilities							
ENHANCEMENT AND PROTECTION OF FRESHWATER BODIES AND WETLANDS							
Provision of buffer zones for wetlands and surface waters							
PROPER PROJECT SITING AND SITE DESIGN							
Maximizing habitat restoration potential through siting and site design							
RIPARIAN HABITAT ENHANCEMENT							
Reconfiguration of a channel to active stream							
Use of multi-layered diverse vegetation				■			
Proper plant selection				■			
Installation of appropriate soil type fills							

CLIMATE CHANGE ADAPTATION							BIODIVERSITY ACTION											CROSS-CUTTING		
CCA3.2	CCA3.3	CCA3.4	CCA3.5	CCA3.6	CCA3.7	CCA3.8	BIO3.2	BIO3.3. Use prevention, abatement, and control measures against pressures on biodiversity (other than NbS)										CCA3.9	CCA3.10	CCA4.1
Optimize project siting and site design to avoid threats	Implement strategies to maximize resourcefulness (materials, water, land)	Implement strategies to maximize durability	Implement strategies to maximize adaptability	Implement strategies to maximize redundancy	Implement strategies to maximize integration	Implement strategies to optimize system recovery	Implement nature-based solutions (NbS)	Land/freshwater/sea change	Resource Exploitation	Water pollution	Air pollution	Soil pollution	Waste pollution	Noise pollution	Light pollution	Climate change and related impacts	Introduction of Invasive species	Develop local skills and capabilities	Promote inclusivity	Implement strategies to promote reflective capacity
■			■				■	■								■				
■								■								■				
			■				■	■								■				
■	■		■				■	■		■		■				■				
■			■				■	■		■						■				
■		■	■			■										■				
										■										
■			■																	
■		■	■				■									■				
■			■				■									■				
	■		■						■	■						■				
	■								■							■				
	■								■											
■			■				■	■								■				
■			■							■	■	■		■		■				
■			■				■			■						■				
■			■				■									■				
			■				■		■							■				
			■													■				

PROJECT STRATEGIES IN RELATION TO THE INTEGRATED CLIMATE-BIODIVERSITY RESPONSE CRITERIA (CONTINUED)

CLIMATE CHANGE MITIGATION

	CCM3.2 Implement energy efficiency strategies	CCM3.3 Implement decarbonization strategies	CCM3.4 Implement electrification strategies	CCM3.5 Implement carbon capture & storage strategies	CCM3.6 Optimize material sourcing practices	CCM3.7 Implement resource circularity strategies	CCM3.8 Optimize project-related transport and logistics
MINIMIZATION OF TEMPORARY DISTURBANCE ON WILDLIFE DURING CONSTRUCTION							
Application of construction period limits							
Installation of wildlife exclusion fencing during construction							
CONTROL AGAINST THE INTRODUCTION OF INVASIVE SPECIES							
Invasive species abatement before construction							
Control of imported soils during construction							
Post-construction invasive species controls							
WATER POLLUTION PREVENTION							
Runoff reduction and treatment onsite							
Stormwater Pollution Prevention Plan (SWPPP) for spill and leak prevention during construction							
Avoidance of herbicide and pesticide use							
MONITORING							
Monitoring of energy generation and use	■	■					
Monitoring of water use							
Monitoring of bioretention basins' performance							
Upper Penitencia Creek improvement project 10-year monitoring							

CLIMATE CHANGE ADAPTATION							BIODIVERSITY ACTION											CROSS-CUTTING		
CCA3.2 Optimize project siting and site design to avoid threats	CCA3.3 Implement strategies to maximize resourcefulness (materials, water, land)	CCA3.4 Implement strategies to maximize durability	CCA3.5 Implement strategies to maximize adaptability	CCA3.6 Implement strategies to maximize redundancy	CCA3.7 Implement strategies to maximize integration	CCA3.8 Implement strategies to optimize system recovery	BIO3.2 Implement nature-based solutions (NbS)	BIO3.3. Use prevention, abatement, and control measures against pressures on biodiversity (other than NbS)										CCA3.9 Develop local skills and capabilities	CCA3.10 Promote inclusivity	CCA4.1 Implement strategies to promote reflective capacity
								Land/freshwater/sea change	Resource Exploitation	Water pollution	Air pollution	Soil pollution	Waste pollution	Noise pollutio	Light pollution	Climate change and related impacts	Introduction of Invasive species			
								■												
								■												
			■													■	■			
			■													■	■			
			■													■	■			
	■		■				■		■	■		■				■				
									■	■		■								
									■	■		■								
																■				□
	■								■							■				□
			■	■			■			■										□
			■				■										■			□

NOTES

1. The Silicon Valley Corridor is generally defined as the I-880 and I-680 transportation corridor extending from the City of Fremont in south Alameda County through the cities of Milpitas, San José and Santa Clara. The Corridor is geographically constrained by the San Francisco Bay to the west and Diablo Mountain Range to the east.
2. "VTA's BART Silicon Valley Phase II," Santa Clara Valley Transportation Authority, accessed November 7, 2025, https://www.vta.org/projects/bart-sv/phase-ii.
3. "Berryessa Transit Center & Berryessa/North San José BART Station," Santa Clara Valley Transportation Authority, accessed November 27, 2024, https://www.vta.org/projects/bart-sv/phase-i/berryessa.
4. "Project Benefits," Santa Clara Valley Transportation Authority, accessed November 7, 2025, https://vtabart.org/project-benefits/.
5. Santa Clara Valley Transportation Authority, "ZPH Session: Berryessa Transit Center, Envision Platinum—Assessing Sustainability and Transferring Lessons Learned to Transportation Projects," presentation by Ann Calnan and Rachael Keish at the Zofnass Program virtual sessions, March 24, 2022.
6. Two trails: One runs along Coyote Creek and another along the south side of Upper Penitencia Creek, parallel to Berryessa Road. An existing trail along Upper Penitencia Creek runs in close proximity to the Berryessa Station Campus, and it currently stops at King Road. This trail is slated to connect with Berryessa Station as part of the project.
7. Santa Clara Valley Transportation Authority, "ZPH Session: Berryessa Transit Center, Envision Platinum—Assessing Sustainability and Transferring Lessons Learned to Transportation Projects," 2022.
8. United States Federal Register, *Notices*, Vol. 67, no. 25 (February 6, 2002): 5636-5637, https://www.govinfo.gov/content/pkg/FR-2002-02-06/pdf/02-2796.pdf.
9. Ibid.
10. David Ackerly et al., *San Francisco Bay Area Summary Report, in California's Fourth Climate Change Assessment* (Sacramento, CA: State of California, 2018), Publication No. CCCA4-SUM-2018-005.
11. Ibid.
12. Ibid.
13. Ibid., 7.
14. A 100-year flood event has a 1 percent chance of occurrence in any year. A 500-year flood event has a 0.2 percent chance of occurrence in any year.
15. "About," Santa Clara Valley Transportation Authority, accessed November 25, 2025, https://www.vta.org/about.
16. Santa Clara Valley Transportation Authority, S*ustainability Plan 2020*, 2020, 2, https://www.vta.org/sites/default/files/2020-06/SustainabilityPlan2020_Accessible_0.pdf.
17. Ibid. 4-5.
18. Santa Clara Valley Transportation Authority, *Board Memorandum*, January 27, 2020, https://www.vta.org/sites/default/files/2020-02/Climate_Emergency_Declaration-Thursday%2CFeb_6_%202020%20%281%29_0.pdf.
19. Santa Clara Valley Transportation Authority, *VTA Climate Action and Adaptation Plan, Volume I: Final CAAP*, January 2024, https://www.vta.org/sites/default/files/2024-01/VTA_CAAP_Final_508web.pdf
20. "Sustainability," Bay Area Rapid Transit, accessed November 7, 2025, https://www.bart.gov/about/projects/cars/sustainability#:~:text=BART%20challenged%20Bombardier%20to%20build,that%20the%20trains%20currently%20use.
21. California Air Resources Board, *Climate Change Scoping Plan: A Framework for Change* (State of California, December 2008), https://ww2.arb.ca.gov/sites/default/files/classic/cc/scopingplan/document/adopted_scoping_plan.pdf.
22. City of Santa Clara, *Santa Clara Climate Action Plan* (December 3, 2013), https://www.santaclaraca.gov/home/showpublisheddocument/10170/635713044859030000.
23. Santa Clara Valley Transportation Authority, *Valley Transportation Plan for 2030* (VTA, February 2005), https://www.vta.org/sites/default/files/2023-06/VTP-2030.pdf.
24. City of San Jose, *Envision San José 2040: General Plan* (adopted November 1, 2011, as amended on October 24, 2024), https://www.sanjoseca.gov/home/showpublisheddocument/22359/637928744399330000.
25. "Stormwater Discharges from Transportation Sources," United States Environmental Protection Agency, accessed November 7, 2025, https://www.epa.gov/npdes/stormwater-discharges-transportation-sources.
26. "Sustainability," Bay Area Rapid Transit, accessed November 23, 2025, https://www.bart.gov/sustainability

G. McNeilus

Climate-Biodiversity Action in a

Wind Energy Project

Chapter 9

PROJECT NAME:	UPLAND PRAIRIE WIND FARM
LOCATION:	POWESHIEK COUNTY, IOWA, US
AREA:	47,000,000 SQ M (11,640 ACRES)
OWNER/OPERATOR:	INTERSTATE POWER AND LIGHT (ALLIANT ENERGY'S IOWA UTILITY SUBSIDIARY)
PROJECT DEVELOPMENT:	APEX CLEAN ENERGY AND INTERSTATE POWER AND LIGHT
PROJECT CONSTRUCTION & OPERATION:	ALLIANT ENERGY
OWNER'S ENGINEER:	BLACK & VEATCH
WIND TURBINE VENDOR:	GENERAL ELECTRIC (GE)
ARCHITECTS:	OPN ARCHITECTS
CONSTRUCTION CONTRACTOR:	WHITE CONSTRUCTION INC

UPLAND PRAIRIE WIND PROJECT

Integration of a wind generation facility with ongoing agricultural activity

PROJECT OVERVIEW

Project Description

The Upland Prairie wind farm is a 300-megawatt project located in Clay and Dickinson counties in northwest Iowa, in close proximity to the city of Everly.

Alliant Energy Corporation[1] is a regulated, investor-owned public utility holding company that provides electric and natural gas service in Iowa, Wisconsin, and Minnesota.[2] To move towards a cleaner energy future and expand its renewable energy portfolio, Alliant Energy's Iowa subsidiary, Interstate Power and Light (IPL),[3] purchased the Upland Prairie farm in 2017 from Apex[4] Clean Energy Holdings,[5] a company that develops utility-scale renewable energy facilities. The Upland Prairie project was developed in a single phase by Apex Clean Energy and IPL. Following its development, the project was constructed in 2018, commissioned in 2019, and is currently operated by Alliant Energy.[6] GE Renewable Energy was selected as the turbine supplier for the wind power project.

The wind project includes 121 wind turbines with an installed capacity of 300,000 MW that can power 108,000 homes in Iowa. It is a distributed project that extends within an agricultural area of approximately 30,000 acres. It is the largest individual wind farm developed by Alliant Energy,[7] with a cost of $430 million, representing a significant economic development for the city of Everly.

The wind turbines are situated within approximately 88 parcels located on parcels leased from landowners in the project area. They occupy a small portion of each parcel in order to allow continuation of the farming activity. The overall layout is therefore carefully designed to minimize the impact on the ongoing agricultural activity. In addition, the turbines are sited to optimize performance with setbacks to protect existing structures and natural elements. Also, for safety reasons, all turbines are situated 1,200 feet or more away from residential dwellings.

The project has been perceived positively by the local communities, who have watched revenue flowing in not only through property taxes but also through landowner payments. The resulting additional income is invested back into the community, thus contributing to the growth of local businesses. Both Apex and, later on, Alliant Energy have invested in developing close relationships with the local communities through regular meetings, open houses, and relevant presentations in order to promote public engagement and reinforce inclusivity.

Project Purpose

The Upland Prairie Wind project serves Alliant Energy's purpose of delivering cost-competitive clean energy to its Iowa customers, thereby contributing to

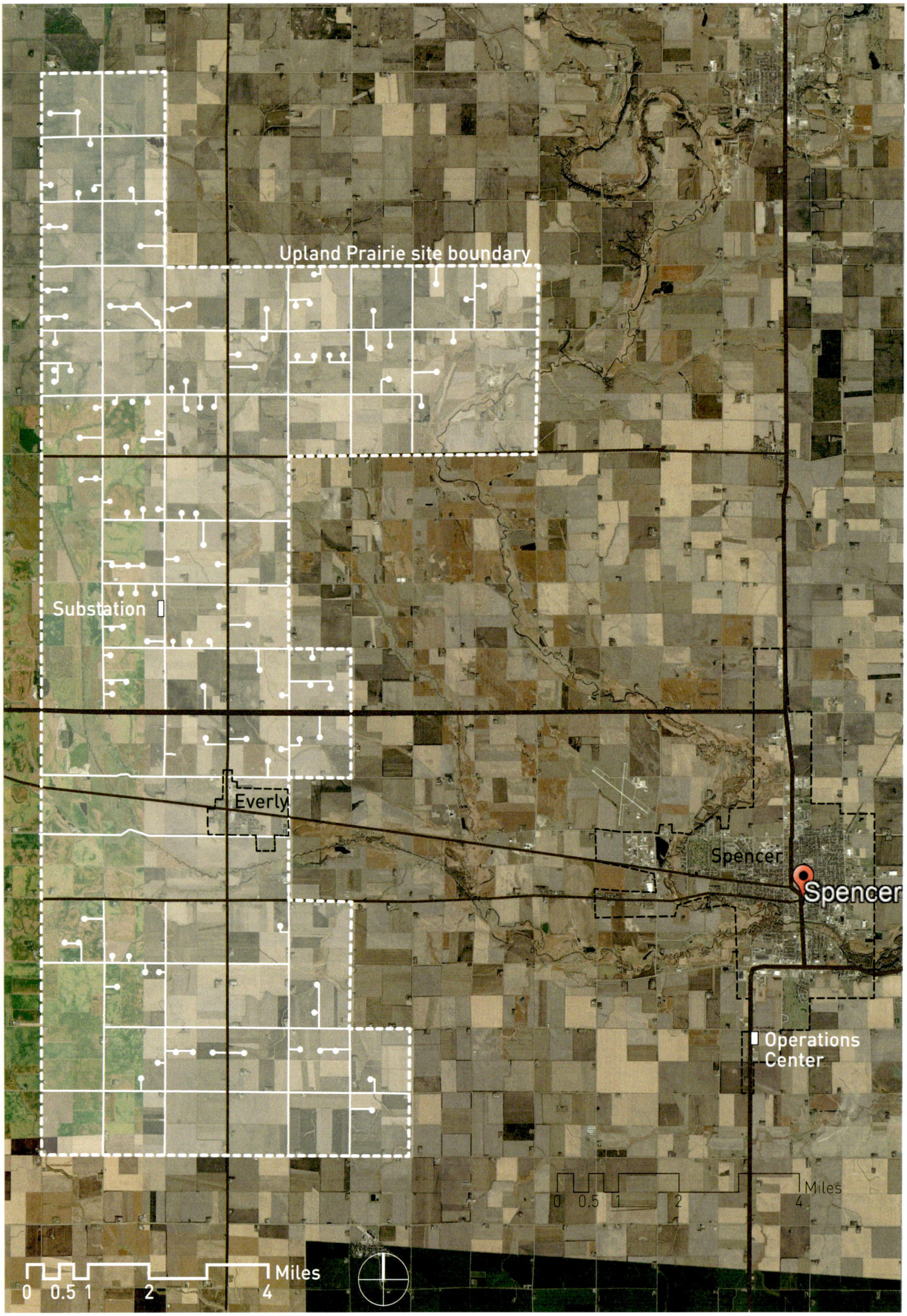

Aerial map of the site showing the location of the wind turbines, the Spicer Substation, and the Spencer Operations Center

the company's goal of increasing its renewable energy capacity from 20% in 2018 to 33% by 2024.

According to Terry Kouba, the President of Alliant Energy's Iowa energy company, the company is continuing to transition its generating fleet to cleaner, more cost-effective sources, which will promote a 40% reduction in carbon emissions by 2030 and an 80% reduction by 2050.[8]

The company has invested roughly $1.8 billion in wind energy generation in Iowa, adding up to 1,000 MW, which is enough to power up to 430,000 Iowa homes a year. The wind energy, generated by all the relevant projects combined together, was expected to represent one third of the energy provided to the Iowan customers by 2020.[9]

The Upland Prairie's annual energy output in particular, is enough to supply the energy for about 130,000 households a year. The project was expected to result in more than $40 million in local tax benefits over its life as well as approximately $45 million in landowner payments for the next 25 years.[10]

Apart from delivering clean energy to the customers, the project also seeks to keep landowners satisfied, and to provide economic benefits to the communities with the minimum impacts on the agricultural activities, the environment and the natural habitat.

MAIN PROJECT COMPONENTS

The Upland Prairie Wind Farm project consists of the following components:

- 121 wind turbines (and appurtenant facilities)
- The Spencer Operations Center
- Access roads
- The Spicer Substation
- Meteorological evaluation towers

Wind Turbines

The project is equipped with a total of 121 state-of-the-art wind turbines, based on GE's proven 2 MW platform. These include three types of GE turbines: the 2.3-116, the 2.5-116, and GE's newly introduced 2.5-127 model.[11] The 127-meter rotor, combined with the 2 MW electrical system, achieves a high capacity factor and increased annual energy production.

These types of turbines were selected by the project team for their high performance, durability, and noise reduction potential. The wind turbines can withstand higher wind speeds, have a lower cut-in wind speed, and have a high capacity factor. The high performance of the turbines results in using fewer units per project, minimizing thus disturbance to ecologically sensitive areas, like wetlands, on the project site.[12]

The Spencer Operations Center

For the operation of the Upland Prairie Wind Farm, a new operations center was constructed about 5 miles east of the project site, in Spencer, Iowa.

The operations center was designed according to Leadership in Energy and Environmental Design (LEED) principles in order to minimize its energy and water use. It is the only component with substantial energy needs. It has used electricity from the grid only in 2010 for its initial operation. Since then, the power needed was partially provided by the project's own wind power

Aerial image of the Spencer Operations Center

Example of wind turbines sited within agricultural parcels

Street views of the Upland Prairie site

generation. Overall, it has been designed to be energy efficient and is expected to consume annually 26% less energy compared to a facility using typical components.

Access Roads

For the development, construction, operation, maintenance and decommissioning of the project, a significant impact on several existing roads in Clay and Dickinson Counties occurred. The project developer agreed to the use, repair, and improvement of the designated roads. When access to farm fields was not possible, additional access roads to crop fields were constructed. The location of the new roads was decided in consultation with the landowner.

The Spicer Substation

The project has its own substation and interconnection to the electric grid. The substation is centrally located within the site. It was studied through the Midcontinent Independent System Operator (MISO), the Regional Transmission Organization, to ensure that the project's electrical output can be accommodated on the grid.

Aerial image of the Spicer Substation

NOTABLE SUSTAINABLE PRACTICES

Land-use Synergies

One of the project's most distinctive practices was its successful integration with existing agricultural land use.

The project maintained the use of agriculture while introducing a whole new activity, that of wind energy generation. The wind turbines were strategically distributed across agricultural land plots so as not to interfere with crop production and freshwater habitats. Of the 51,810 acres of prime farmland, only 0.16% was affected by the placement of wind towers and associated infrastructure. This limited disturbance ensures that over 99% of the farmland remains intact. This approach was made possible through fair lease agreements with landowners, which in turn generated additional income for the community.

Project Stakeholder Engagement

IPL focused on addressing community needs and fostering strong stakeholder relationships since acquiring the project. After taking over from Apex following CUP approval, IPL ensured continuous engagement with landowners and the community.

Landowners were involved throughout the project, particularly in decisions regarding turbine siting, road locations, and infrastructure planning. IPL adjusted layouts based on their feedback to minimize disruptions to agricultural activities. Transparency was maintained through community presentations, public meetings, and a project office where concerns could be raised. Regular communication was ensured via dedicated landowner liaisons and weekly open houses like "Coffee with Cliff," keeping the public informed at every stage.

In addition to engaging stakeholders, IPL prioritized local workforce development, hiring from local union halls, and providing specialized wind energy construction training, enhancing workers' skills for future opportunities. IPL also worked with local emergency services, providing tower rescue training and coordinating with response teams to ensure preparedness for potential emergencies. Environmental concerns were addressed through wildlife analyses in collaboration with the US Fish and Wildlife Service and Iowa Department of Natural Resources, with voluntary measures like setbacks from bat nesting areas. The project also improved local infrastructure, including roads and water systems, and designed the operations center as a community hub with meeting spaces and public trails.

Overall, IPL integrated community input, minimized impacts, and built long-term relationships, ensuring both project success and lasting benefits for the area.

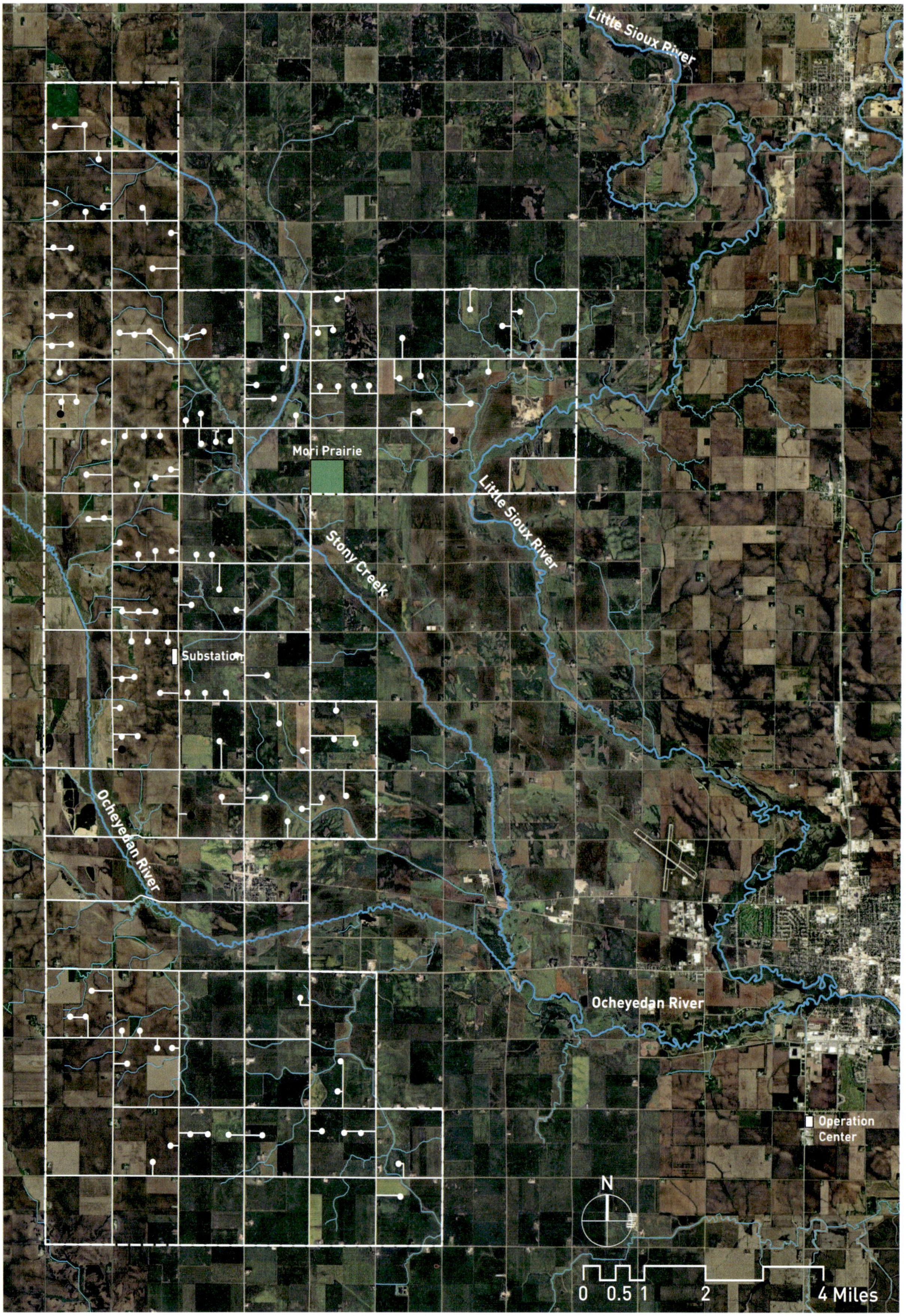

Water bodies in the site area

ANALYSIS BASED ON THE INTEGRATED CRITERIA

ASSESS

Existing Conditions

The Upland Prairie project required extensive assessments to measure the environmental status of the region and the status of communities where the wind farm would be located. That process involved, as appropriate, preliminary site evaluation, site characterization, and field studies to document wildlife and habitat and to predict project impacts.

Prior to the siting of the turbines, a third-party consultant was hired to perform a Desktop Analysis survey of the Upland Prairie Wind Project to assess the presence of wetlands and other waters of the United States. The analysis used National Wetlands Inventory (NWI) data from the United States Fish and Wildlife Service (USFWS), United States Geological Survey (USGS) stream and river data, Natural Resource Conservation Service (NRCS) hydric soil data, and aerial imagery from Google Earth Pro. This analysis concluded that there are 1,632 areas within the project boundary characterized as wetland or waters of the United States.

Following the Desktop Analysis, a Wetland Delineation Report and an updated Jurisdiction Delineation Report were conducted in order to determine whether or not wetlands existed within an area of investigation around proposed project facilities.

The project is situated on agricultural lands that are drained through the use of extensive networks of subsurface drainage tiles in order to support farming. The natural topography of the land consists of gently rolling plains, which do not include high-risk hillsides or steep slopes. The project area is located in a generally flat region with the largest increase in elevation being 130 feet and occurring gradually over 2 miles.

There are two major aquifers underlying the project site: the Dakota Aquifer and the Cambrian-Ordovician Aquifer. The Upland Prairie Wind site includes three notable habitat corridor areas that are likely conducive to wildlife movement. These habitat corridors occur along the Ocheyedan River through the central-south portion of the site, the Stony Creek corridor across the northern portion of the site, and the Little Sioux River that slightly overlaps the northeast portion of the site.

The preliminary site evaluation also indicated the potential existence of a threatened species, the Northern long-eared bat, and a species of special concern, the bald eagle. Western EcoSystems Technology, Inc. (WEST) was hired as a third-party consultant to perform several field surveys on these species.

The rivers and streams offer cover and shelter in the form of some small forested areas, prairies, and parks that are not developed or cultivated and that have native and/or taller vegetation areas where animals can hide. Trees, bushes or shrubs, and tall grass can be used for perching and provide nesting materials, and certain tree species offer sheltering locations for bats under their bark. Older, open barns in agricultural areas are also used by bat species such as the Northern long-eared bat. Eagles typically occur near large rivers where they can find food sources; the Ocheyedan River corridor is the site of a bald eagle nest. Birds and bats forage for insects in and over forested and prairie areas, especially near water, where they can also drink, and where certain prey insects such as mosquitoes may congregate. Birds and bats may also forage over parts of agricultural fields and parks, but often return to river or stream corridors to continue movement through the landscape.

At the time of the project, the Northern long-eared bat was classified as threatened species under the federal Endangered Species Act. It is worth highlighting that it was reclassified as endangered in 2022 due to a widespread disease that affected its population. The bald eagle, an endangered species in the 1960s, was delisted from the Endangered Species Act in 2007, due to its significant recovery; however, it is still protected under the Migratory Bird Treaty Act and the Bald and Golden Eagle Protection Act.

Finally, within the project site, there is a 40-acre preserve of rare, unplowed, tallgrass prairie plant community, owned and managed by the Nature Conservancy, the Mori Prairie. It is an outstanding and very rare example of remnant prairies in Northern Iowa. Tallgrass prairies historically covered vast regions of

OVERVIEW OF THE STATE OF BIODIVERSITY AT THE PROJECT SITE

		Wind farm site	Operation center site	Substation site
Site of high ecological value		No	No	No
Proximity to ecosystem		Small habitats, river and stream corridors south of the project area		Land surrounding the substation is intensively cropped agricultural fields with occasional residences
Pre-existing site condition		Intensive row crop agricultural production	Small industrial park in the interface between the edge of Grinnell town and the surrounding rural, agricultural area	Agricultural production
STATE OF BIODIVERSITY (BEFORE THE PROJECT)	Species	• Small habitats, river and stream corridors south of the project area are particularly important for wildlife movement • Rivers and streams in and around the project area offer cover and shelter for eagles • Birds and bats forage for insects in and over forested and prairie areas, especially near water • Older, open barns in agricultural areas are also used by bat species such as the Northern long-eared bat for forage	No data available	No data available
	Ecosystem	Degraded soil in agricultural fields due to the application of fertilizers and pesticides		

Iowa, and now less than one percent remains. Remnant prairies, like the Mori Prairie, provide habitat for avian species of conservation concern and offer a rich diversity of plant species.

Overall, since the project site is used for intensive agricultural production, the small habitats and river and stream corridors are particularly important for wildlife movement, as there is almost no functional habitat or shelter for wildlife remaining in the agricultural fields.

Assessing Drivers of Climate Change and Biodiversity Loss

To identify how the project would potentially impact the existing conditions, the following studies were conducted:

- **A Life Cycle Analysis (LCA) calculation** of the whole project. It estimates the greenhouse gas emissions generated during construction and operations in carbon dioxide equivalents. The annual fossil fuel emissions offset at Upland Prairie is estimated at 592,627 tons of CO_2 equivalent. Over the 40-year project life, it is expected to have a net negative impact of over 23.3 million tons of GHG emissions.
- **A cradle-to-grave LCA of the 2.5xl wind turbine:** This was run by GE, the wind turbine manufacturer, to assess the climate change or global warming impacts, measured in carbon dioxide equivalents (CO_2e), across the lifecycle of the 2.5xL wind turbine. It included raw materials acquisition and preprocessing, manufacturing, site prep and installation, operation and maintenance, and end of life. The result of the LCA showed that 65% of lifecycle GHG emissions came from supply chain activities, with 24% from distribution (site preparations and installation of the wind turbine), 6% from use (operation and maintenance), 5% from end-of-life, and <1% from production.
- **IPL's Avian Protection Plan** (included in the Alliant Energy Sustainability Report): Aligned with US Fish and Wildlife Service (USFWS) Land-Based Wind Energy Guidelines (WEG), this plan determines the highest risk areas for birds (relevant to poles, wires, wind farms, and other infrastructure). Based on this plan, IPL could take enhanced action in those regions to protect birds.

The main pressure on biodiversity that existed on the site before the project development was the application of fertilizers and pesticides on the land related to the agricultural activity.

Standard traps identified by the project team include impacts to the environment due to the placement of the turbines if only the minimum county-required setbacks are used, with no other environmental, safety, or technical considerations or best practices. The

DEPENDENCIES ON ECOSYSTEM SERVICES

ECOSYSTEM SERVICES			DEPENDENCIES	
			project	community
PROVISIONING SERVICES	Biomass provisioning	Food crops		■
		Grazed biomass provisioning		■
		Livestock provisioning		
		Aquaculture provisioning		
		Wood provisioning		
	Terrestrial and aquatic plants as energy source			
	Other raw material provisioning			
	Genetic material			
	Water supply	Potable water		
		Non-potable water for processes	■	■
		Non-potable water for irrigation	■	■
REGULATING AND MAINTENANCE SERVICES	Global climate regulation		■	■
	Rainfall pattern regulation (at sub-continental scale)			
	Local (micro and meso) climate regulation		■	■
	Air filtration		■	■
	Soil quality regulation		■	
	Soil and sediment retention	Soil erosion control	■	
		Landslide mitigation		
	Solid waste remediation			
	Water purification (water quality amelioration)	Retention and breakdown of nutrients	■	
		Retention and breakdown of other pollutants	■	
	Water flow regulation	Baseline flow maintenance	■	
		Peak flow mitigation	■	
	Flood mitigation	Coastal protection		
		River flood mitigation	■	
	Storm mitigation			
	Noise attenuation		■	■
	Pollination			■
	Biological control	Pest control		
		Disease control		
	Nursery population and habitat maintenance			■
CULTURAL SERVICES	Recreation-related			■
	Visual amenity			■
	Education, scientific and research			
	Spiritual, artistic and symbolic			
	Ecosystem and species appreciation			■

appropriate setbacks from wetlands, water bodies, bat habitat, eagle nest areas, grasslands, and other environmentally sensitive features should be maintained in order to minimize the risks of noise and air pollution, and the overall risk of species disturbance.[13]

Additionally, the turbine lighting at night may cause light pollution. However, it is hard to fully reduce it as nighttime lighting is needed for aviation safety in accordance with the Federal Aviation Administration's (FAA) requirements.

There is also a risk of spills when precipitation levels are above the average level or during storms. The risk of water pollution is high if stormwater, from entraining sediment or other pollutants, is transferred to the nearest water body.

One of the most essential community dependencies is the ongoing food production that takes place across the site, which was historically used as agricultural land. The project respects the community's needs and maintains a balance by preserving the agricultural land use in combination with the wind farm.

Assessing Climate Change-Driven Threats and Risks

The main climate change-related shocks that may affect the project are increased precipitation and storms, and the associated severity and duration of flooding, as well as higher winds.

In order to assess climate change-related threats and risks, the project team has developed, among others, the following studies:

- **Climate Impact Assessment and Adaptation Plan**: It provides a comprehensive overview of the potential climate change impacts to the project and the associated design and operational mitigation strategies.
- **Hydrology Study:** IPL further assessed and avoided potential flood risks to turbines and access roads through detailed hydrological modeling of the project site, incorporating accurate elevation data into the model and siting turbines and other elements outside the forecasted flood elevations, even when those were outside the FEMA floodplain.

SUMMARY OF EXISTING AND PROJECT-DRIVEN PRESSURES ON BIODIVERSITY

PRESSURES ON BIODIVERSITY		EXISTING	PROJECT DRIVEN
Land change, freshwater, sea change			Land change, freshwater change
Pollution	Water		Runoff
	Air		During construction
	Soil	Application of fertilizers and pesticides on agricultural land	
	Waste		During construction, maintenance and end of life
	Noise		Operational noise and vibration
	Light		Turbine lighting as night airspace obstruction for safety
Introduction of invasive species		No data available	No data available

CLIMATE CHANGE-RELATED SHOCKS AND RISKS

CLIMATE CHANGE-RELATED ACUTE SHOCKS AND CHRONIC STRESSORS	CLIMATE CHANGE-RELATED RISKS
Increased precipitation/Number and intensity of storm events	Increased risk of spills and stormwater contamination
	Risk of damage in wind turbines due to flooding
	Roads damaged
Higher winds or adverse weather (i.e. icing)	Risk of damage in wind turbines due to high intensity wind events
	Safety risk for workers and public

The project is committed to several sustainability targets that align with national and state policies as well as with best practice guidelines. IPL worked with Black & Veatch, its engineer, during Project development to create a policy to avoid wetlands, water bodies, bat habitat, eagle nest areas, grasslands, and other environmentally sensitive features.

Commitment to Renewable Energy

Iowa has, in recent years, been one of the most progressive states in the United States in terms of the installation of renewable energy, and Alliant Energy has been a part of this growth effort from the beginning.

Iowa became the first state that adopted a RPS (Renewable Portfolio Standard) in 1983, therefore establishing a goal for renewable energy development. The goal was initially to generate 105 MW of wind energy within the state, a target that was eventually easily surpassed thanks to the progressive energy policies of Iowa. Since then, the state has stayed committed to its renewable energy adoption goals, achieving by 2012 the displacement of 84.7 million metric tons of CO_2 through wind energy generation. In 2021, the Iowa Environmental Council announced its vision for Iowa to reach 100% clean power generation by 2035.[14]

Iowa currently produces the highest percentage of electricity by wind of any state (64.7%). It generates more electricity from wind than any other single source.[15]

Similarly, Alliant Energy is committed to providing its customers with clean, renewable energy by reducing its reliance on fossil fuels in its service territory.

As a result, the Upland Prairie Project contributes to reducing fossil fuel dependence, supporting Iowa's position as a top state for wind generation as a percentage of electricity production. This initiative stems from Alliant's broader commitment to sustainable energy and goes beyond any federal, state, or local regulatory requirement for clean energy.

Commitment to Minimizing Environmental Impact

According to Alliant Energy's Corporate Sustainability standards, the project was committed to following best practices regarding setbacks, wind turbines and equipment siting as well as responsible material sourcing. This commitment is reflected in Alliant Energy's and GE's sustainability reports, which highlight the project's focus on reducing ecological impacts, preserving biodiversity and ensuring environmentally responsible development across all project phases.

Commitment to Local Habitat and Wildlife Conservation

Common environmental concerns associated with the Upland Prairie Wind Farm include the impacts on birds and bats. To align with federal wildlife conservation goals, the project adheres to the US Fish and Wildlife Service (USFWS) Land-Based Wind Energy Guidelines (WEG). The WEG provides a structured process for working with USFWS to site wind projects and turbines to avoid areas where wildlife impacts are likely to be high and difficult or costly to mitigate. The project thus commits to reducing risks to protected species of migratory birds and bats, especially those species protected under the Endangered Species Act.[16]

In this effort, several biological studies were conducted, including avian and raptor nest surveys and bat habitat assessments, to inform turbine siting and

ALLIANT ENERGY'S TARGET FOR THE TRANSITION IN ENERGY RESOURCES

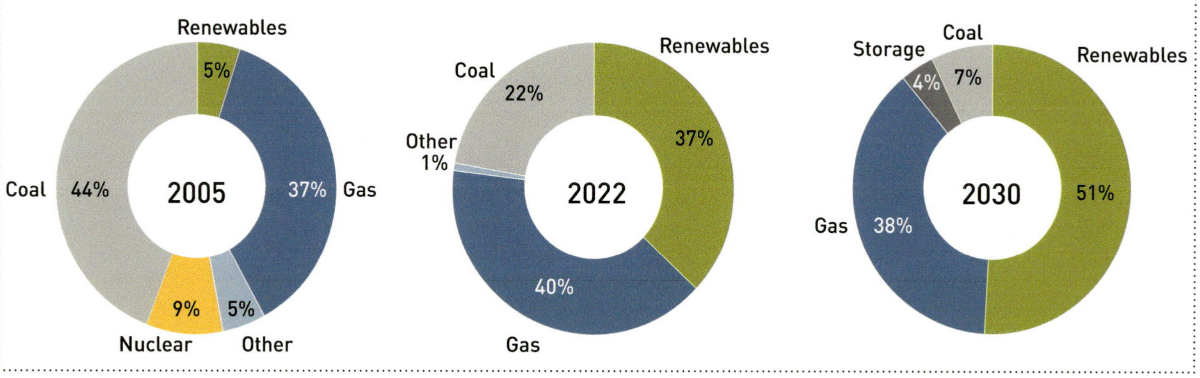

minimize impacts on protected species. A Bird and Bat Conservation Strategy was developed to comply with conservation laws, emphasizing impact monitoring and mitigation throughout the project lifecycle.

In addition, several surveys[17] required by the USFWS Eagle Conservation Plan Guidance were performed. They identified minimal impact on local raptor populations, including eagles. The findings guided Alliant in siting turbines to protect raptors. Setback policies were also applied to create buffer zones around critical eagle habitats.

Through discussions with the Iowa Department of Natural Resources (IDNR) and the United States Fish and Wildlife Service (USFWS), as well as through the completion of several biological studies by third parties, it was determined that development of the Upland Prairie Wind project avoids land that is determined to be prime habitat. The risk assessment completed for the project is in accordance with the USFWS 2012 Land-based Wind Energy Guidelines and the 2013 Eagle Conservation Plan Guidance.

Alliant's project development policy is to avoid wetlands, water bodies, and related features by locating project components a minimum of 300 feet from any of these natural features. At the Upland Prairie site, setbacks were applied based on coordination with the USFWS, best practices established on previous project experience, and Alliant's commitment to environmental responsibility and sustainability. Their efforts surpass local zoning requirements and emphasize habitat preservation.

Commitment to Wetland and Waterway Protection

Although Clay County zoning ordinances do not require any setbacks between wind energy facilities and wetlands or water bodies, Alliant Energy voluntarily committed to adopting stricter measures to protect environmentally sensitive areas. Aligning with the stricter requirements of Dickinson County, where a 660-foot setback is required for lands and waters owned or managed by the Iowa Department of Natural Resources (IDNR), Alliant Energy applied similar or greater setbacks within Clay County.

Alliant Energy worked closely with IDNR to identify sensitive areas and then applied the company's best practice setbacks. All field-delineated wetlands are fully protected, with no temporary or permanent impacts allowed during IPL wind projects. Water crossings and other direct impacts to wetlands and water bodies are avoided to the maximum extent possible, preserving natural hydrologic connections and sediment transport.

As a result, the project meets and exceeds county regulations, minimizing impacts on wetlands and waterways and preserving prime habitat.

Commitment to Stakeholder Engagement

The previous commitments demonstrate Alliant Energy's alignment with federal, state, and local sustainability policies while advancing renewable energy with minimal environmental disruption.

Apart from the biodiversity- and environmental-related project targets, the project teams have uncovered several community needs, goals, and issues through research, private, and public consultations. The project takes place in two counties, Clay and Dickinson, both part of the NWIPDC Region III Planning District. Among the Commission's goals, as noted in the report from 2015, the goals "promote an emphasis on industries that enhance the region" and "promote and encourage net regional job growth" speak directly to the Upland Prairie project. IPL invested approximately $1 billion to expand its wind operations in Iowa and therefore advance clean energy to support Iowa's vision for growing renewable energy and enhance the region. Being a wind project, not only does it create construction jobs with good wages, but it also benefits supporting industries.

The project teams made significant efforts to align the project's vision and goals to the needs and goals of the host and affected communities through continued active stakeholder engagement.

RESPOND

To respond to its commitments, the project was designed in accordance with the Envision methodology for sustainable infrastructure, and as a result, it received the Envision Platinum Award. The following strategies were implemented as part of its sustainable approach. Although the project development was not strictly led by climate change and biodiversity criteria, it is clear that even typical sustainable strategies respond by default to multiple climate change and biodiversity impacts and risks. This section describes the project's main sustainability strategies and highlights the relevance of each sustainable strategy to specific climate change and biodiversity criteria by mentioning the corresponding code numbers at the end of each description.

LIST OF SUSTAINABLE PROJECT STRATEGIES

GHG EMISSIONS REDUCTION

Generation of renewable energy (300 MW of wind power)
The 2.5xl wind turbine reduces net carbon emissions by replacing a portion of grid electricity that would otherwise be generated through the use of fossil fuels (high carbon emissions) with renewable energy (minimal carbon emissions). Over the 40-year project life, it has a net negative impact of over 23.3 million tons of greenhouse gas emissions.
[CCM1.2] [CCM3.3]

Solar infrastructure provision at the operations center
The Spencer Operations Center will include provisions for a roof-mounted or steel armature-mounted PV array and reserve onsite space for a future ground-mounted solar array to support the center's power needs during future operations.
[CCM1.2] [CCM3.3]

No significant dependence on petroleum fuels
The project's use of petroleum products will be limited to fuel for construction vehicles and equipment for approximately 1 year, followed by a 40-year operating life during which it will use relatively small quantities of oil and lubricants in the nacelle of each of the 121 turbines.
[CCM1.1] [CCM1.3] [CCM3.3]

Construction Scope 1 emissions reduction
The construction equipment used on the project is factory equipped with idling-limiters, and all vehicles and equipment are outfitted with devices that meet the latest EPA criteria with respect to emissions.
[CCM1.1] [CCM3.2]

ENERGY USE REDUCTION

LEED-aligned energy-efficient features for the operations center
The energy consumed by the Upland Prairie Wind Farm is limited to the energy demands of the Spencer Operations Center, which is designed and specified for energy efficiency. The use of the following LEED-aligned features results in a reduction of energy usage by 26.0% annually at the operations center compared to a similar conventional facility. This equates to an energy use reduction of 2,345 MWh over the 40-year life of the project.[18]

Energy-efficient features at the operations center:

- High-efficiency, long-life LED fixtures
[CCM1.2] [CCM3.2]
- Occupancy sensor controls to reduce energy when spaces are no longer occupied. Timers automatically turn off lights after 15 minutes of space vacancy
[CCM1.2] [CCM3.2]
- Daylight sensors to automatically dim fixtures in spaces with sufficient daylight (light level is maintained at 30 foot-candles at +30 inches)
[CCM1.2] [CCM3.2]
- Diffused, translucent light into the center's garage area to provide more natural light
[CCM1.2] [CCM3.2]
- Photocell-enabled control and countdown timer that switches on exterior lighting to reduce energy consumption and light pollution
[CCM1.2] [BIO1.6] [CCM3.2]
- A geothermal heat pump HVAC system for more energy savings over conventional HVAC systems.
[CCM1.1] [CCM1.2] [CCM3.3]
- High-efficiency, low environmental impact Dyson Airblade hand dryers are installed in restrooms
[CCM1.2] [CCM3.2]

Merging three operations centers into one energy-efficient building
IPL used to operate two old, non-energy-efficient operations centers in Rolfe, IA, and Peterson, IA. The Spencer Operations Center was built to combine the three facilities into a new, energy-efficient, sustainable building. Instead of constructing a third building (non-energy-efficient), IPL has opted to combine the three operations centers (two existing and the one needed for the project) into one modern, energy-efficient building. It was the most energy-efficient option for the project, but this strategy also reduces overall energy consumption by IPL operations centers in the area.
[CCM1.2] [CCM3.2] [CCA3.7]

High SRI value for the majority of the hardscape:
Over 90% of the hardscape has a high solar reflectance index (SRI) value of 29 or higher. Hardscapes include the concrete foundations of the wind turbines and the MET weather stations, which will use light-colored concrete, the project substation, and the Spencer Operations Center's parking lot and courtyard. More specifically: SRI values for concrete are estimated at 38, for Portland cement concrete (PCC) paving at 40, and for the aggregates (gravel) at 46 to 55. In addition, tree planting around the operations center entrance will shade portions of the outdoor surfaces and the building walls.
[CCM1.2] [CCM3.2] [BIO3.3]

USE OF HIGH-PERFORMANCE COMPONENTS

High-performance GE 2.5-127 turbines
The project includes 121 high-performance GE 2.5-127 wind turbines, which offer a best-in-class capacity factor and increased Annual Energy Production.[19] These turbines also feature a lower cut-in speed,[20] allowing electricity generation to begin at lower wind speeds.
[CCM1.2] [CCA1.4] [CCM3.2] [CCM3.3] [CCA3.3] [CCA3.5]

Windfree technology for reactive power support
Windfree, a GE feature used in the project, provides reactive power support to the grid when there is no wind. This innovative system maintains voltage stability and overall grid reliability while reducing the need for additional, energy-intensive stabilization sources. It thus allows the wind turbines to contribute to a steady and efficient energy infrastructure, enhancing the project's sustainability and lowering the carbon footprint associated with conventional energy sources.
[CCM1.1] [CCA1.2] [CCM3.2] [CCA3.5]

Incorporation of a feeder bay to minimize downtimes
The integration of the feeder bay into the project's substation provides efficient management of energy from the wind turbines while minimizing the likelihood of future downtime.
[CCM1.1] [CCA1.2] [CCM3.2] [CCA3.5] [CCA3.8]

PROJECT USEFUL LIFE EXTENSION

Advanced analytics for extended operating life
The wind analytics program relies on monitoring data in order to design, operate, and maintain wind generation facilities for optimal performance. This strategy aims to extend operating life to 40 years, significantly surpassing the industry standard of 20 years. The in-house data analysis program performs monthly analysis of operational data with software tools, such as Pi Historian and MATLAB. Using statistics and pattern-matching algorithms it can detect operational anomalies.
[CCM1.1] [CCM1.3] [CCM3.7] [CCA3.4] [CCA4.1]

Condition monitoring for predictive maintenance
The project utilizes GE's Condition Monitoring System to provide advanced warnings of potential issues in wind turbines, helping to minimize secondary damage and defects.
A computer system continuously monitors and analyzes the performance of the Alliant Energy and GE wind turbines. The in-house data analysis program is designed to predict potential issues before they occur, allowing for predictive maintenance that replaces traditional scheduled maintenance. By using real-time data and advanced analytics, it helps ensure that turbines are maintained only when necessary, improving efficiency and reducing unnecessary downtime. This system enables the team to conduct preventive maintenance on turbines that may not be performing optimally, improving overall reliability and operational efficiency.
[CCM1.1] [CCM1.3] [CCA1.3] [CCA3.4] [CCA4.1]

Strategic service contracting for long-term maintenance
A strategic service and maintenance contracting approach provides comprehensive maintenance services for wind turbines over a 20-year period. This approach incentivizes maintenance and service practices that support an extended operating life. Upon the conclusion of the initial agreement, a new service and maintenance agreement will be necessary for the subsequent 20 years of operation.
[CCM1.1] [CCM1.2] [CCM1.3] [CCA1.4] [CCM3.7] [CCA3.4] [CCA4.1] [BIO3.3]

Enhanced foundations for wind turbines
At certain turbine locations closer to potential flood elevations, turbines required a deeper pier foundation to ensure their stability. These pier foundations extend approximately 21 feet below ground, still significantly above the groundwater level in the area.
[CCM3.7] [CCA3.2] [CCA3.4]

Durable construction materials and components for the extended lifespan of the operations center
To achieve a long service life, the operations center was constructed with materials that are resilient to the various weather conditions the facility will face over its lifespan. This includes weathering steel panels, which oxidize to create a protective layer around the panel, and precast concrete sandwich panels.
[CCM1.1] [CCM1.3] [CCA1.4] [CCM3.7] [CCA3.4] [BIO3.3]

Digital replica for enhanced performance
The wind farm utilizes a digital replica of the project to analyze and improve performance. This technology allows for comprehensive monitoring and analysis, enabling the project team to predict and respond to customer issues effectively. As a result, it can enhance the operational and financial performance of the wind turbines, the plant, or the entire fleet.
[CCM1.1] [CCA1.4] [CCA4.1]

Frequent turbine data collection for adaptability to changing conditions
Turbine data will be collected and analyzed frequently to adapt controls to changing environmental conditions.
[CCM1.1] [CCA1.3.] [CCA1.4] [CCA3.5] [CCA4.1]

Scheduled grease replacement to reduce failure rates
Grease replacement is carried out during bearing inspections on a scheduled basis to extend the project's useful life by reducing failure rates. This maintenance strategy for main bearing inspection and grease replacement is "above and beyond" the industry standard recommendations.
[CCM1.1] [CCM1.3] [CCA1.3] [CCA1.4.] [CCM3.7] [CCA3.4]

Power curve testing by an independent engineer
The Wind Turbine Supply Agreement specifies that an independent engineer perform a Power Curve Test during commissioning. This test verifies that the turbine's predicted power output at varying wind speeds aligns with its actual performance during operation.
[CCM3.2]

MATERIAL INPUT REDUCTION

Recycled steel in wind turbine fabrication
The project incorporates a high percentage of recycled steel in the fabrication of its wind turbines. Each turbine contains 57% recycled plate steel. The percentage of recycled material in the turbines, measured by weight, is approximately equivalent to the percentage of recycled material by cost. The plate steel used by GE contains 95% or more recycled scrap metal, depending on the specific steel mill supplying the material in the US. According to the United Nations Environment Programme, recycling metals is two to ten times more energy-efficient than producing metals from virgin materials, while the use of reclaimed steel results in approximately 75% fewer GHG emissions.
[CCM1.3] [CCA1.4] [CCM3.7] [CCA3.3] [CCA3.6] [BIO3.3]

Optimized power grid strategy for less need of substations and transmission lines
The project team has strengthened the power grid by incorporating higher voltage lines. Since power can travel further at higher voltages, fewer substations are required, minimizing the overall infrastructure footprint. These higher voltage lines are placed underground, thus not exposed to weather extremes, improving reliability of power transmission while lowering costs.
[CCM1.3] [BIO1.3] [BIO1.4] [CCM3.7] [CCA3.3] [BIO3.3]

Use of fly ash (byproduct) in concrete mixes
Fly ash, a byproduct found in coal-fired power plants across the state, is incorporated into the project's concrete mixes to minimize waste and reduce overall project costs.
[CCM1.3] [CCA1.4] [BIO1.4] [CCM3.7] [CCA3.3] [BIO3.3]

Locations of sourced materials

LOCALLY SOURCED MATERIALS

Use of locally and regionally sourced materials

The project utilizes locally and regionally sourced materials, including aggregate, concrete, seed, rebar, embedment rings, and crane mats, all obtained from within a 300-mile radius. The project team was guided by Envision's preferable distances for sourcing key types of materials (50 miles for soils and aggregate, 100 miles for concrete, 250 miles for plants, and 500 miles for all other materials).[21]

More than 96% of the total cost of materials used in the project comes from regional sources.

More specifically:

- Fill (accounting for 3.2 percent of project materials' cost) was sourced within 50 miles (80 km)
- White Construction found no quarries located within the required 50 mile range of the project that could supply the necessary aggregate (accounting for 32 percent of materials' cost), so it was sourced from outside the 50 mile range. However, in order to reduce the impacts from transporting the materials, the aggregate will first be brought by rail to the center of the project area, and then by truck to the specific locations within the project site where it is needed
- Concrete (estimated 32 percent of material costs) was sourced from a batch plant located within 100 miles
- Plants and seed for restoration of disturbed land (0.5 percent of materials' cost) were purchased from a seller within the project area
- Rebar (accounts for 27 percent of materials' cost) was manufactured in Polo, MO (258 miles) or Kankakee, IL (458 miles)
- Embedment rings (3.2 percent of materials' cost) were manufactured in Minonk, IL (359 miles)
- Anchor bolts (3.2 percent of materials' cost) were sourced from a known manufacturer outside the 500 mile radius (Pueblo, CO, 700 miles) because of the need to ensure quality

[CCM1.3] [CCA3.8]

Use of a byproduct of local agricultural activities

By sourcing the straw locally (an agriculture byproduct obtained from the dry stalks of cereal plants after the grain and chaff have been removed), the project will reduce transportation costs and increase community involvement.

[CCM1.3] [CCA1.4] [CCA3.7] [CCM3.8]

Optimized material transport logistics

For materials that could not be sourced locally, multi-modal transportation methods (i.e. the material is transported in large quantities by rail, then transferred by truck to Everly, Iowa) will reduce carbon emissions caused by heavy truck traffic.
[CCM1.3] [CCM3.8]

SUSTAINABLE PROCUREMENT PRACTICES

Sustainable procurement of fill materials

The procurement of sand, gravel, and fill material from Dave's Sand and Gravel met construction requirements, supporting the project's sustainability assessment criteria.
[CCM1.3] [BIO1.3] [BIO1.4] [BIO1.5] [BIO1.6] [CCM3.6] [BIO3.3]

Use of approved vendor lists to ensure sustainable procurement

The project ensures sustainable procurement by relying on approved vendor lists, which require suppliers of materials and structures, including wind turbines, to demonstrate commitments to improved performance, human rights protection, and environmental stewardship.
[CCM1.3] [BIO1.3] [BIO1.4] [BIO1.5] [BIO1.6] [CCM3.6] [BIO3.3]

Triple-bottom-line supplier vetting strategy for sustainable procurement

The project team conducts a vetting process for suppliers using questionnaires focusing on their procurement practices from a triple bottom line perspective, reviewing social/community, environmental, and economic aspects of the supply chain. The construction contractor, White Construction utilizes a Pre-Qualification Application to exclude vendors who do not meet the company's minimum requirements.
[CCM1.3] [BIO1.3] [BIO1.4] [BIO1.5] [BIO1.6] [CCM3.6] [BIO3.3]

SOLID WASTE DIVERSION

Reuse of excess aggregate for community and project needs

Excess aggregate, a byproduct generated by the project and stored onsite, will be reused to repair roads and address the needs of both the project and the community. This byproduct will also be utilized in the temporary laydown yard and for constructing temporary turning radii on county roads.
[CCM1.3] [BIO1.6] [CCM3.7] [BIO3.3]

Recycling of operational waste

78% of the waste materials generated by Upland Prairie Wind Farm's operation (Spencer Operations Center waste) are diverted from the landfill. IPL contracts OSI Environmental to collect and manage the majority of the operations center waste material (i.e. batteries, used oil filters, used oil, light ballasts, capacitors, cardboard, coolant, and plastic containers etc.). All of these materials are all recycled and diverted from landfill disposal.
[CCM1.3] [BIO1.6] [CCM3.7] [BIO3.3]

Reduction of excavated material taken off site

The project was designed to reuse 100 percent of the excavated materials from each turbine site, with all excavated material being retained at the location it originated:

- Excavated materials were utilized to obtain the 2% grade required at each turbine site to ensure water drainage away from the turbine foundation
- 100% of the material excavated for trenching activities was used to backfill the trenches
- Any required fill related to the turbine elevation changes or the substation and switchyard grading was imported from Dave's Sand and Gravel, located at the center of the site

[CCM1.3] [BIO1.6] [CCM3.7] [BIO3.3]

END-OF-LIFE REPURPOSING

Recyclable wind turbine towers

A large portion of the wind turbine structure is designed for easy separation into major components, facilitating efficient recycling or reuse. 90% of the steel, copper, aluminum, brass, non-ferro, and cast-iron materials and 100% of the battery (lead-acid) material will be recycled. Additionally, 100% of the blades will be recycled in the United States. The recycled blades will be repurposed for beneficial uses including manhole covers, pallets, and composite panels.
[CCM1.3] [CCM3.7] [CCA3.5]

PROTECTION AGAINST EXTREME EVENTS

Digital solutions for adaptive response

The implementation of a digital solutions system enables responsive actions to changes in operating conditions, such as higher extreme winds (maximum 50-year 10-minute wind speeds and maximum 50-year 3-second wind gusts), stronger storms, or adverse weather. This approach enhances operational resilience and safety under varying environmental conditions.
[CCM1.3] [CCM3.7] [CCA3.5] [CCA3.8] [CCA4.1]

Wind turbine selection for local weather extremes

As part of the turbine selection process for the project, it was required that GE approves the operational and survival suitability of the turbine model used against the local atmospheric extremes. These turbines are capable of handling a maximum sustained gust of up to 85 mph, significantly higher than any measured wind speeds in the area. Each turbine model is designed and certified to withstand certain winds by shutting off when the wind reaches a certain speed, and will not turn back on until the four-minute rolling average wind speed decreases to a certain speed.
This design feature protects both the equipment and the project's surroundings from potential damage during high wind speed events such as tornadoes, severe storms, straight-line winds, etc.
[CCM1.1] [CCA1.2] [CCA1.3] [CCA1.4] [CCM3.7] [CCA3.4] [CCA3.5]

Safety measures (fail safely)

Project siting has followed county ordinances and best practices for safety. Upland Prairie has been designed so that all wind turbines are a minimum of 1.1 times the total height (which is 534 feet in this case), from any public street, overhead utilities, or adjacent property lines. Private property owners can, however, provide written consent to allow a turbine to be placed closer to their property lines, and that consent has been obtained in all necessary cases. All turbines are 1,200 feet or more from residential dwellings.
[CCA3.4]

Use of remote weather sensing and warning systems

Alliant Energy uses Weather Sentry Online, a software monitoring system that utilizes remote sensing and networks to generate warnings. This system enables wind project operators to respond quickly to any potentially hazardous weather, and take appropriate measures to ensure the safety of the workers and the public.
[CCA1.4] [CCA3.8] [CCA4.1]

STORMWATER AND FLOOD CONTROL

Avoidance of floodplains

Through detailed hydrological modeling incorporating higher resolution elevation data, potential flood risks are assessed and mitigated, ensuring that all project elements are sited outside forecasted flood elevations, even in areas not indicated as floodplains by FEMA (FEMA 100-year floodplain). The project ensures that all permanent features, such as wind turbines and access roads, are located in areas where FEMA-designated 100-year floodplains are avoided.[22] The final grading of each turbine is located above the 100-year floodplain flow depth with a minimum setback of 150 feet from the floodplain designation.
The installation of underground collection lines in certain locations is the only infrastructure activity that occurred within the 100-year floodplain and the area was restored immediately after installation to its pre-construction condition, thus causing negligible impacts to the floodplain.
[CCA1.2] [CCA1.3] [CCA1.4] [CCA3.2]

Effective runoff and infiltration control practices to not exacerbate the existing flooding risk in the wider area

Due to the nature and size of a wind project, the disturbed areas are small and noncontiguous, so stormwater will be managed primarily through infiltration and drainage measures already in place in the project area. Since the project area generally consists of nearly flat to gently rolling agricultural lands, stormwater runoff is managed via overland flow and field conveyances (tiles) into various drainage channels and tributaries. The project limits the use of impervious surfaces to allow groundwater infiltration to continue largely unaffected compared to pre-project conditions.
As such, runoff and infiltration in all areas of the project, including floodplains, will remain almost unchanged after the project is constructed. The pre- and post-construction runoff coefficients were calculated at 0.30 and 0.35, respectively, and no one area will have a significant effect on the overall stormwater flow of the site or surrounding area.
[CCA1.2] [CCA1.3] [CCA1.4] [BIO1.3] [BIO1.5] [CCA3.2] [CCA3.5] [BIO3.3]

Detention pond construction

A detention pond was employed on the northern end of the substation area to ensure that the water storage capacity is maintained at the original levels, in

Flood hazard information according to FEMA FIRMs (Flood Insurance Rate Map), 2021

accordance with Iowa Department of Natural Resources Permit requirements. According to these requirements, projects that disturb 10 or more acres require a detention pond with a minimum capacity of 3,600 cubic feet per acre disturbed.

In addition, a drainage swale and stormwater detention pond was constructed at the Spencer Operations Center to manage runoff associated with up to a 100 year storm event, thereby maintaining and enhancing water quality.[23] The detention pond ensures that the water storage capacity is maintained at the original levels, preventing flooding in surrounding areas. Regular inspections are also conducted to ensure functionality.
[CCA1.2] [CCA1.3] [CCA1.4] [BIO1.5] [BIO1.6] [CCA3.2] [CCA3.5] [BIO3.3]

Turbine placement for improved drainage
Wind turbines were strategically sited on elevated terrain whenever feasible to mitigate drainage challenges during periods of heavy rainfall. This approach minimizes potential water accumulation and ensures consistent operational performance, even in extreme weather conditions.
[CCA1.2] [CCA1.3] [CCA1.4] [BIO1.4] [CCM3.7] [CCA3.2] [CCA3.4]

Avoidance of large stream crossings during construction
Rerouting the crane path for the project to eliminate large stream (and accompanying floodplain) crossings so that no culverts or backfill would need to be placed in Stony Creek, the large stream that runs through the site.
[BIO1.3] [BIO3.3]

MAXIMIZED REDUNDANCY

Provision of a backup battery powered by the turbines
Some project components that rely on energy from the distribution network to operate in the event of an energy shortage will utilize a backup battery powered by the turbines to maintain continued operation.
[CCA1.2] [CCA3.6]

WORKFORCE TRAINING PROGRAMS

Training programs for employee development
Alliant Energy provided meter training for all onsite employees and offered development training to salaried staff. Gearbox (GBX) and hydraulics training were also available to some technicians to enable proper operations and maintenance. In addition, GE provided Condition Monitoring System (CMS) training to their employees to introduce the basic system architecture, hardware, and software, as well as how to prepare monthly reports and diagnostics information in detail.
[CCA1.4] [CCA3.9]

POTABLE WATER USE REDUCTION

Use of native plantings to eliminate outdoor water usage
The project utilizes native species landscaping around the Spencer Operations Center to eliminate the need for outdoor irrigation and water demands during operation. This approach has achieved a 100% reduction in outdoor water usage.
[CCA1.4] [BIO1.4] [CCA3.3] [BIO3.3]

Selection of plumbing fixtures for maximum water conservation
All of the plumbing fixtures for the Grinnell Operations Center have been specified on the basis of achieving maximum water conservation and meeting regulatory minimum flow requirements.
[CCA1.4] [BIO1.4] [CCA3.3] [BIO3.3]

Water-free wind energy generation
Wind energy requires no water for electricity generation, unlike thermoelectric systems. This project essentially removes the dependence on water in the generation supply chain, thereby protecting existing water resources.
[CCA1.4] [BIO1.4] [CCA3.3] [BIO3.3]

PROPER PROJECT SITING

Strategic siting to preserve agricultural use

The project facilities are sited to enable the continued use of surrounding farmland, allowing landowners to maintain normal agricultural operations during and after construction.
[BIO1.3] [CCA3.2] [BIO3.3]

Minimization of project footprint

The project design focused on minimizing the footprint of its facilities to reduce disturbance to prime agricultural land. Of the 51,810 acres of prime farmland, only 0.16% was impacted by turbine placement and infrastructure. This limited disturbance ensures that over 99% of the farmland, including areas of statewide importance, prime farmland, and prime farmland if drained on the site, remains available for ongoing agricultural use. Detailed micrositing visits were conducted, adjusting many access roads to limit disturbance.
[BIO1.3] [BIO3.3]

PRESERVATION OF SITES OF HIGH ECOLOGICAL VALUE

Maintaining adequate setbacks from sensitive wildlife habitats

Setbacks were implemented from areas that provide shelter for wildlife and serve as movement corridors, including rivers, streams, forests, bat and bird habitats, wetlands, prairies/Conservation Reserve Program (CRP) land, parks, and eagle nesting areas. To protect key wildlife habitats, the following buffer zones have been established between wind turbines and natural areas:

- A 1,000-foot setback from suitable Northern long-eared bat habitat, with turbine blade feathering adjusted to the manufacturer's cut-in speed. In accordance with the Endangered Species Act, the project conducted a Northern Long-Eared Bat Habitat Assessment and Acoustic Survey to protect this species. Using USFWS standards, turbines were sited 1,000 feet away from suitable bat habitats, ensuring minimal interference with their roosting and foraging areas
- A 1.6-mile setback from the onsite bald eagle nest. During the Avian Use Survey and Raptor Nest Survey work by Western EcoSystems Technology, Inc. (WEST), it was noted that there is an active bald eagle nest within the project boundary in the area west of Everly
- A 656-foot setback from wetlands that provide critical wildlife habitat
- A 656-foot setback from grassland areas

[BIO1.3] [BIO3.2] [BIO3.3]

Avoidance of forested and wetland areas for crane paths

Crane paths are routed to non-forested areas and small, non-wetland water bodies with gentle banks. Surface water crossings are thus minimized and large streams or rivers are avoided whenever possible. As a result, the need for culvert installations and gravel in streams is eliminated. This approach allows the crane to cross the area only once, creating a temporary impact that can recover naturally or be restored. In addition, existing tree clearings are minimized, further reducing impacts on potential habitats, and ensuring the conservation of both terrestrial and freshwater ecosystems.
[CCA1.3] [BIO1.3] [CCA3.2] [BIO3.3]

Strategic siting of wind turbines to minimize impacts of the project on avian species

Siting of the turbines was determined based on special studies in order to minimize impacts of the project on avian species, particularly raptors, including eagles. Data gathered during raptor nesting habitat surveys[24] informed the siting of turbines and helped assess potential risks to raptors, including eagles, from the Upland Prairie Wind Project.
[BIO1.3] [BIO3.3]

Implementation of a noise and shadow flicker buffer

A 1-mile buffer around the Mori Prairie[25] parcel, owned by the Nature Conservancy, was established to mitigate noise and shadow flicker effects on human enjoyment, as well as to reduce the potential for blade strikes on sensitive grassland avian species.
[BIO1.3] [BIO1.6] [BIO3.3]

Monitoring of raptor nest activities

Alliant Energy enlisted WEST to conduct ongoing monitoring of eagle nesting activities and flight paths to and from the nest. This monitoring aims to better understand and minimize potential risks to eagles.
[BIO1.3] [BIO1.6] [BIO3.3] [CCA4.1]

PROVISION OF WETLAND AND WATER BODIES BUFFERS

Maintaining adequate setbacks from aquatic habitat features

Buffer zones have been established between turbines and aquatic habitat features (e.g., streams, lakes, ponds) to ensure protection of aquatic habitats and their ecosystems:

- A 50-foot setback from small streams or drainages that do not provide habitat
- A 656-foot setback from larger streams without associated habitat
- A 1,312-foot setback from rivers with riparian habitats

[BIO1.3] [BIO1.5] [BIO3.2] [BIO3.3]

Avoidance of permanent infrastructure on large water bodies

Permanent infrastructure, such as culvert crossings and gravel, is avoided on large water bodies to prevent lasting environmental impacts.

[BIO1.3] [BIO1.6] [CCA3.2] [BIO3.3]

PROTECTION OF SOIL HEALTH

Implementation of native prairie landscaping

The Spencer Operations Center integrates sustainable landscaping by transforming the surrounding land from agriculture to native prairie plantings. The project restored approximately nine out of the total 12 acres of the site with native plantings, increasing the storage capacity to levels above pre-development conditions. The area surrounding the substation will be fully restored to its previous condition and planted with native prairie vegetation, aiding soil stabilization and rebuilding soil structure to full functionality in the long term.

Native prairie is prioritized as the main landscaping feature across all areas outside buildings, parking, and equipment zones. This provides multiple benefits to the natural world while significantly reducing the need for IPL to maintain, fertilize, water, or use pesticides on the large prairie area of the operations center and the substation's surroundings. This shift not only supports local biodiversity but also significantly reduces phosphorus application by approximately 1,176 pounds per year,[26] contributing to healthier soil and water quality in the area. In addition, this approach has reduced operational water demand, achieving a 100% reduction in outdoor water usage at the operations center. The space also creates new pollinator habitat, another key sustainable focus area for IPL.

[CCA1.4] [BIO1.3] [BIO1.6] [CCA3.3] [BIO3.3]

Disconnection of subsurface drainage tiles

The existing subsurface drainage tiles that were used previously for the farmland are now disconnected to divert runoff away from groundwater. This action reduces nutrient export, slows runoff, and promotes crop development by restoring natural hydrology.[27]

[BIO1.5] [BIO3.3]

Restoration of disturbed soils

All disturbed soils from construction were restored to their pre-project condition and functions (row crop agriculture). This involved:

- Reusing excavated soil for foundations and backfilling at their respective excavations
- Replacing topsoil removed during excavation to restore the site to pre-project soil structure and conditions

[CCA1.4] [BIO1.3] [BIO1.6] [CCA3.3]

Minimization of impermeable surfaces

The project minimized impermeable surfaces associated with structures through thoughtful design, allowing for a more natural level of infiltration to occur, such as via permeable gravel paving.

[BIO1.3] [BIO1.6] [CCA3.2] [BIO3.2]

PRESERVATION OF WATER QUALITY

Avoidance of direct connections between impervious facilities and waterways

To protect water quality and manage runoff, the project ensures that no direct connections are established between impervious facilities and waterways, thereby minimizing the risk of pollution and preserving the natural hydrology of the area.

[BIO1.6] [BIO3.3]

Use of existing roads and bridges to avoid water crossings

The project team coordinated with county authorities to authorize the use of short stretches of county roads or bridges when it helped to avoid crane crossings of large streams or rivers. Routing the construction crane path to minimize surface water crossings and avoid large streams or rivers whenever possible eliminates

the need for culvert installations and gravel in streams. This approach allows the crane to cross the area only once, creating a temporary impact that can recover or be restored. Additionally, by preserving the natural landscape of the area, it contributes to stormwater management.
[CCM1.3] [CCA1.3] [CCA1.4] [BIO1.3] [BIO1.4] [CCM3.7] [CCA3.2] [CCA3.3] [CCA3.7] [BIO3.3]

Construction above groundwater level
The foundation footings for the turbines are designed to be at most 10-15 feet below grade, ensuring that construction activities and final facilities remain above the area's groundwater depth.[28]
[BIO1.6] [BIO3.3]

Runoff management in facility siting
The facilities at the wind farm have been strategically sited to prevent direct runoff into any drainage.
[BIO1.6] [BIO3.3]

Implementation of stormwater containment structures
To manage stormwater events, the substation includes impermeable concrete containment structures under large power transformers, adhering to Spill Prevention Control and Countermeasure Plan requirements.
[CCA1.2] [CCA1.3] [CCA3.2] [BIO3.3]

Use of straw bales for erosion and sediment control during construction works
Spencer Livestock provided straw bales to support White's erosion and sediment control needs during land disturbance activities. This practice aligns with federal, state, and local stormwater regulations, helps manage sediment runoff during construction, and prevents water quality impacts. The use of straw bales also reduces project costs, minimizes waste, supports local businesses, and fosters long-term community relationships.
[CCA1.2] [CCA1.3] [CCM3.7] [CCA3.3] [BIO3.3]

Minimization of temporary impacts from collection lines
The project minimized temporary impacts by combining the installation of underground electrical collection lines with the construction crane path. These areas will be restored to pre-project conditions and will not inhibit floodplain functions.
[CCA1.2] [CCA1.3] [CCM3.7] [CCA3.3] [BIO3.3]

Weekly Inspections and Monitoring Protocol for the construction site
Inspections and monitoring protocols for the site are conducted weekly and following rainfall events during construction.[29]
[BIO1.6] [BIO3.3] [CCA 4.1]

Regular site monitoring for environmental compliance during construction
A construction environmental observer will visit the site regularly during the construction period to document and address any necessary repairs or upgrades to pollution control and erosion control measures.
[BIO1.6] [BIO3.3] [CCA 4.1]

AIR POLLUTION REDUCTION

Dust and carbon emissions mitigation measures
Dust mitigation is achieved by watering the site access roads, public roads, and earthmoving activities on an as-needed basis. In some instances, and where practical, the use of calcium chloride or similar dust control measures was implemented. Dust-generating activities such as aggregate production, installation, and hauling were expected to be minimized since the road design selected for the project utilizes a cement treated sub-base which reduces the amount of aggregate required.
[BIO1.6] [BIO3.3]

NOISE AND VIBRATION REDUCTION

Selection of low noise turbines
Low Noise Trailing Edge (LNTE) turbine upgrades were selected, with collaboration between the turbine manufacturer and an acoustical consultant who surveyed baseline noise levels and model sound dispersion in the area.
[BIO1.6] [BIO3.3]

Automatic monitoring for potential vibration issues
Automatic monitoring systems used for predictive maintenance can also identify and prevent any potential vibration issues.[30]
[BIO1.6] [BIO3.3]

LIGHT EMISSIONS REDUCTION

Minimization of turbine lighting

Efforts were made to minimize the required turbine lighting, primarily at night for airspace obstruction, resulting in the removal of lighting requirements from one turbine following FAA review based on the project's layout.[31]
[BIO1.6] [BIO3.3]

Reduction of light spill at the operations center

Light spillage at the operations center and substation was minimized through:

- Use of exterior lighting fixtures with full cutoff lenses that guarantee no light is above 90 degrees
- Use of the minimum number of fixtures and the lowest height necessary to provide adequate lighting coverage for onsite safety and security

[BIO1.6] [BIO3.3]

SUMMARY OF HOW PROJECT RESPONSES ALIGN WITH THE MITIGATION HIERARCHY (AS ADAPTED FOR CLIMATE CHANGE MITIGATION AND ADAPTATION WITHIN THE CRITERIA)

The following tables summarize how the implemented project strategies align with the steps of the mitigation hierarchy, as adapted for climate change mitigation, adaptation, and biodiversity loss within the criteria. Efforts were made to avoid, reduce impacts and exposure to location- and project-specific threats and risks.

			RESPONSES BASED ON THE GHG MANAGEMENT HIERARCHY STEPS					
			AVOID/ ELIMINATE	REDUCE	SUBSTITUTE/ REPLACE	COMPENSATE/ OFFSET offsite	COMPENSATE/ OFFSET onsite	TRANSFORM
GHG EMISSIONS	Scope 1 GHG emissions	Existing			■			
		Project-driven	■	■				
	Scope 2 GHG emissions	Existing						
		Project-driven	■	■				
	Scope 3 GHG emissions (upstream)	Project-driven	■	■				
	Scope 3 GHG emissions (downstream)	Project-driven	■	■				

	RISKS DRIVEN BY THE IDENTIFIED CLIMATE-RELATED THREATS (FLOODING, EXTREME WIND)		RESPONSES BASED ON THE RESILIENCE HIERARCHY STEPS				
			AVOID THREAT	REDUCE EXPOSURE	ABSORB/ MINIMIZE IMPACT	RESTORE/ RECOVER	TRANSFORM
CLIMATE RISKS	Service continuity risk		■	■	■		
	Physical asset risk		■	■	■		
	Resource availability risk	Water		■			
		Materials		■			
		Land		■			
		Workforce		■			
	Supply chain continuity risk			■			

SUMMARY OF HOW PROJECT RESPONSES ALIGN WITH THE MITIGATION HIERARCHY (AS ADAPTED FOR BIODIVERSITY WITHIN THE CRITERIA)

				RESPONSES BASED ON THE MITIGATION HIERARCHY STEPS					
				AVOID	MINIMIZE	RESTORE	OFFSET offsite	OFFSET onsite	TRANSFORM
PRESSURES ON BIODIVERSITY	Land change		Existing			■			
	Land change		Project-driven	■	■	■			
	Freshwater change		Existing		■				
	Freshwater change		Project-driven	■					
	Resource exploitation								
	Pollution	Water	Existing			■	N/A		
		Water	Project-driven		■		N/A		
		Air	Existing				N/A		
		Air	Project-driven				N/A		
		Soil	Existing			■	N/A		
		Soil	Project-driven		■		N/A		
		Waste	Existing				N/A		
		Waste	Project-driven		■		N/A		
		Noise	Existing				N/A		
		Noise	Project-driven		■		N/A		
		Light	Existing				N/A		
		Light	Project-driven		■		N/A		
	Climate change								
	Introduction of invasive species		Existing				N/A		
	Introduction of invasive species		Project-driven	■			N/A		

To summarize the Upland Prairie Wind Farm's contribution to integrated climate-biodiversity action, the selected project strategies are presented in relation to the drivers of climate change, climate-related risks, and pressures on biodiversity that were identified within the integrated action criteria. The following matrix table is a project-specific version of the generic Opportunity Matrix presented as part of the integrated action criteria.

Similarly to the generic matrix, this table illustrates the opportunities of most of the strategies to address more than one impact at a time and, by extension, contribute to the three types of action simultaneously.

The matrix particularly underscores the potential for integrated action of various groups of sustainable project strategies, such as:

- Material input reduction
- Extension of project's useful life
- Use of locally and regionally sourced materials
- Sustainable procurement strategies
- Stormwater and flood control
- Protection of freshwater bodies and wetlands through adequate buffer zones
- Proper project siting and site design
- Protection of water quality

Several of these integrated potential strategies fall under the category of nature-based solutions (NbS). These strategies are then presented in a separate table in relation to their potential to support ecosystem services both for the project and the adjacent community. These ecosystem services mirror, to a high degree, nature's responses to the impact drivers and risks of climate change and biodiversity loss, and, within the present framework, represent project's and community's dependencies on biodiversity.

PROJECT STRATEGIES IN RELATION TO IMPACT DRIVERS AND RISKS OF CLIMATE CHANGE AND BIODIVERSITY LOSS

	DRIVERS OF CLIMATE CHANGE			
	Scope 1 emissions	Scope 2 emissions	Scope 3 emissions	
			Upstream	Downstream
GHG EMISSIONS REDUCTION				
Generation of renewable energy	■			
Solar infrastructure provision at the operations center	■			
No significant dependence on petroleum fuels	■		■	
Construction Scope 1 emissions reduction	■			
ENERGY USE REDUCTION				
High-efficiency, long-life LED fixtures		■		
Occupancy sensor controls		■		
Daylight sensors to automatically dim fixtures when daylight is sufficient		■		
Diffused, translucent light into the center's garage area to provide more natural light		■		
Photocell enabled control and countdown timer		■		
A geothermal heat pump HVAC system	■	■		
High-efficiency, low environmental impact Dyson Airblade hand dryers in restrooms		■		
Merging three operations centers into one energy-efficient building		■		
High SRI value for the majority of the hardscapes		■		
USE OF HIGH-PERFORMANCE COMPONENTS				
High-performance GE 2.5-127 turbines	■			
Windfree technology for reactive power support	■			
Incorporation of a feeder bay to minimize downtimes	■			
PROJECT USEFUL LIFE EXTENSION				
Advanced analytics for extended operating life	■			■
Condition monitoring system for preventive maintenance	■			■
Strategic service contracting for long-term wind turbine maintenance	■			■
Enhanced foundations for wind turbines				
Durable construction materials and components for extended lifespan	■			■
Digital replica for enhanced wind turbine performance	■			
Frequent turbine data collection for adaptability to changing conditions	■			
Scheduled grease replacement to reduce failure rates	■			■
Power curve testing by independent engineer	■			
MATERIAL INPUT REDUCTION				
Recycled steel in wind turbine fabrication			■	
Optimized power grid strategy for less need of substations and transmission lines			■	
Use of fly ash (byproduct) in concrete mixes			■	
LOCALLY SOURCED MATERIALS				
Use of locally and regionally sourced materials			■	
Use of a byproduct of local agricultural activities			■	

CLIMATE-RELATED RISKS							PRESSURES ON BIODIVERSITY									
Service continuity	Physical asset	Resource availability				Supply chain	Land/freshwater/ sea change	Resource exploitation	Pollution						Climate change	Introduction of invasive species
		Water	Material	Land	Workforce				Water	Air	Soil	Waste	Noise	Light		
															■	
															■	
															■	
															■	
			■									■			■	
															■	
															■	
															■	
														■	■	
															■	
															■	
			■	■											■	
															■	
															■	
■															■	
■															■	
								■				■			■	
	■							■				■			■	
			■					■							■	
	■														■	
	■		■					■							■	
■															■	
■	■														■	
■	■		■					■							■	
■															■	
			■												■	
			■				■	■							■	
			■					■							■	
						■									■	
			■			■		■				■			■	

PROJECT STRATEGIES IN RELATION TO IMPACT DRIVERS AND RISKS OF CLIMATE CHANGE AND BIODIVERSITY LOSS (CONTINUED)

	DRIVERS OF CLIMATE CHANGE			
	Scope 1 emissions	Scope 2 emissions	Scope 3 emissions	
			Upstream	Downstream
Optimized material transport logistics			■	
SUSTAINABLE PROCUREMENT STRATEGIES				
Sustainable procurement of fill materials			■	
Use of approved vendor lists to ensure that the procurement process is sustainable			■	
Triple-bottom-line supplier vetting strategy for sustainable procurement			■	
SOLID WASTE DIVERSION				
Reuse of excess aggregate for community and project needs				■
Recycling of operational waste				■
Reduction of excavated material taken off site				■
END-OF-LIFE REPURPOSING				
Recyclable wind turbine towers				■
PROTECTION AGAINST EXTREME EVENTS				
Digital solutions for adaptive response				■
Wind turbine selection for weather extremes	■			
Safety measures (fail safely)				
Use of remote weather sensing and warning systems				
STORMWATER AND FLOOD CONTROL				
Avoidance of floodplains				
Effective runoff and infiltration control practices				
Detention pond construction				
Turbine placement for improved drainage				
Avoidance of large stream crossings (Stony Creek) during construction				
MAXIMIZED REDUNDANCY				
Provision of a backup battery powered by the turbines				
WORKFORCE TRAINING PROGRAMS				
Training programs for employee development				
POTABLE WATER USE REDUCTION				
Use of native plantings to eliminate outdoor water usage				
Selection of plumbing fixtures for maximum water conservation				
Water-free wind energy generation				
PROPER PROJECT SITING				
Strategic siting to preserve agricultural use				
Minimization of project footprint				
PRESERVATION OF SITES OF HIGH ECOLOGICAL VALUE				
Maintaining adequate setbacks from sensitive wildlife habitats				
Avoidance of forested and wetland areas for crane paths				

CLIMATE-RELATED RISKS							PRESSURES ON BIODIVERSITY									
Service continuity	Physical asset	Resource availability				Supply chain	Land/freshwater/ sea change	Resource exploitation	Pollution						Climate change	Introduction of invasive species
		Water	Material	Land	Workforce				Water	Air	Soil	Waste	Noise	Light		
															■	
							■	■	■	■	■	■			■	
							■	■	■	■	■	■			■	
							■	■	■	■	■	■			■	
								■				■			■	
												■			■	
								■				■			■	
			■					■				■				
■	■		■													
■	■		■													
■	■															
■					■											
■	■		■													
■	■		■	■					■		■					
■	■		■	■					■		■					
■	■		■	■				■								
	■						■									
■																
					■											
		■						■								
		■						■								
		■						■								
				■			■									
				■			■									
							■									
	■						■									

PROJECT STRATEGIES IN RELATION TO IMPACT DRIVERS AND RISKS OF CLIMATE CHANGE AND BIODIVERSITY LOSS (CONTINUED)

	DRIVERS OF CLIMATE CHANGE			
	Scope 1 emissions	Scope 2 emissions	Scope 3 emissions	
			Upstream	Downstream
Strategic siting of wind turbines to minimize impacts of the project on avian species				
Implementation of a noise and shadow flicker buffer				
Monitoring of raptor nest activities				
PROVISION OF WETLAND AND WATER BODIES BUFFERS				
Maintaining adequate setbacks from aquatic habitat features				
Avoidance of permanent infrastructure on large water bodies				
PROTECTION OF SOIL HEALTH				
Implementation of native prairie landscaping				
Disconnection of subsurface drainage tiles				
Restoration of disturbed soils				
Minimization of impermeable surfaces				
PRESERVATION OF WATER QUALITY				
Avoidance of direct connections between impervious facilities and waterways				
Use of existing roads and bridges to avoid water crossings			■	
Construction above groundwater level				
Runoff management in facility siting				
Implementation of stormwater containment structures				
Use of straw bales for erosion and sediment control during construction works			■	■
Minimization of temporary impacts from collection lines				
Weekly inspections and monitoring protocol for the site				
Regular site monitoring for environmental compliance				
AIR POLLUTION REDUCTION				
Dust and carbon emissions mitigation measures				
NOISE AND VIBRATION REDUCTION				
Selection of low noise turbines				
Automatic monitoring to detect any potential vibration issues				
LIGHT EMISSIONS REDUCTION				
Minimization of turbine lighting		■		
Reduction of light spill at the operations center				

CLIMATE-RELATED RISKS | PRESSURES ON BIODIVERSITY

Service continuity	Physical asset	Resource availability				Supply chain	Land/freshwater/ sea change	Resource exploitation	Pollution						Climate change	Introduction of invasive species
		Water	Material	Land	Workforce				Water	Air	Soil	Waste	Noise	Light		
							■									
							■						■			
							■									
							■		■	■	■	■				
		■					■		■	■	■	■				
		■					■		■		■					
									■							
							■				■					
							■		■		■					
		■							■							
			■				■		■							
									■		■					
									■							
■	■															
			■						■		■	■				
			■				■		■		■					
									■		■					
									■	■	■					
										■						
													■			
													■			
														■		
														■		

IMPLEMENTED NATURE-BASED SOLUTIONS IN RELATION TO THEIR POTENTIAL CONTRIBUTIONS (ECOSYSTEM SERVICES)

	ECOSYSTEM SERVICES (available to the project and/or community)										
	PROVISIONING SERVICES										
	Biomass provisioning					Terrestrial and aquatic plants as energy source	Other raw material provisioning	Genetic material	Water supply		
IMPLEMENTED NATURE-BASED SOLUTIONS	Food crops	Grazed biomass provisioning	Livestock provisioning	Aquaculture provisioning	Wood provisioning				Potable water	Non-potable water for processes	Non-potable water for irrigation
Maintaining adequate setbacks from sensitive wildlife habitats											
Maintaining adequate setbacks from aquatic habitat features									■		
Avoidance of permanent infrastructure on large water bodies											
Minimization of project footprint to preserve farmland	■										
Disconnection of subsurface drainage tiles									■		
Implementation of native prairie landscaping											
Minimization of impermeable surfaces											

ECOSYSTEM SERVICES (available to the project and/or community)																								
REGULATING & MAINTENANCE SERVICES																				CULTURAL SERVICES				
Global climate regulation	Rainfall pattern regulation	Local (micro and meso climate) regulation	Air filtration	Soil quality regulation	Soil and sediment retention		Solid waste remediation	Water purification		Water flow regulation		Flood mitigation		Storm mitigation	Noise attenuation	Pollination	Biological control		Nursery population and habitat maintenance	Recreation-related	Visual amenity	Education, scientific and research	Spiritual, artistic and symbolic	Ecosystem and species appreciation
					Soil erosion control	Landslide mitigation		Retention and breakdown of nutrients	Retention and breakdown of other pollutants	Baseline flow maintenance	Peak flow mitigation	Coastal protection	River flood mitigation				Pest control	Disease control						
																			■					■
		■	■						■	■	■		■						■		■			■
									■	■	■		■								■			■
		■																	■		■			■
								■		■	■													■
■		■		■	■			■	■	■	■						■		■		■			■
		■								■	■		■											■

TRACK

The Upland Prairie Wind Farm employs multiple strategies to promote reflective capacity and ensure that operational and environmental performance are consistently tracked throughout the life of the project. The project integrates a digital replica system to enhance turbine performance and enable real-time tracking of operational data. Frequent turbine data collection further supports adaptability, allowing the project to respond efficiently to changing conditions.

Alliant Energy's wind analytics program relies heavily on monitoring data in order to design, operate, and maintain wind generation facilities for optimal performance over a 40-year operating life, which is substantially longer than the industry standard of 20 years. The existing in-house data analysis program utilizes monthly analysis of operational data with software tools such as Pi Historian and MATLAB. This analysis uses statistics and pattern-matching algorithms of turbine operational data to detect operational anomalies. An operational anomaly may provide early warning of equipment failure and allow for faster repair time. When an anomaly is found, the operation and maintenance staff are alerted to investigate and resolve the issue. In addition, existing in-house software EtaPRO automates more of this analysis. EtaPRO has non-statistical pattern recognition-based anomaly detection and diagnostics to provide early warning of equipment failure.

In addition to the in-house data analysis program, Alliant Energy utilizes commercially available tools for collecting, analyzing, and reacting to monitoring and performance data. One of the platforms implemented, GE's Renewable Energy Digital, exemplifies extending monitoring beyond industry standards. The GE digital wind farm technology suite collects, analyzes, and responds to large volumes of data. These tools enable efficient operation and maintenance and asset management of the facilities. The PREDIX internet platform collects and monitors large quantities of industrial monitoring data. These data improve operational performance of their power generation by making real-time decisions. Condition Monitoring System (CMS) from PulsePOINT™ is an advanced monitoring and diagnostic service, which detects the faults in their early state and performs preventive repair.

The company also uses a computer system to assist with long-term monitoring and analysis of both Vestas and GE wind turbines. The goals of this system are to develop predictive maintenance practices to replace current scheduled maintenance and help in business decisions by providing a more accurate prediction of major component lifespans.

The maintenance plan for Upland Prairie Wind Farm (UPL) is housed within the IBM Maximo Asset Management platform, an enterprise asset management software solution used to track the operation, maintenance, and disposal of assets. The turbine Preventive Maintenance (PM) work orders are generated in Maximo based on elapsed time from prior PM completion and results of prior inspections. Maximo generates work requirements, checklists, capture results, and track labor and material consumed. These data are used as inputs to the analysis program to assist with turbine asset management.

At the Spencer Operations center a drainage swale and stormwater detention pond is constructed to manage runoff associated with up to a 100-year storm event, thereby maintaining and enhancing water quality. This ensures that the water storage capacity is maintained at the original levels, providing effective stormwater management by preventing flooding in surrounding areas. To ensure functionality, inspections of the detention pond are regularly conducted.

By integrating multiple systems—including in-house performance analytics, commercial digital monitoring systems, onsite inspections, maintenance management, and environmental monitoring—the Upland Prairie Wind Farm establishes a comprehensive framework for continuous tracking and adaptive response. This approach demonstrates how renewable energy infrastructure can sustain long-term operational efficiency while protecting ecosystems and responding to the challenges of climate change.

KEY TAKEAWAYS

This section seeks to provide additional examples of implementing the criteria in infrastructure projects. The following table presents the Upland Prairie Wind Farm's strategies in relation to the response criteria—the generic strategies that are proposed within the integrated action criteria. This process provides examples of real-world solutions as implemented in a renewable energy generation project and links them with the response criteria they relate to.

Most of the project's sustainable strategies are relevant to more than one generic strategy. This alternative reading aligns with the analysis of project strategies in relation to the impact drivers and risks of climate change and biodiversity loss they address, as presented in the project-specific Opportunity Matrix in the previous section. Again, it is the same groups of sustainable strategies that stand out as more relevant to the criteria, several of which coincide with nature-based solutions:

- Material input reduction
- Extension of project's useful life
- Use of locally and regionally sourced materials
- Sustainable procurement strategies
- Stormwater and flood control
- Protection of freshwater bodies and wetlands through adequate buffer zones
- Proper project siting and site design
- Protection of water quality

That is no coincidence since in the present framework, the criteria are interlinked and designed in such a way that the proposed response criteria (strategies) address specific impact drivers and risks.

Moreover, the strategies that the following table highlights as relevant to all types of action are related to the four key strategies for integrated action presented in the integrated action criteria:

- Resource circularity
- Optimized material sourcing
- Resourcefulness
- Nature-based solutions

Despite the key role of the above strategies and the opportunities they present for all types of action, all strategies listed under the climate change mitigation, adaptation, and biodiversity action response criteria, along with all project strategies presented, are indispensable components of climate and biodiversity action. Representative examples are strategies that involve prevention, abatement, and control measures that fall under the category of "doing everything else better." These are typical sustainable actions to minimize environmental impact. When properly implemented, following best practices, they can have a significant positive effect on biodiversity action, especially in the current context of changing climate conditions and the need to ensure that ecosystems maintain their capacity to adapt.

In general, the analysis performed for this wind energy project reaffirms that sustainable project strategies are integral to the criteria for climate and biodiversity action.

PROJECT STRATEGIES IN RELATION TO THE INTEGRATED CLIMATE-BIODIVERSITY RESPONSE CRITERIA

CLIMATE CHANGE MITIGATION

	CCM3.2 Implement energy efficiency strategies	CCM3.3 Implement decarbonization strategies	CCM3.4 Implement electrification strategies	CCM3.5 Implement carbon capture & storage strategies	CCM3.6 Optimize material sourcing practices	CCM3.7 Implement resource circularity strategies	CCM3.8 Optimize project-related transport and logistics
GHG EMISSIONS REDUCTION							
Generation of renewable energy		■					
Solar infrastructure provision at the operations center		■					
No significant dependence on petroleum fuels		■					
Construction Scope 1 emissions reduction	■						
ENERGY USE REDUCTION							
High-efficiency, long-life LED fixtures	■						
Occupancy sensor controls	■						
Daylight sensors to automatically dim fixtures when daylight is sufficient	■						
Diffused, translucent light into the center's garage area to provide more natural light	■						
Photocell enabled control and countdown timer	■						
A geothermal heat pump HVAC system	■	■					
High-efficiency, low environmental impact Dyson airblade hand dryers in restrooms	■						
Merging three operations centers into one energy-efficient building	■						
High SRI value for the majority of the hardscapes	■						
USE OF HIGH-PERFORMANCE COMPONENTS							
High-performance GE 2.5-127 turbines	■	■					
Windfree technology for reactive power support	■						
Incorporation of a feeder bay to minimize downtimes	■						
PROJECT USEFUL LIFE EXTENSION							
Advanced analytics for extended operating life						■	
Condition monitoring system for preventive maintenance							
Strategic service contracting for long-term wind turbine maintenance						■	
Enhanced foundations for wind turbines							
Durable construction materials and components for extended lifespan						■	
Digital replica for enhanced wind turbine performance							
Frequent turbine data collection for adaptability to changing conditions							
Scheduled grease replacement to reduce failure rates						■	

CLIMATE CHANGE ADAPTATION

BIODIVERSITY ACTION

CROSS-CUTTING

CCA3.2 Optimize project siting and site design to avoid threats	CCA3.3 Implement strategies to maximize resourcefulness (materials, water, land)	CCA3.4 Implement strategies to maximize durability	CCA3.5 Implement strategies to maximize adaptability	CCA3.6 Implement strategies to maximize redundancy	CCA3.7 Implement strategies to maximize integration	CCA3.8 Implement strategies to optimize system recovery	BIO3.2 Implement nature-based solutions (NbS)	BIO3.3. Use prevention, abatement, and control measures against pressures on biodiversity (other than NbS): Land/freshwater/sea change	BIO3.3. Resource exploitation	BIO3.3. Water pollution	BIO3.3. Air pollution	BIO3.3. Soil pollution	BIO3.3. Waste pollution	BIO3.3. Noise pollution	BIO3.3. Light pollution	BIO3.3. Climate change and related impacts	BIO3.3. Introduction of invasive species	CCA3.9 Develop local skills and capabilities	CA3.10 Promote inclusivity	CCA4.1 Implement strategies to promote reflective capacity
					■															
	■		■																	
			■																	
			■			■														
		■																		□
		■																		□
		■							■				■							□
■		■																		
		■							■				■							
																				□
			■																	□
		■																		

PROJECT STRATEGIES IN RELATION TO THE INTEGRATED CLIMATE-BIODIVERSITY RESPONSE CRITERIA (CONTINUED)

	CLIMATE CHANGE MITIGATION						
	CCM3.2 Implement energy efficiency strategies	CCM3.3 Implement decarbonization strategies	CCM3.4 Implement electrification strategies	CCM3.5 Implement carbon capture & storage strategies	CCM3.6 Optimize material sourcing practices	CCM3.7 Implement resource circularity strategies	CCM3.8 Optimize project-related transport and logistics
Power curve testing by independent engineer	■						
MATERIAL INPUT REDUCTION							
Recycled steel in wind turbine fabrication						■	
Optimized power grid strategy for less need of substations and transmission lines						■	
Use of fly ash (byproduct) in concrete mixes						■	
LOCALLY SOURCED MATERIALS							
Use of locally and regionally sourced materials							■
Use of a byproduct of local agricultural activities						■	■
Optimized material transport logistics							■
SUSTAINABLE PROCUREMENT STRATEGIES							
Sustainable procurement of fill materials					■		
Use of approved vendor lists to ensure that the procurement process is sustainable					■		
Triple- bottom-line supplier vetting strategy for sustainable procurement					■		
SOLID WASTE DIVERSION							
Reuse of excess aggregate for community and project needs						■	
Recycling of operational waste						■	
Reduction of excavated material taken off site						■	
END-OF-LIFE REPURPOSING							
Recyclable wind turbine towers						■	
PROTECTION AGAINST EXTREME EVENTS							
Digital solutions for adaptive response							
Wind turbine selection for local weather extremes						■	
Safety measures (fail safely)							
Use of remote weather sensing and warning systems							
STORMWATER AND FLOOD CONTROL							
Avoidance of floodplains							
Effective runoff and infiltration control practices							
Detention pond construction							

CLIMATE CHANGE ADAPTATION							BIODIVERSITY ACTION											CROSS-CUTTING		
CCA3.2	CCA3.3	CCA3.4	CCA3.5	CCA3.6	CCA3.7	CCA3.8	BIO3.2	BIO3.3. Use prevention, abatement, and control measures against pressures on biodiversity (other than NbS)										CCA3.9	CA3.10	CCA4.1
Optimize project siting and site design to avoid threats	Implement strategies to maximize resourcefulness (materials, water, land)	Implement strategies to maximize durability	Implement strategies to maximize adaptability	Implement strategies to maximize redundancy	Implement strategies to maximize integration	Implement strategies to optimize system recovery	Implement nature-based solutions (NbS)	Land/freshwater/sea change	Resource exploitation	Water pollution	Air pollution	Soil pollution	Waste pollution	Noise pollution	Light pollution	Climate change and related impacts	Introduction of invasive species	Develop local skills and capabilities	Promote inclusivity	Implement strategies to promote reflective capacity
																■				
	■			■												■				
	■								■							■				
	■								■							■				
																■				
									■				■			■				
																■				
								■	■	■	■	■	■			■				
								■	■	■	■	■	■			■				
								■	■	■	■	■	■			■				
													■			■				
													■			■				
													■			■				
			■		■															
			■			■														□
		■	■																	
		■																		
						■														□
■																				
■										■		■								
■			■							■		■								

PROJECT STRATEGIES IN RELATION TO THE INTEGRATED CLIMATE-BIODIVERSITY RESPONSE CRITERIA (CONTINUED)

CLIMATE CHANGE MITIGATION

	CCM3.2 Implement energy efficiency strategies	CCM3.3 Implement decarbonization strategies	CCM3.4 Implement electrification strategies	CCM3.5 Implement carbon capture & storage strategies	CCM3.6 Optimize material sourcing practices	CCM3.7 Implement resource circularity strategies	CCM3.8 Optimize project-related transport and logistics
Turbine placement for improved drainage						■	
Avoidance of large stream crossings (Stony Creek) during construction							
MAXIMIZED REDUNDANCY							
Provision of a backup battery powered by the turbines							
WORKFORCE TRAINING PROGRAMS							
Training programs for employee development							
POTABLE WATER USE REDUCTION							
Use of native plantings to eliminate outdoor water usage							
Selection of plumbing fixtures for maximum water conservation							
Water-free wind energy generation							
PROPER PROJECT SITING							
Strategic siting to preserve agricultural use							
Minimization of project footprint							
PRESERVATION OF SITES OF HIGH ECOLOGICAL VALUE							
Maintaining adequate setbacks from sensitive wildlife habitats							
Avoidance of forested and wetland areas for crane paths							
Strategic siting of wind turbines to minimize impacts of the project on avian species							
Implementation of a noise and shadow flicker buffer							
Monitoring of raptor nest activities							
PROVISION OF WETLAND AND WATER BODIES BUFFERS							
Maintaining adequate setbacks from aquatic habitat features							
Avoidance of permanent infrastructure on large water bodies							
PROTECTION OF SOIL HEALTH							
Implementation of native prairie landscaping				■			
Disconnection of subsurface drainage tiles							
Restoration of disturbed soils							
Minimization of impermeable surfaces							

CLIMATE CHANGE ADAPTATION							BIODIVERSITY ACTION											CROSS-CUTTING		
CCA3.2	CCA3.3	CCA3.4	CCA3.5	CCA3.6	CCA3.7	CCA3.8	BIO3.2	BIO3.3. Use prevention, abatement, and control measures against pressures on biodiversity (other than NbS)										CCA3.9	CA3.10	CCA4.1
Optimize project siting and site design to avoid threats	Implement strategies to maximize resourcefulness (materials, water, land)	Implement strategies to maximize durability	Implement strategies to maximize adaptability	Implement strategies to maximize redundancy	Implement strategies to maximize integration	Implement strategies to optimize system recovery	Implement nature-based solutions (NbS)	Land/freshwater/sea change	Resource exploitation	Water pollution	Air pollution	Soil pollution	Waste pollution	Noise pollution	Light pollution	Climate change and related impacts	Introduction of invasive species	Develop local skills and capabilities	Promote inclusivity	Implement strategies to promote reflective capacity
■		■																		
								■												
				■																
																		□		
	■								■											
	■								■											
	■								■											
■								■												
								■												
							■	■												
								■												
								■												
								■						■						
								■												□
							■	■												
■								■												
	■						■	■		■		■								
										■										
								■												
■							■	■												

PROJECT STRATEGIES IN RELATION TO THE INTEGRATED CLIMATE-BIODIVERSITY RESPONSE CRITERIA (CONTINUED)

CLIMATE CHANGE MITIGATION

	CCM3.2 Implement energy efficiency strategies	CCM3.3 Implement decarbonization strategies	CCM3.4 Implement electrification strategies	CCM3.5 Implement carbon capture & storage strategies	CCM3.6 Optimize material sourcing practices	CCM3.7 Implement resource circularity strategies	CCM3.8 Optimize project-related transport and logistics
PRESERVATION OF WATER QUALITY							
Avoidance of direct connections between impervious facilities and waterways							
Use of existing roads and bridges to avoid water crossings						■	
Construction above groundwater level							
Runoff management in facility siting							
Implementation of stormwater containment structures							
Use of straw bales for erosion and sediment control during construction works						■	
Minimization of temporary impacts from collection lines							
Weekly inspections and monitoring protocol for the construction site							
Regular site monitoring for environmental compliance during construction							
AIR POLLUTION REDUCTION							
Dust and carbon emissions mitigation measures							
NOISE AND VIBRATION REDUCTION							
Selection of low noise turbines							
Automatic monitoring to detect any potential vibration issues							
LIGHT EMISSIONS REDUCTION							
Minimization of turbine lighting							
Reduction of light spill at the operations center							

CLIMATE CHANGE ADAPTATION

BIODIVERSITY ACTION

CROSS-CUTTING

CCA3.2	CCA3.3	CCA3.4	CCA3.5	CCA3.6	CCA3.7	CCA3.8	BIO3.2	BIO3.3. Use prevention, abatement, and control measures against pressures on biodiversity (other than NbS)										CCA3.9	CA3.10	CCA4.1
Optimize project siting and site design to avoid threats	Implement strategies to maximize resourcefulness (materials, water, land)	Implement strategies to maximize durability	Implement strategies to maximize adaptability	Implement strategies to maximize redundancy	Implement strategies to maximize integration	Implement strategies to optimize system recovery	Implement nature-based solutions (NbS)	Land/freshwater/sea change	Resource exploitation	Water pollution	Air pollution	Soil pollution	Waste pollution	Noise pollution	Light pollution	Climate change and related impacts	Introduction of invasive species	Develop local skills and capabilities	Promote inclusivity	Implement strategies to promote reflective capacity
										■		■								
■	■				■			■		■										
										■		■								
										■		■								
■										■		■								
	■									■		■	■							
								■				■					■			
										■										□
										■		■								□
											■									
														■						
														■						□
															■					
															■					

NOTES

1. Alliant Energy is a Midwest company that delivers energy solutions with a portfolio of electric generating facilities operating on a diverse mix of fuels, including coal, natural gas and renewable resources.
2. "Our Sponsors," Alliant Energy Center, accessed October 29, 2025, https://www.alliantenergycenter.com/about-us/sponsors.
3. IPL is a wholly owned subsidiary of Alliant Energy. It is a utility company that engages in the generation and distribution of electricity. It is also involved in the distribution and transportation of natural gas.
4. Alliant Energy Corporation, "Alliant Energy announces new Upland Prairie Wind Farm," *PR Newswire*, September 29, 2017, https://www.prnewswire.com/news-releases/alliant-energy-announces-new-upland-prairie-wind-farm-300528521.html.
5. "Power plant profile: Upland Prairie Wind, US," Power Technology, last modified October 21, 2024, https://www.power-technology.com/marketdata/upland-prairie-wind-us. Apex Clean Energy Holdings LLC (Apex), formerly Apex Wind Energy is a renewable energy company that offers building utility-scale generation facilities.
6. Michelle Froes, "Apex sells Upland Prairie Wind project to Alliant Energy Corporation subsidiary," *Windpower Engineering & Development*, October 2, 2017, https://www.windpowerengineering.com/apex-sells-upland-prairie-wind-project-alliant-energy-corporation-subsidiary/.
7. "GE Renewable Energy and Alliant Energy Add 470 MW Wind Capacity in Iowa," General Electric (GE), May 8, 2018, https://www.ge.com/news/press-releases/ge-renewable-energy-and-alliant-energy-add-470-mw-wind-capacity-iowa.
8. "Alliant Energy's English Farms and Upland Prairie Wind Farms Earn Envision Platinum Award for Sustainable Infrastructure," Alliant Energy, July 22, 2019, https://investors.alliantenergy.com/News--Presentations/news/news-details/2019/Alliant-Energys-English-Farms-and-Upland-Prairie-Wind-Farms-Earn-Envision-Platinum-Award-for-Sustainable-Infrastructure/default.aspx.
9. "Alliant Energy to build English Farms Wind Farm," Alliant Energy, December 19, 2017, https://alliantenergy.wordpress.com/2017/12/19/alliant-energy-to-build-english-farms-wind-farm/.
10. Alliant Energy Corporation, "Alliant Energy announces new Upland Prairie Wind Farm," *PR Newswire*, September 29, 2017, https://www.prnewswire.com/news-releases/alliant-energy-announces-new-upland-prairie-wind-farm-300528521.html.
11. The first number indicates nominal power (MW), and the second is rotor diameter (m); larger rotors capture more wind and generate more energy at lower speeds.
12. Alliant Energy Corporation, "Alliant Energy's English Farms and Upland Prairie Wind Farms Earn Envision Platinum Award for Sustainable Infrastructure," *PR Newswire*, July 22,2019, https://www.prnewswire.com/news-releases/alliant-energys-english-farms-and-upland-prairie-wind-farms-earn-envision-platinum-award-for-sustainable-infrastructure-300887839.html.
13. Construction of the project, including noise, presence of construction workers, and large vehicle movements will likely cause some wildlife to avoid construction areas and remain closer to the river and stream corridors, inhibiting movement between corridors somewhat during the construction timeframe. During operation, only the small areas occupied by the turbine towers, turbine blade rotor swept area, and substation would have to be physically avoided by wildlife. Turbine noise may also cause some species to travel farther away from turbine locations. Most birds and bats would be expected to slightly alter their flight patterns over the project to avoid colliding with turbines, but they would still be able to move through the area above and below the rotor-swept area of the turbines.
14. "Council sets goal to reach 100% clean power in Iowa by 2035," Iowa Environmental Council, August 26, 2021, https://www.iaenvironment.org/newsroom/energy-news/council-sets-goal-to-reach-100-clean-power-in-iowa-by-2035.
15. "Iowa Wind Fact Sheet," Iowa Environmental Council, August, 2024, https://www.iaenvironment.org/webres/File/Wind%20Fact%20Sheet%202024.pdf.
16. The Endangered Species Act of 1973 is the primary US law for protecting endangered and threatened species and their habitats.
17. The Avian Use Survey is a requirement of the USFWS Eagle Conservation Plan Guidance and was performed in order to compile a comprehensive survey of all birds using the project area. This study documented all birds seen or heard within certain distances of the survey points and within certain timeframes depending on the size and type of bird. The dataset developed during this survey was used to inform siting of the turbines to avoid or minimize impacts to avian species. Summaries of these datasets were also used to gather input from the USFWS about the potential impacts of the Upland Prairie Wind project on bald eagles, avian species of concern, and bats.
18. The only substantial demands are associated with the Spencer Operations Center, located about 5 miles east of the project site. It is designed in accordance with LEED principles to minimize its energy and water use.
19. "GE Renewable Energy and Alliant Energy to add 470 MW of wind capacity in Iowa," GE, May 8, 2018, https://www.ge.com/news/press-releases/ge-renewable-energy-and-alliant-energy-add-470-mw-wind-capacity-iowa.
20. "Wind speed, cut-in and cut-out," Build, accessed October 27, 2025, https://build.com.au/wind-speed-cut-and-cut-out. A turbine's cut-in speed is the point at which the turbine starts generating electricity from turning. The cut-out point denotes how fast the turbine can go before wind speeds get so fast that it risks damage from further operation. The primary safety issue with wind turbines comes from overspeeding, so some sort of stall or brake mechanism is needed to shut down the turbine before it reaches this danger zone.
21. Institute for Sustainable Infrastructure, *Envision Rating System for Sustainable Infrastructure Manual Version 2*, 2015, 80.
22. The model showed that there was flood potential at higher elevations than shown by FEMA floodplain data; therefore, Alliant Energy moved several turbines to ensure that the flood elevations were avoided.
23. Stormwater management and downstream water quality at the substation site will improve as a result of the detention pond in that it will slow down and capture sediment and other pollutants out of the runoff water while maintaining the connection with upstream and downstream areas.
24. The Aerial Raptor Nest Survey is a requirement of the USFWS Eagle Conservation Plan Guidance and was performed in order to understand the raptor nesting habitat surrounding the project area. The study recorded raptor nests in or within 5 miles of the project boundary and documented the nests by species as well as by occupancy status and condition.
25. The Nature Conservancy's Mori Prairie, is a 40 acre preserve of rare, unplowed, tall grass prairie plant community in Northern Iowa.
26. Restoring native prairie vegetation to these areas owned by IPL provides multiple benefits to the natural world while significantly

reducing the need for IPL to maintain, fertilize, water, or use pesticides on the large prairie area of the operations center and the substation surroundings. In addition, the space creates new pollinator habitat, another key sustainable focus area for IPL.

27. The current drainage method used by landowners in the area depends on tile line installation under crop fields that rapidly drain water away, causing stormwater to be almost immediately released to streams.
28. The estimated minimum depth to the top of the shallowest groundwater aquifer is 300 feet below ground surface.
29. An environmental observer (Black & Veatch) will verify that construction stormwater management controls are installed and performing correctly and that other environmental protection commitments and plans are being implemented successfully.
30. Alliant Energy Corporation, "Alliant Energy's English Farms and Upland Prairie Wind Farms Earn Envision Platinum Award for Sustainable Infrastructure," *PR Newswire*, July 22, 2019, https://www.prnewswire.com/news-releases/alliant-energys-english-farms-and-upland-prairie-wind-farms-earn-envision-platinum-award-for-sustainable-infrastructure-300887839.html.
31. Federal Aviation Administration (FAA) requires that turbines on the edges of rows or those more than a half-mile from other turbines be lit, with lighting consisting of synchronized flashing red lights.

BIBLIOGRAPHY AND SOURCES

Ackerly, David, Andrew Jones, Mark Stacey, and Bruce Riordan. *San Francisco Bay Area Summary Report*. In *California's Fourth Climate Change Assessment*. State of California, 2018. Publication No. CCCA4-SUM-2018-005.

Alcamo, Joseph et al. *Ecosystems and Human Well-Being: A Framework for Assessment*. Island Press, 2003. https://wedocs.unep.org/bitstream/handle/20.500.11822/8768/Ecosystem_and_human_well_being_a_framework_for_assessment.pdf?sequence.

Alderton, Matt. "Could carbon fiber be the superhero of composite building materials?" *Autodesk*. December 10, 2021. https://www.autodesk.com/design-make/articles/carbon-fiber-building?

Allen & Overy and Climate Policy Initiative (CPI). *How big is the Net Zero financing gap?* September 2023. https://www.climatepolicyinitiative.org/wp-content/uploads/2023/09/How-big-is-the-Net-Zero-financing-gap-2023.pdf.

Alliant Energy. "Alliant Energy to build English Farms Wind Farm." December 19, 2017. https://alliantenergy.wordpress.com/2017/12/19/alliant-energy-to-build-english-farms-wind-farm/.

Alliant Energy. "Alliant Energy's English Farms and Upland Prairie Wind Farms Earn Envision Platinum Award for Sustainable Infrastructure." July 22, 2019. https://investors.alliantenergy.com/News--Presentations/news/news-details/2019/Alliant-Energys-English-Farms-and-Upland-Prairie-Wind-Farms-Earn-Envision-Platinum-Award-for-Sustainable-Infrastructure/default.aspx.

Alliant Energy Center. "Our Sponsors." Accessed October 29, 2025. https://www.alliantenergycenter.com/about-us/sponsors.

Alliant Energy Corporation. "Alliant Energy announces new Upland Prairie Wind Farm." *PR Newswire*, September 29, 2017. https://www.prnewswire.com/news-releases/alliant-energy-announces-new-upland-prairie-wind-farm-300528521.html.

Armstrong McKay, David I., Arie Staal, Jesse F. Abrams, et al. "Exceeding 1.5°C Global Warming Could Trigger Multiple Climate Tipping Points." *Science* 377, no. 6611 (2022): eabn7950. https://doi.org/10.1126/science.abn7950.

Ash, Neville, Hernán Blanco, Clair Brown, et al. "Conceptual Frameworks for Ecosystem Assessment: Their Development, Ownership, and Use Chapter 3." In *Ecosystems and Well-being: A manual for assessment practitioners*. Island Press, 2010. https://portals.iucn.org/library/efiles/documents/2010-029.pdf.

Bay Area Rapid Transit. "Sustainability." Accessed November 7, 2025. https://www.bart.gov/about/projects/cars/sustainability#:~:text=BART%20challenged%20Bombardier%20to%20build,that%20the%20trains%20currently%20use.

Bennun, L., J. van Bochove, C. Ng, C. Fletcher, D. Wilson, N. Phair, and G. Carbone. *Mitigating Biodiversity Impacts Associated with Solar and Wind Energy Development: Guidelines for Project Developers*. Gland, Switzerland: IUCN; Cambridge, UK: The Biodiversity Consultancy, 2021. https://doi.org/10.2305/IUCN.CH.2021.04.en.

Booth, Hollie et al. "Operationalizing Transformative Change for Business in the Context of Nature Positive." *One Earth* 7, no. 7 (2024): 1235–49. https://doi.org/10.1016/j.oneear.2024.06.003.

Build. "Wind speed, cut-in and cut-out." Accessed October 27, 2025. https://build.com.au/wind-speed-cut-and-cut-out.

Bradshaw, Corey J. A., Paul R. Ehrlich, Andrew Beattie, et al. "Underestimating the Challenges of Avoiding a Ghastly Future." *Frontiers in Conservation Science* 1 (January 2021): 615419. https://doi.org/10.3389/fcosc.2020.615419.

Burgess, Neil D., Natasha Ali, Jacob Bedford, et al. "Global Metrics for Terrestrial Biodiversity." *Annual Review of Environment and Resources* 49, no. 1 (October 18, 2024): 673–709. https://doi.org/10.1146/annurev-environ-121522-045106.

California Air Resources Board. *Climate Change Scoping Plan: A Framework for Change*. Sacramento: State of California, December 2008. https://ww2.arb.ca.gov/sites/default/files/classic/cc/scopingplan/document/adopted_scoping_plan.pdf.

Cambridge Conservation Initiative and Natural Capital Coalition. *Biodiversity and Natural Capital* (flyer, May 2017). https://www.cambridgeconservation.org/wp-content/uploads/2017/05/Flyer_May2017_draft_v3_LD.pdf.

Campaign for Nature. "Why 30%." Accessed November 20, 2025. https://www.campaignfornature.org/why-30-1.

City of San Jose. *Envision San José 2040: General Plan*. Adopted November 1, 2011, as amended on October 24, 2024. https://www.sanjoseca.gov/home/showpublisheddocument/22359/637928744399330000.

City of Santa Clara. *Santa Clara Climate Action Plan*. December 3, 2013. https://www.santaclaraca.gov/home/showpublisheddocument/10170/635713044859030000.

Clifford Chance. "Nature's Paris? Key Outcomes and Measures of the Kunming-Montreal Global Biodiversity Framework (COP15)." January 20, 2023. https://www.cliffordchance.com/briefings/2023/01/nature-s-paris--key-outcomes-and-measures-of-the-kunming-montrea0.html.

Climate Action Coalition. "About the Climate Action Coalition." Accessed November 25, 2025. https://coalition.climateaction.org/about/.

Climate Adaptation Platform. "Data Show Global Cities Impact of 3°C Rise Trajectory." October 10, 2024. https://climateadaptationplatform.com/data-show-global-cities-impact-of-3c-rise-trajectory/.

Climate Disclosure Standards Board (CDSB). *CDSB Framework for Reporting Environmental & Social Information*. January 2022. https://www.cdsb.net/what-we-do/reporting-frameworks.

Climate Disclosure Standards Board (CDSB). *CDSB Framework: Application Guidance for Biodiversity-Related Disclosures*. November 2021. https://www.cdsb.net/biodiversity.

Convention on Biological Diversity (CBD). *Decision Adopted by the Conference of the Parties to the Convention on Biological Diversity: CBD/COP/DEC/15/4*. December 19, 2022. https://www.cbd.int/doc/decisions/cop-15/cop-15-dec-04-en.pdf.

Convention on Biological Diversity (CBD). "2030 Targets." Accessed November 20, 2025. https://www.cbd.int/gbf/targets.

Convention on Biological Diversity (CBD). "Target 7: Reduce Pollution to Levels That Are Not Harmful to Biodiversity." Accessed November 2, 2025. https://www.cbd.int/gbf/targets/7.

Convention on Biological Diversity (CBD) and United Nations Environment Programme (UNEP). *Updated Glossary for the Draft Post-2020 Global Biodiversity Framework*. Montreal: Open-Ended Working Group on the Post-2020 Global Biodiversity Framework, Fifth Meeting. December 3–5, 2022. https://www.cbd.int/doc/c/c3ab/388d/950ddc02586468a814120acf/wg2020-05-04-en.pdf.

De Lamo X., Martin Jung, Piero Visconti, Guido Schmidt-Traub, Lera Miles and Valerie Kapos. *Strengthening synergies: how action to achieve post-2020 global biodiversity conservation targets can contribute to mitigating climate change*. UNEP-WCMC, 2020. https://www.unep.org/resources/report/strengthening-synergies-achieve-biodiversity-goals.

Endangered Wildlife Trust. *The Biological Diversity Protocol (BD Protocol)*. National Biodiversity and Business Network – South Africa, 2020. https://nbbnbdp.org/wp-content/uploads/2022/05/bdp_final_080321.pdf.

Ermgassen, Sophus O.S.E. Zu, Michael Howard, Leon Bennun, et al. "Are Corporate Biodiversity Commitments Consistent With Delivering 'Nature-positive' Outcomes? A Review of 'Nature-positive' Definitions, Company Progress and Challenges." *Journal of Cleaner Production* 379 (October 21, 2022): 134798. https://doi.org/10.1016/j.jclepro.2022.134798.

European Environment Agency. *EEA Technical Report No 12/2015: Exploring Nature-based Solutions: The role of green infrastructure in mitigating the impacts of weather- and climate change-related natural hazards*. EEA, 2015. https://www.eea.europa.eu/publications/exploring-nature-based-solutions-2014.

European Union. "Climate Action." *EUR-Lex Glossary of European Union Policies*. Accessed November 25, 2025. https://eur-lex.europa.eu/EN/legal-content/glossary/climate-action.html.

Froes, Michelle. "Apex sells Upland Prairie Wind project to Alliant Energy Corporation subsidiary." *Windpower Engineering & Development*, October 2, 2017. https://www.windpowerengineering.com/apex-sells-upland-prairie-wind-project-alliant-energy-corporation-subsidiary/.

General Electric. "GE Renewable Energy and Alliant Energy Add 470 MW Wind Capacity in Iowa." Press release. May 8, 2018. https://www.ge.com/news/press-releases/ge-renewable-energy-and-alliant-energy-add-470-mw-wind-capacity-iowa

Global Reporting Initiative (GRI). "Biodiversity crisis emphasizes need for corporate transparency." June 4, 2021. https://www.globalreporting.org/news/news-center/biodiversity-crisis-emphasizes-need-for-corporate-transparency/.

Global Sustainability Standards Board (GSSB). *Consolidated Set of the GRI Standards*. Amsterdam, GSSB, 2024.

Greenhouse Gas (GHG) Protocol. *Scope 3 Frequently Asked Questions*. June 2022. https://ghgprotocol.org/sites/default/files/2022-12/Scope%203%20Detailed%20FAQ.pdf.

GRESB Infrastructure. *2022 Asset Assessment: 2022 GRESB Infrastructure Fund & Asset Assessment Pre-Release*. Internal report, 2022.

Haines-Young, Roy, and Marion Potschin. *Common International Classification of Ecosystem Services (CICES) V5.1 and Guidance on the Application of the Revised Structure*. Fabis Consulting Ltd., 2018. https://cices.eu/content/uploads/sites/8/2018/01/Guidance-V51-01012018.pdf.

Hein, Lars, Ken Bagstad, Neville Crossman, et al. *SEEA Experimental Ecosystem Accounting: Towards a definition and classification of ecosystem services for SEEA*. Final Report, 2018. Discussion paper. https://seea.un.org/sites/seea.un.org/files/lg_24_bg_ecosystem_services_classification.pdf.

Infrastructure Australia. *Resilience Principles: Infrastructure Australia's approach to resilience*. June 2022. https://www.infrastructureaustralia.gov.au/sites/default/files/2022-06/Resilience%20Principles%20-%20Infrastructure%20Australia%27s%20approach%20to%20resilience.pdf.

Institute for Sustainable Infrastructure (ISI). "About Envision." Accessed November 11, 2025, https://sustainableinfrastructure.org/envision/about/.

Institute for Sustainable Infrastructure (ISI). *Envision Rating System for Sustainable Infrastructure Manual Version 2*. ISI, 2015. https://dpw.lacounty.gov/contracts/resources/doc/EnvisionGuidanceManual2015.pdf.

Institute for Sustainable Infrastructure (ISI). *Envision: Sustainable Infrastructure Framework Version* 3. ISI, 2018. https://sustainableinfrastructure.org/wp-content/uploads/EnvisionV3.9.7.2018.pdf.

Institute of Environmental Management and Assessment (IEMA). *Pathways to Net Zero: Using the IEMA GHG Management Hierarchy*. IEMA, 2020. https://www.iema.net/resources/reading-room/2020/11/26/pathways-to-net-zero-using-the-iema-ghg-management-hierarchy-november-2020.

Intergovernmental Panel on Climate Change (IPCC), ed. "Investment and Finance." In *Climate Change 2022 - Mitigation of Climate Change: Working Group III Contribution to the Sixth Assessment Report of the Intergovernmental Panel on Climate Change*, 1547–1640. Cambridge University Press, 2023. https://doi.org/10.1017/9781009157926.017.

Intergovernmental Panel on Climate Change (IPCC). "About the IPCC." Accessed November 3, 2025. https://www.ipcc.ch/about/.

Intergovernmental Panel on Climate Change (IPCC). "Annex I: Glossary." Edited by R. van Diemen. In *Climate Change and Land: An IPCC Special Report on Climate Change, Desertification, Land Degradation, Sustainable Land Management, Food Security, and Greenhouse Gas Fluxes in Terrestrial Ecosystems*. Edited by P. R. Shukla, et al., 2019. https://doi.org/10.1017/9781009157988.010.

Intergovernmental Panel on Climate Change (IPCC). "IPCC – Intergovernmental Panel on Climate Change: History." Accessed November 1, 2025. https://archive.ipcc.ch/organization/organization_history.shtml.

Intergovernmental Panel on Climate Change (IPCC). "IPCC Glossary Search." Accessed November 12, 2025. https://apps.ipcc.ch/glossary/.

Intergovernmental Panel on Climate Change (IPCC). *Climate Change 2007: Impacts, Adaptation and Vulnerability. Contribution of Working Group II to the Fourth Assessment Report of the Intergovernmental Panel on Climate Change*. Edited by M.L. Parry, O.F. Canziani, J.P. Palutikof, P. J. van der Linden and C.E. Hanson. Cambridge University Press, 2007. https://www.ipcc.ch/site/assets/uploads/2018/03/ar4_wg2_full_report.pdf.

Intergovernmental Panel on Climate Change (IPCC). *Climate Change 2022: Impacts, Adaptation and Vulnerability. Contribution of Working Group II to the Sixth Assessment Report of the Intergovernmental Panel on Climate Change*. Edited by H.-O. Pörtner, et al. Cambridge University Press, 2022. https://doi.org/10.1017/9781009325844.

Intergovernmental Panel on Climate Change (IPCC). *Global Warming of 1.5°C: An IPCC Special Report on the Impacts of Global Warming of 1.5°C above Pre-industrial Levels and Related Global Greenhouse Gas Emission Pathways, in the Context of Strengthening the Global Response to the Threat of Climate Change, Sustainable Development, and Efforts to Eradicate Poverty*. Edited by V. Masson-Delmotte, P. Zhai, H.-O. Pörtner et al. Cambridge University Press, 2022. https://doi.org/10.1017/9781009157940.

Intergovernmental Panel on Climate Change (IPCC). *Managing the Risks of Extreme Events and Disasters to Advance Climate Change Adaptation. A Special Report of Working Groups I and II of the Intergovernmental Panel on Climate Change*. Edited by C.B. Field, et al. Cambridge University Press, 2012.

Intergovernmental Panel on Climate Change (IPCC). "Summary for Policymakers." In *Climate Change 2023: Synthesis Report. Contribution of Working Groups I, II and III to the Sixth Assessment Report of the Intergovernmental Panel on Climate Change*. Edited by Core Writing Team, H. Lee, and J. Romero. Geneva: IPCC, 2023. https://doi.org/10.59327/IPCC/AR6-9789291691647.001.

Intergovernmental Panel on Climate Change (IPCC). *The Evidence is Clear: The Time for Action is Now. We Can Halve Emissions by 2030*. April 2022. Accessed October 29, 2024. https://www.ipcc.ch/2022/04/04/ipcc-ar6-wgiii-pressrelease/.

Intergovernmental Science-Policy Platform on Biodiversity and Ecosystem Services (IPBES). "Glossary." Accessed November 25.2025. https://www.ipbes.net/node/42052#.

Intergovernmental Science-Policy Platform on Biodiversity and Ecosystem Services (IPBES). "IPBES-IPCC Co-Sponsored Workshop on Biodiversity and Climate Change." June 10, 2021. https://www.ipbes.net/events/ipbes-ipcc-co-sponsored-workshop-biodiversity-and-climate-change.

Intergovernmental Science-Policy Platform on Biodiversity and Ecosystem Services (IPBES). *Global Assessment Report on Biodiversity and Ecosystem Services of the Intergovernmental Science-Policy Platform on Biodiversity and Ecosystem Services*. Edited by E.S. Brondizio, Joseph Settele, Sandra Diaz, et al. Bonn, Germany: IPBES secretariat, 2019.

Intergovernmental Science-Policy Platform on Biodiversity and Ecosystem Services (IPBES). *Summary for Policymakers of the Global Assessment Report on Biodiversity and Ecosystem Services of the Intergovernmental Science-Policy Platform on Biodiversity and Ecosystem Services*. Edited by S. Diaz, J. Settele, E.S. Brondizio, et al. Bonn, Germany: IPBES secretariat, 2019. https://zenodo.org/records/3553579.

Intergovernmental Science-Policy Platform on Biodiversity and Ecosystem Services (IPBES). *Summary for Policymakers of the Thematic Assessment Report on Invasive Alien Species and Their Control of the Intergovernmental Science-Policy Platform on Biodiversity and Ecosystem Services*. Edited by Helen E. Roy, Ana Pauchard, Peter Stoett, et al. Bonn: IPBES Secretariat, 2023. https://doi.org/10.5281/zenodo.7430692.

Intergovernmental Science-Policy Platform on Biodiversity and Ecosystem Services (IPBES). *The IPBES Regional Assessment Report on Biodiversity and Ecosystem Services for Asia and the Pacific*. Secretariat of the Intergovernmental Science-Policy Platform on Biodiversity and Ecosystem Services, 2018. https://doi.org/10.5281/zenodo.3237373.

International Energy Agency (IEA). "Electrification." Accessed November 21, 2025. https://www.iea.org/energy-system/electricity/electrification.

International Energy Agency (IEA). "Energy Efficiency." Accessed November 25, 2025. https://www.iea.org/energy-system/energy-efficiency-and-demand/energy-efficiency.

International Federation of Consulting Engineers. *Infrastructure and Climate Change: Sustainable Infrastructure Series—A Global Leadership Forum Discussion Paper*. FIDIC, 2023. https://issuu.com/fidic/docs/infrastructure_and_climate_change_july_2023_?fr=xKAE9_zU1NQ.

International Union for Conservation of Nature (IUCN). *Defining Nature-based Solutions*. Resolution of the World Conservation Congress at its session in Hawaii. United States of America, September 1-10, 2016. https://portals.iucn.org/library/sites/library/files/resrecfiles/WCC_2016_RES_069_EN.pdf.

International Union for Conservation of Nature (IUCN). *Global Standard for Nature-based Solutions: A User-friendly Framework for the Verification, Design and Scaling Up of NbS, First edition*. Gland, Switzerland: IUCN, 2020. https://portals.iucn.org/library/sites/library/files/documents/2020-020-En.pdf.

Iowa Environmental Council. "Council sets goal to reach 100% clean power in Iowa by 2035." August 26, 2021. https://www.iaenvironment.org/newsroom/energy-news/council-sets-goal-to-reach-100-clean-power-in-iowa-by-2035.

Iowa Environmental Council. "Iowa Wind Fact Sheet." August, 2024. https://www.iaenvironment.org/webres/File/Wind%20Fact%20Sheet%202024.pdf.

KPMG IMPACT. *The time has come: The KPMG Survey of Sustainability Reporting 2020*. KPMG, 2020. https://assets.kpmg.com/content/dam/kpmg/my/pdf/time-has-comes.pdf.

Ladislaw, Sarah, and Stephen J. Naimoli. "Climate Solutions Series: Deep Decarbonization Pathways." *Center for Strategic and International Studies (CSIS)*. March 10, 2020. https://www.csis.org/analysis/climate-solutions-series-deep-decarbonization-pathways.

Levin, Kelly, Taryn Fransen, Clea Schumer, Chantal Davis and Sophie Boehm. "What Does 'Net-Zero Emissions' Mean?" *World Resources Institute*. March 20, 2023. https://www.wri.org/insights/net-zero-ghg-emissions-questions-answered#.

Mauchaza, Boipelo Tshwene, Matthew Harris, Maria Antonova, Filipa Ramalheiro and Aarty Iyengar. *What Is the State of Evidence on How Nature-Based Solutions Promote Transformative Change for the Sustainable Use and Management of Biodiversity in Socio-Ecological Systems?* Biodiversa+, European Biodiversity Partnership, and UNEP-WCMC, March 2024. https://www.biodiversa.eu/wp-content/uploads/2024/07/Biodiversa_D4.8_Desk-Study-NBS-Transformative-change.pdf.

McElwee, Pamela. "A Tale of Two Panels: Learning and Coordinating across IPCC, IPBES, and Other Science-Policy Interfaces." *Climatic Change* 178, no. 3 (2025): 45. https://doi.org/10.1007/s10584-025-03869-9.

Millennium Ecosystems Assessment. "Overview of the Millennium Ecosystem Assessment." Accessed November 24. 2025. https://www.millenniumassessment.org/en/About.html#1.

National Academies of Sciences, Engineering, and Medicine. *NCHRP Report 750: Strategic Issues Facing Transportation. Volume 7: Preservation, Maintenance, and Renewal of Highway Infrastructure*. The National Academies Press, 2020. https://doi.org/10.17226/25757.

Natural England. *Biodiversity Net Gain: An introduction to the benefits*. Accessed November 24, 2025. https://naturalengland.blog.gov.uk/wp-content/uploads/sites/183/2022/04/BNG-Brochure_Final_Compressed-002.pdf.

Organisation for Economic Co-operation and Development (OECD). *Resilience Strategies and Approaches to Contain Systemic Threats*. September 2019. https://one.oecd.org/document/SG/NAEC (2019)5/en/pdf.

PA Future. "Can ESG awakening end the biodiversity crisis?" September 29, 2020. https://future.portfolio-adviser.com/can-esg-awakening-end-the-biodiversity-crisis/.

Palomo, Ignacio, Bruno Locatelli, Iago Otero, et al. "Assessing Nature-based Solutions for Transformative Change" *One Earth*, vol. 4, no. 5, May 1, 2021, pp. 730–41, https://doi.org/10.1016/j.oneear.2021.04.013.

Pörtner H.O., Robert J Scholes, John Agard, et al. *Scientific outcome of the IPBES-IPCC co-sponsored workshop on biodiversity and climate change*; IPBES secretariat. Zenodo, 2021. DOI:10.5281/zenodo.4659158.

Power Technology. "Power plant profile: Upland Prairie Wind, US." Last modified October 21, 2024. https://www.power-technology.com/marketdata/upland-prairie-wind-us.

Principles for Responsible Investment. "Integrating Biodiversity Considerations into Infrastructure." August 1, 2023. https://public.unpri.org/infrastructure/integrating-biodiversity-considerations-into-infrastructure/11611.article.

Protected Planet. "Protected Planet Report 2024." Accessed November 25, 2025. https://digitalreport.protectedplanet.net/.

Pultarova, Tereza. "Pollutant or Resource? [Carbon Dioxide]" *Engineering and Technology* 14, no 2 (March 2019), https://doi.org/10.1049/et.2019.0204.

Reisinger, Andy, Annette L. Cowie, Oliver Geden, et al."Science-Based Targets Miss the Mark," *Communications Earth & Environment* 5, no 383. July 23, 2024. https://doi.org/10.1038/s43247-024-01535-z.

Santa Clara Valley Transportation Authority. "About." Accessed November 25, 2025. https://www.vta.org/about.

Santa Clara Valley Transportation Authority. "Berryessa Transit Center & Berryessa/North San José BART Station." Accessed November 27, 2024. https://www.vta.org/projects/bart-sv/phase-i/berryessa.

Santa Clara Valley Transportation Authority. *Board Memorandum*. January 27, 2020. https://www.vta.org/sites/default/files/2020-02/Climate_Emergency_Declaration-Thursday%2CFeb_6_%202020%20%281%29_0.pdf.

Santa Clara Valley Transportation Authority. "Project Benefits." Accessed November 7, 2025. https://vtabart.org/project-benefits/.

Santa Clara Valley Transportation Authority. *Sustainability Plan 2020*. 2020. https://www.vta.org/sites/default/files/2020-06/SustainabilityPlan2020_Accessible_0.pdf.

Santa Clara Valley Transportation Authority. *Valley Transportation Plan for 2030* (VTA, February 2005). https://www.vta.org/sites/default/files/2023-06/VTP-2030.pdf.

Santa Clara Valley Transportation Authority. *VTA Climate Action and Adaptation Plan, Volume I: Final CAAP*. January 2024. https://www.vta.org/sites/default/files/2024-01/VTA_CAAP_Final_508web.pdf.

Santa Clara Valley Transportation Authority. "VTA's BART Silicon Valley Phase II." Accessed November 7, 2025. https://www.vta.org/projects/bart-sv/phase-ii.

Santa Clara Valley Transportation Authority. "ZPH Session: Berryessa Transit Center, Envision Platinum—Assessing Sustainability and Transferring Lessons Learned to Transportation Projects." Presentation by Ann Calnan and Rachael Keish at the Zofnass Program virtual sessions, March 24, 2022.

Science Based Targets Network. *Science-Based Targets for Nature: Initial Guidance for Business*. Science Based Targets Network, 2020. https://sciencebasedtargetsnetwork.org/wp-content/uploads/2020/11/Science-Based-Targets-for-Nature-Initial-Guidance-for-Business.pdf.

Science Based Targets Network. *Step 4: Act*, SBTN, 2023. https://sciencebasedtargetsnetwork.org/companies/take-action/act/.

Science for Environment Policy. *Ecosystem Services and the Environment, In-depth Report 1*. Produced for the European Commission. DG Environment by the Science Communication Unit. UWE Bristol, 2015. https://uwe-repository.worktribe.com/output/834982.

Science-based Targets Initiative. "What Are Science-Based Targets?" Accessed October 30, 2025. https://sciencebasedtargets.org/how-it-works.

Secretariat of the Convention on Biological Diversity. *Biodiversity and the 2030 Agenda for Sustainable Development: Technical Note*. Secretariat of the Convention on Biological Diversity, 2016. https://www.cbd.int/development/doc/biodiversity-2030-agenda-technical-note-en.pdf.

Secretariat of the Convention on Biological Diversity. *Global Biodiversity Outlook 5 – Summary for Policy Makers*. Montréal: Secretariat of the Convention on Biological Diversity, 2020. https://www.cbd.int/gbo/gbo5/publication/gbo-5-spm-en.pdf.

Secretariat of the Convention on Biological Diversity. *Interlinkages between Biological Diversity and Climate Change: Advice on the Integration of Biodiversity Considerations into the Implementation of the United Nations Framework Convention on Climate Change and Its Kyoto Protocol*. Montreal, SCBD, 2003. CBD Technical Series no. 10. https://www.cbd.int/doc/publications/cbd-ts-10.pdf.

Secretariat of the Convention on Biological Diversity and United Nations Environment Programme. *Convention on Biological Diversity (CBD). Text and Annexes*. Secretariat of the Convention on Biological Diversity, 2011. https://www.cbd.int/doc/legal/cbd-en.pdf.

Society for Ecological Restoration. "What Is Ecological Restoration?" Accessed November 24, 2025. https://www.ser-rrc.org/what-is-ecological-restoration/.

Stephenson, P.J. and G. Carbone. *Guidelines for planning and monitoring corporate biodiversity performance*. IUCN, 2021. https://doi.org/10.2305/iucn.ch.2021.05.en.

Task Force on Climate-related Financial Disclosures (TCFD). *Recommendations of the Task Force on Climate-related Financial Disclosures: Final Report*. June, 2017. https://assets.bbhub.io/company/sites/60/2021/10/FINAL-2017-TCFD-Report.pdf.

Task Force on Climate-related Financial Disclosures (TCFD). *Technical Supplement: The Use of Scenario Analysis in Disclosure of Climate-Related Risks and Opportunities*. June 2017. https://assets.bbhub.io/company/sites/60/2020/10/FINAL-TCFD-Technical-Supplement-062917.pdf.

Taskforce on Nature-related Financial Disclosures (TNFD). "About us: Mission." Accessed November 27, 2025. https://tnfd.global/about/#mission.

Taskforce on Nature-related Financial Disclosures (TNFD). *Glossary Version 1.0*. September 2023. https://tnfd.global/wp-content/uploads/2023/09/Glossary_of_key_terms_v1.pdf?v=1702506695.

Taskforce on Nature-related Financial Disclosures (TNFD). *Proposed Technical Scope Recommendations for the TNFD*. June 2021. https://tnfd.global/wp-content/uploads/2021/07/TNFD-%E2%80%93-Technical-Scope-3.pdf.

Taskforce on Nature-related Financial Disclosures (TNFD). *Recommendations of the Task Force on Nature-related Financial Disclosures*. September 2023. https://tnfd.global/publication/recommendations-of-the-taskforce-on-nature-related-financial-disclosures/.

Taskforce on Nature-related Financial Disclosures (TNFD). *The TNFD Nature-related Risk & Opportunity Management and Disclosure Framework Beta v0.1 release*. TNFD, 2023. https://www.learningfornature.org/wp-content/uploads/2023/05/23-23882-TNFD_v0.4_Short_Summary_v5.pdf.

TEEB for National and International Policy Makers. *The Economics of Ecosystems and Biodiversity*. TEEB, 2009. Chapter 3. https://www.cbd.int/financial/values/g-accounting-teeb.pdf.

TEEB. *Mainstreaming the Economics of Nature: A Synthesis of the Approach, Conclusions and Recommendations of TEEB*. TEEB, 2010. https://www.academia.edu/14081672/The_economics_of_ecosystems_and_biodiversity_mainstreaming_the_economics_of_nature_a_synthesis_of_the_approach_conclusions_and_recommendations_of_TEEB.

Thacker, S., D. Adshead, C. Fantini, et al. *Infrastructure for climate action*. UNOPS, 2021. https://content.unops.org/publications/Infrastructure-for-climate-action_EN.pdf.

The Biodiversity Consultancy. *A cross-sector guide for implementing the Mitigation Hierarchy*. CSBI, 2015. https://www.thebiodiversityconsultancy.com/fileadmin/user_upload/A_cross-sector_guide_for_implementing_the_Mitigation_Hierarchy.pdf.

The CEO Water Mandate. *Driving Harmonization of Water Stress, Scarcity, and Risk Terminology: Discussion Paper*. January 2014. https://ceowatermandate.org/files/Driving_Harmonization_of_Water_Terminology_draft.pdf.

The Rockefeller Foundation and ARUP. *City Resilience Framework*. April 2014; updated December, 2015. https://www.rockefellerfoundation.org/wp-content/uploads/City-Resilience-Framework-2015.pdf.

UK National Infrastructure Commission. *Anticipate, React, Recover: Resilient Infrastructure Systems*. May 2020. https://nic.org.uk/app/uploads/Anticipate-React-Recover-28-May-2020.pdf.

UN Decade on Restoration. "What Is Ecosystem Restoration?." Accessed November 24. 2025.

https://www.decadeonrestoration.org/what-ecosystem-restoration.

UN Global Compact. *Water Action Hub's NbS Benefits Explorer*. Accessed November 26, 2025. https://nbsbenefitsexplorer.net/tool#.

United Nations, *Paris Agreement*. December 12, 2015, in *United Nations Treaty Collection*. https://unfccc.int/sites/default/files/english_paris_agreement.pdf.

United Nations. "1.5°C: what it means and why it matters." Accessed November 24, 2025. https://www.un.org/en/climatechange/science/climate-issues/degrees-matter.

United Nations. "Climate change recognized as 'threat multiplier', UN Security Council debates its impact on peace." January 25, 2019. https://news.un.org/en/story/2019/01/1031322.

United Nations. "Jim Skea: Every Fraction of a Degree of Warming Matters United Nations." November 15, 2022. https://www.un.org/en/climatechange/voices-from-cop27/jim-skea.

United Nations. "Tackling Biodiversity & Climate Crises Together and Their Combined Social Impacts." Press release. Issued by the Secretariat of the Intergovernmental Science-Policy Platform on Biodiversity and Ecosystem Services (IPBES). June 10, 2021. https://www.un.org/sustainabledevelopment/blog/2021/06/tackling-biodiversity-climate-crises-together-and-their-combined-social-impacts/.

United Nations. "The Climate Crisis – A Race We Can Win." Accessed November 20, 2025, https://www.un.org/en/un75/climate-crisis-race-we-can-win.

United Nations. *United Nations Framework Convention on Climate Change*. New York: United Nations, 1992.

United Nations Climate Change. "Background - Cooperation with the IPCC." Accessed November 21, 2025. https://unfccc.int/topics/science/workstreams/cooperation-with-the-ipcc/background-cooperation-with-the-ipcc.

United Nations Development Programme (UNDP). "What Is Climate Change Adaptation and Why Is It Crucial?" January 2024. https://climatepromise.undp.org/news-and-stories/what-climate-change-adaptation-and-why-it-crucial.

United Nations Environment Programme (UNEP). "Slow climate adaptation threatening lives and economies." Press release. October 29, 2025. https://www.unep.org/news-and-stories/press-release/slow-climate-adaptation-threatening-lives-and-economies#:~:text=A%20worrying%20gap&text=International%20public%20adaptation%20finance%20flows%20to%20developing%20countries%20were%20US,as%20much%20as%20current%20flows.

United Nations Environment Programme (UNEP). "The Sectoral Solution to Climate." Accessed November 5, 2025. https://www.unep.org/interactive/sectoral-solution-climate-change/.

United Nations Environment Programme (UNEP). *Adaptation Gap Report 2023: Underfinanced. Underprepared. Inadequate investment and planning on climate adaptation leaves world exposed*. Nairobi: UNEP, 2023. https://doi. org/10.59117/20.500.11822/43796.

United Nations Environment Programme (UNEP). *Adaptation Gap Report 2024: Come hell and high water — As fires and floods hit the poor hardest, it is time for the world to step up adaptation actions*. Nairobi: UNEP, 2024. https://doi.org/10.59117 /20.500.11822/46497.

United Nations Environment Programme (UNEP). *Adaptation Gap Report 2025: Running on Empty. The World Is Gearing Up for Climate Resilience — Without the Money to Get There*. Edited by H. Neufeldt et al. Nairobi: UNEP, 2025. https://wedocs.unep.org/20.500.11822/48798.

United Nations Environment Programme (UNEP). *Emissions Gap Report 2019*. Nairobi: UNEP, 2019. https://wedocs.unep.org/rest/api/core/bitstreams/30f3e088-bd80-4c7e-a338-3840156783db/content.

United Nations Environment Programme (UNEP). *Emissions Gap Report 2021: The Heat Is On – A World of Climate Promises Not Yet Delivered*. Nairobi: UNEP, 2021. https://www.unep.org/emissions-gap-report-2021.

United Nations Environment Programme (UNEP). *Emissions Gap Report 2022: The Closing Window — Climate crisis calls for rapid transformation of societies*. Nairobi: UNEP, 2022. https://www.unep.org/emissions-gap-report-2022.

United Nations Environment Programme (UNEP). *Emissions Gap Report 2023: Broken Record – Temperatures hit new highs, yet world fails to cut emissions (again)*. Nairobi: UNEP, 2023. https://doi.org/10.59117/20.500.11822/43922.

United Nations Environment Programme (UNEP). *Emissions Gap Report 2024: No more hot air ... please! With a massive gap between rhetoric and reality, countries draft new climate commitments*. Nairobi: UNEP, 2024. https://doi.org/10.59117/20.500.11822/46404.

United Nations Environment Programme (UNEP). *Expert Input to the Post-2020 Global Biodiversity Framework: Transformative Actions on All Drivers of Biodiversity Loss Are Urgently Required to Achieve the Global Goals by 2050*. CBD/WG2020/3/INF/11. Note by the Executive Secretary (14 January 2022). https://www.cbd.int/doc/c/16b6/e126/9d46160048cfcf74cadcf46d/wg2020-03-inf-11-en.pdf.

United Nations Environment Programme (UNEP). *GEO-6 Key Messages*. 2019. Accessed November 20, 2025. https://wedocs.unep.org/rest/api/core/bitstreams/87ef3956-dfc6-4acf-b102-2c6f81951d85/content.

United Nations Environment Programme (UNEP). *The Long View: Exploring Product Lifetime Extension*. 2017. https://wedocs.unep.org/20.500.11822/22394.

United Nations Environment Programme (UNEP). *UNEP Emissions Gap Report 2024 Key Messages*. UNEP, 2024. https://wedocs.unep.org/rest/api/core/bitstreams/e7d6bc1f-f83b-492c-8fb4-01f574a1adfd/content.

United Nations Environment Programme (UNEP). *UNEP Emissions Gap Report 2025 Key Messages*. UNEP, 2025. https://wedocs.unep.org/rest/api/core/bitstreams/162130ba-17df-4f8d-80be-fed7384c8442/content.

United Nations Environment Programme Finance Initiative (UNEP FI). "Governments Adopt First Global Strategy to Finance Biodiversity: Implications for Financial Institutions." Press release. March 3, 2025. https://www.unepfi.org/themes/ecosystems/governments-adopt-first-global-strategy-to-finance-biodiversity-implications-for-financial-institutions/.

United Nations Environment Programme World Conservation Monitoring Centre (UNEP-WCMC). "Research Reveals Benefits of Joint Action on Climate and Nature." November 2020. https://www.unep-wcmc.org/en/news/research-reveals-major-benefits-of-joint-action-on-climate-and-nature.

United Nations et al. *System of Environmental Economic Accounting—Ecosystem Accounting (SEEA EA)*. White cover publication, pre-edited text subject to official editing, 2021.

United Nations Framework Convention on Climate Change (UNFCCC). *Matters Relating to Adaptation: Glasgow–Sharm el-Sheikh Work Programme on the Global Goal on Adaptation Referred to in Decision 7/CMA.3*. December 2023. https://unfccc.int/sites/default/files/resource/cma2023_L18_adv.pdf.

United Nations Global Compact (UNGC) and International Union for Conservation of Nature (IUCN). *A framework for corporate action on biodiversity and ecosystem services*. UNGC, IUCN, 2012. https://portals.iucn.org/library/sites/library/files/documents/2012-036.pdf.

United Nations Office for Disaster Risk Reduction (UNDRR). *How to make infrastructure resilient: The Handbook for Implementing the Principles for Resilient Infrastructure*. UNDRR, 2023. https://www.undrr.org/media/87213/download?startDownload=20251205.

United Nations Office for Disaster Risk Reduction (UNDRR). *Principles for Resilient Infrastructure*. UNDRR, 2022. https://www.undrr.org/publication/principles-resilient-infrastructure.

United Nations Office for Disaster Risk Reduction (UNDRR). *Sendai Framework for Disaster Risk Reduction 2015-2030*. UNDRR, 2015. https://www.undrr.org/media/16176/.

United States Environmental Protection Agency (US EPA) and Green Power Partnership. *Offsets and RECs: What's the difference?* February 2018. https://www.epa.gov/sites/default/files/2018-03/documents/gpp_guide_recs_offsets.pdf.

United States Environmental Protection Agency (US EPA). "Stormwater Discharges from Transportation Sources." Accessed November 7, 2025. https://www.epa.gov/npdes/stormwater-discharges-transportation-sources.

United States Environmental Protection Agency (US EPA). *National Ecosystem Services Classification System (NESCS): Framework Design and Policy Application*. EPA-800-R-15-002. Washington, DC: United States Environmental Protection Agency, 2015. Accessed November 12, 2025https://www.epa.gov/sites/default/files/2015-12/documents/110915_nescs_final_report_-_compliant_1.pdf.

United States Federal Register. *Notices*. Vol. 67, no. 25 (February 6, 2002): 5636-5637. https://www.govinfo.gov/content/pkg/FR-2002-02-06/pdf/02-2796.pdf.

Vincent, K. *Climate Adaptation and Its Measurement: Challenges and Opportunities*. Washington, DC: produced for the Resilience Evaluation, Analysis and Learning (REAL) Associate Award. 2024. https://climatelinks.org/sites/default/files/asset/document/2024-05/AdaptationMeasurement_508_final_0.pdf.

World Bank Group. "The Road to Rio+20 and Beyond: 20 Years on From the Earth Summit." November 16, 2011. https://www.worldbank.org/en/news/feature/2011/11/16/road-to-rio-twenty-years-after-earth-summit.

World Bank Group. "Water in Circular Economy and Resilience (WICER)." September 15, 2021. https://www.worldbank.org/en/topic/water/publication/wicer.

World Bank Group. *Biodiversity Offsets: A User Guide*. World Bank Group, October 2016. https://documents1.worldbank.org/curated/en/344901481176051661/pdf/110820-WP-BiodiversityOffsetsUserGuideFinalWebRevised-PUBLIC.pdf.

World Business Council for Sustainable Development (WBCSD) and Net Zero Initiative. *Guidance on Avoided Emissions: Helping Business Drive Innovations and Scale Solutions toward Net Zero*. March 2023. https://www.wbcsd.org/wp-content/uploads/2023/09/Climate-Avoided-Emissions-guidance_WBCSD.pdf.

World Business Council for Sustainable Development (WBCSD) and World Resources Institute (WRI). *Greenhouse Gas (GHG) Protocol: Required Greenhouse Gases in Inventories: Accounting and Reporting Standard Amendment*. WBCSD and WRI, 2013. https://ghgprotocol.org/sites/default/files/2022-12/Required%20gases%20and%20GWP%20values_0.pdf.

World Business Council for Sustainable Development (WBCSD) and World Resources Institute (WRI). *The Greenhouse Gas Protocol: A Corporate Accounting and Reporting Standard, 2004*. http://pdf.wri.org/ghg_protocol_2004.pdf.

World Business Council for Sustainable Development (WBCSD). "Nature Positive Initiative Launches to Promote the Integrity and Implementation of the Global Goal for Nature." September 6, 2023. https://www.wbcsd.org/news/nature-positive-initiative/.

World Business Council for Sustainable Development (WBCSD). *Circular Transition Indicators V4*.WBCSD, 2023. https://www.wbcsd.org/wp-content/uploads/2023/09/Circular_Transition_Indicators_v4.pdf.

World Commission on Environment and Development. "Chapter 2: Towards Sustainable Development." In *Our Common Future*. United Nations, 1987. http://www.un-documents.net/ocf-02.htm#I.

World Economic Forum (WEF). *Bridging the Gap: How to Finance the Net-Zero Transition WHITE PAPER*. January 2025.

https://reports.weforum.org/docs/WEF_Bridging_the_Gap_How_to_Finance_the_Net_Zero_Transition_2025.pdf.

World Economic Forum. *Measuring Stakeholder Capitalism: Towards Common Metrics and Consistent Reporting of Sustainable Value Creation*. Accessed November 11, 2025. https://www.weforum.org/publications/measuring-stakeholder-capitalism-towards-common-metrics-and-consistent-reporting-of-sustainable-value-creation/.

World Economic Forum. *Nature Risk Rising: Why the Crisis Engulfing Nature Matters for Business and the Economy.* WEF and PWC, 2020. https://www3.weforum.org/docs/WEF_New_Nature_Economy_Report_2020.pdf.

World Economic Forum. *The Global Risks Report 2025: Insight Report.* WEF, 2025. https://reports.weforum.org/docs/WEF_Global_Risks_Report_2025.pdf.

World Green Building Council. *Bringing Embodied Carbon Upfront: Coordinated Action for the Building and Construction Sector to Tackle Embodied Carbon.* WGBC, 2019. https://worldgbc.s3.eu-west-2.amazonaws.com/wp-content/uploads/2022/09/22123951/WorldGBC_Bringing_Embodied_Carbon_Upfront.pdf.

World Meteorological Organization. *State of the Global Climate 2024.* No. 1368. WMO, 2025. https://wmo.int/sites/default/files/2025-03/WMO-1368-2024_en.pdf.

World Water Council. *International High Level Panel on Water announces outcomes to alert the world to the severity of global water issues.* March 2018. https://www.worldwatercouncil.org/sites/default/files/2018-03/International_High_Level_Panel_on_Water_announces_outcomes_to_alert_the_world_to_the_severity_of_global_water_issues.pdf.

Wubet, Selamawit. "Adapting to the Inevitable: The Global Goal on Adaptation and the Urgency of Action." *Global Center on Adaptation.* July 9, 2024. Accessed October 30, 2024. https://gca.org/adapting-to-the-inevitable-the-global-goal-on-adaptation-and-the-urgency-of-action/.

WWF, FIDIC, and AECOM (Technical Partner). *A Playbook for Nature-Positive Infrastructure Development: Part of the Sustainable Infrastructure Series.* WWF, FIDIC, and AECOM. October 2023. Accessed November 1, 2025. https://issuu.com/fidic/docs/wwf_playbook_-_single_pages?fr=xKAE9_zU1NQ.

IMAGE CREDITS

P.6: Graph by World Wildlife Fund, "Nature-Based Solutions for Climate Resilient Infrastructure Planning," May 30, 2025, 20, https://www.worldwildlife.org/publications/nature-based-solutions-for-climate-resilient-infrastructure-planning-in-indonesia/. **P.8:** Image by World Wildlife Fund, "Nature-Based Solutions for Climate Resilient Infrastructure Planning," May 30, 2025, 33. **P.14:** Photograph by the United States Geological Survey (USGS), Unsplash, accessed January 27, 2025, https://unsplash.com/photos/9Jvc7VFuo3M. **P.30:** Photograph by Maksims Mordanovs, Pexels, accessed January 27, 2025, https://www.pexels.com/photo/aerial-view-of-frozen-lake-and-forest-landscape-30388212/. **P.46:** Photograph by Mark Stebnicki, Pexels, accessed March 18, 2026, https://www.pexels.com/photo/drone-shot-of-harvester-combine-in-field-26256446/. **P.62:** Graph adapted from International Union for Conservation of Nature and Natural Resources (IUCN), *Net Positive Impact for biodiversity: The conservation case.* (Gland, Switzerland: IUCN, 2015), 4, https://portals.iucn.org/library/sites/library/files/documents/Rep-2015-007.pdf. **P.68:** Photograph by Annie Spratt, Unsplash, accessed October 1, 2024, https://unsplash.com/photos/white-and-black-abstract-painting-d8FjB1_8xHQ. **P.108:** Photograph by Teodor Drobota, Unsplash, accessed October 1, 2024, https://unsplash.com/photos/green-trees-under-white-sky-during-daytime-uyyRJA2an4o. **P.134:** Photograph by Stephen McDaniel Spratt, accessed on December 15, 2024, https://www.pexels.com/photo/road-along-river-in-valley-15397912/. **P.160:** Photograph by Hafizh Maryansyah, Unsplash, accessed January 27, 2025, https://unsplash.com/photos/a-crane-that-is-hanging-from-a-wire-pdogt3a-4e0. **P.164:** Photograph by Shreeja Mani, Unsplash, accessed January 27, 2025, https://unsplash.com/photos/an-aerial-view-of-a-train-track-and-a-river-YyKqBykHM68. **P.168:** (Top left) Map created by the authors using data on administrative boundaries, accessed September 8, 2025, which is made available here under the Open Database License (ODbL).© OpenStreetMap contributors. (Top right) Map created by the authors using data on administrative boundaries, accessed September 8, 2025, which is made available here under the Open Database License (ODbL).© OpenStreetMap contributors. (Bottom) Santa Clara Valley Transportation Authority, *Second Supplemental Environmental Impact Report (SEIR-2), Chapter 2: Introduction*, 3, https://www.vta.org/sites/default/files/documents/2.0_introduction.pdf. **P.169:** (Background image) 3D visualization by Santa Clara Valley Transportation Authority, "ZPH Session: Berryessa Transit Center, Envision Platinum—Assessing Sustainability and Transferring Lessons Learned to Transportation Projects," presentation by Ann Calnan and Rachael Keish at the Zofnass Program virtual sessions, March 24, 2022. (Inset map at upper left) Map adapted from, City of San Jose. *Berryessa BART Urban Village Plan*. San Jose, CA: City of San Jose, amended by on 12-13-2022, 10, https://www.sanjoseca.gov/home/showpublisheddocument/79970/638411065752330000. **P.170:** Historical imagery from Google Earth. Image © Landsat / Copernicus; map data © Google 2023. Overlay layers added by the authors. **P.171:** Photograph by Santa Clara Valley Transportation Authority, *Annual Report 2019*, 9, https://www.vta.org/sites/default/files/2020-04/AnnualReport2019_Accessible.pdf. **P.172:** (Background image) Terrain base map © Google Maps. Map data © Google 2022. Overlay layers added by the authors. **P.173:** (Left and right) Photograph by Santa Clara Valley Transportation Authority, ZPH Session presentation, March 24, 2022. **P.174:** (Left) Historical imagery from Google Earth. Image © 2011 Landsat / Copernicus. Overlay layers added by the authors. (Right) Historical imagery from Google Earth. Image © 2023 Landsat / Copernicus. Overlay layers added by the authors. **P.177:** Base map compiled from FEMA Flood Insurance Rate Map (FIRM) panels for Santa Clara County, California; effective dates 05/18/2009–02/19/2014), obtained from the FEMA Flood Map Service Center (https://msc.fema.gov), accessed October 10, 2025. FEMA basemap panels used unmodified; Overlay layers added by the authors. **P.178:** (All) Photographs by Santa Clara Valley Transportation Authority, ZPH Session presentation, March 24, 2022. **P.187:** (Top and bottom) Adapted from site plans by Santa Clara Valley Transportation Authority (VTA). **P.189:** (Top) Adapted from site plan by Santa Clara Valley Transportation Authority (VTA). (Bottom left and center) Photographs by Santa Clara Valley Transportation Authority, ZPH Session presentation, March 24, 2022. (Bottom right) Photograph by Santa Clara Valley Transportation Authority, "VTA's Berryessa Transit Center," presentation by Ann Calnan and Rachael Keish at the ISI and ZPH Conference: Delivering Merit and Climate Resilience, November 17-18, 2021. **P.191:** (Top left and right) Photographs by Santa Clara Valley Transportation Authority, ZPH Session presentation, March 24, 2022. (Middle) Adapted from site plan by Santa Clara Valley Transportation Authority (VTA).) (Bottom left and right) Photographs by Santa Clara Valley Transportation Authority, ZPH Session presentation, March 24, 2022. **P.192:** Adapted from site plan by Santa Clara Valley Transportation Authority (VTA). **P.195:** (Bottom left and right) Photographs by Santa Clara Valley Transportation Authority, ZPH Session presentation, March 24, 2022. **P.196:** (Bottom) Aerial Photographs by Santa Clara Valley Transportation Authority, ZPH Session presentation, March 24, 2022. **P.197:** Aerial photograph by Santa Clara Valley Transportation Authority (VTA), included in H. T. Harvey & Associates, *Upper Penitencia Creek Improvement Project Year 7 (2019) Monitoring Report*, December 2019, 43. **P.214:** Photograph by Tom Fisk, Pexels, accessed on January 27, 2025, https://www.pexels.com/photo/high-angle-photo-of-wind-turbine-on-field-1605270/. **P.218:** (Background image) Historical imagery from Google Earth. Image © 2020 Landsat / Copernicus. Overlay layers added by the authors. **P.219:** (Top) Google Maps (screenshot). Map data ©2023 Google, TomTom, Vexcel Imaging. Overlay layer added by the authors. (Bottom) Google Earth. Image © 2025 Airbus. Overlay layers added by the authors. **P.220:** Street views, images by Nick_T, KartaView, image capture: October 2023, CC BY 4.0. Images were cropped. **P.221:** Google Earth. Image © 2025 Airbus. Overlay layers added by the authors. **P.222:** (Background image) Historical imagery from Google Earth. Image © 2017 Landsat / Copernicus. Overlay layers added by the authors using data on waterways accessed September 8, 2025, which is made available here under the Open Database License (ODbL).© OpenStreetMap contributors. **P.227:** Adapted from Alliant Energy Corporation, *July 2023 Climate Report (v1.0)*, 11. **P.235:** Base map compiled from FEMA Flood Insurance Rate Map (FIRM) panels for Spencer, Iowa, and surrounding areas; effective dates 11/05/2021–02/22/2024), obtained from the FEMA Flood Map Service Center (https://msc.fema.gov), accessed October 10, 2025. FEMA basemap panels used unmodified; overlay layers added by the author.

RESEARCH TEAM FOR THE DEVELOPMENT OF THE CRITERIA

Spiro N. Pollalis is Professor of Design, Technology, and Management, Emeritus, at the Harvard Design School. He was the founder and Director of the Zofnass Program for the Sustainability of Infrastructure at Harvard (2007-2022), which led to the development of the Envision® Rating System and its transfer to the Institute of Sustainable Infrastructure. Envision is used worldwide for rating the sustainability of infrastructure and for training professionals on infrastructure sustainability. Prof. Pollalis's research at Harvard has focused on sustainability, and in his private practice, he has planned sustainable cities and consulted with governments and multilateral organizations. He taught for 36 years at Harvard and, as a visiting professor at the ETH-Zurich, Switzerland; TU-Delft, Holland; Uni-Stuttgart, Germany; U-Patras, Greece; and has offered joint courses with the Harvard Business School on planning and development. He received his first degree from the University in Athens (EMP) and his Master's and PhD from MIT. His MBA in high technology is from Northeastern University. He has an honorary Master's degree in Architecture from Harvard.

Evgenia Hagistavrou is an architect with expertise in environmental sustainability. She holds a Diploma in Architecture from N.T.U.A (Athens) and a Master's in Landscape Architecture from ETSAB, UPC (Barcelona). Since 2011, she has been part of Prof. Dr. S.N. Pollalis, Inc. and the Zofnass Program for Sustainable Infrastructure at Harvard (2007-2022). Evgenia has contributed as a research associate to two major Zofnass Program projects, resulting in the publication of the books: "Gulf Sustainable Urbanism: The Past" (Hamad Bin Khalifa University Press, 2019) and "Planning Sustainable Cities: An Infrastructure-based Approach" (Routledge, 2016). As a core member of the Zofnass research team, she has worked on ESG systems analysis, sustainability assessment tools for infrastructure, and the integration of sustainability into lifecycle assessments. Her professional experience also includes landscape planning and design, Landscape Integration Impact Assessments, as well as urban planning and sustainable infrastructure planning.

Angela Kouveli is an architect with expertise in urban design and sustainable development. She received her first diploma and a Masters in Architecture from the N.T.U.A (2002, 2004 Athens). She holds a Masters in Advanced Architecture from UPC-IAAC (2006, Barcelona). She has worked at Prof. Dr. S.N. Pollalis, Inc. in the urban design of two projects' masterplans in Pakistan (2010) and South Korea (2006). At the Greek State Company Hellinikon SA (2011-2015 Athens), she worked as team leader for the redevelopment of Athens' former international Airport designing the "Urban Development Model- An integrated sustainable approach." For the Zofnass Program for Sustainable Infrastructure at Harvard (2007-2022), as a research associate she worked in two projects which resulted in the publication of the books: "Gulf Sustainable Urbanism: The Past" (Hamad Bin Khalifa University Press, 2019) and "Planning Sustainable Cities: An Infrastructure-based Approach" (Routledge, 2016).

Dimosthenis Lappas is an architect and urban planner with expertise in sustainable development. He has worked on large-scale urban planning projects and has conducted numerous urban sustainability and resilience studies and assessments on behalf of international organizations and multilateral development banks. Dimosthenis has been a research associate at the Zofnass Program for Sustainable Infrastructure at Harvard University, contributing to research initiatives that led to the publication of the books "Gulf Sustainable Urbanism" (HBKU Press, 2019) and "Planning Sustainable Cities: An Infrastructure-based Approach" (Routledge, 2016). He studied architecture at the Technical University of Athens and received a Master of Architecture from University College London (UCL). He is also an Envision Sustainability Professional (ENV SP), accredited by the Institute for Sustainable Infrastructure.

Eleonora Marinou is an architect specializing in the sustainability of buildings and infrastructure. She holds a Diploma in Architecture from the University of Patras, Greece. Since 2011, she has been part of Prof. Dr. S.N. Pollalis, Inc. and the Zofnass Program for Sustainable Infrastructure at Harvard (2007-2022), working on multiple international urban and infrastructure planning and research projects. She has been a research associate at the Zofnass Program, contributing to research projects that led to the publication of the books: "Gulf Sustainable

Urbanism: The Past" (Hamad Bin Khalifa University Press, 2019) and "Planning Sustainable Cities: An Infrastructure-based Approach" (Routledge, 2016). Her professional experience also includes graphic design, with a focus on publication design.

Olga Tzioti is an architect specializing in sustainable environmental design with experience in both architectural practice and consultancy across a range of project types and scales. Since 2009, as part of Prof. Pollalis' design and research team, she has focused on sustainability at the city scale: planning, infrastructure, and large-scale development. Her work includes international masterplanning projects such as the new sustainable DHA City Karachi and the new administrative town in South Korea. Her research experience includes participation as a research associate in Harvard's "Gulf Sustainable Urbanism" project and the Zofnass Program for Sustainable Infrastructure's "Sustainable Planning Guidelines" project. These collaborations resulted in the publications of "Gulf Sustainable Urbanism: The Past" (Hamad Bin Khalifa University Press, 2019) and "Planning Sustainable Cities: An Infrastructure-based Approach" (Routledge, 2016). In addition, as part of the Zofnass Program research team, she has conducted analyses on ESG systems and worked on the development of KPIs for cities based on the UN SDGs. Olga holds a Diploma in Architecture from the National Technical University of Athens and a Master of Science in Sustainable Environmental Design from the Architectural Association School of Architecture.

CONTRIBUTORS

María Álvarez Mingorance, Chief Sustainability Officer (CSO) at IDOM. With over 20 years of experience, she has led multidisciplinary teams in complex urban and territorial planning projects across Europe, Latin America, and the Middle East. She holds a Master in Architecture and Urban Planning. Currently, at IDOM, she leads the corporate sustainability strategy and the integration of ESG principles into business operations, while fostering sustainability across the company's projects.

Ryan Bartlett directs WWF's program for climate-smart planning and design at their headquarters in Washington. He has worked for WWF for the last 13 years, training staff and partners, developing tools, and tailoring climate and ecosystem science for climate adaptation policy and action. He works with a broad array of stakeholders and partners from major public donors like GCF and GEF, to the private sector and country governments and academia, globally to locally in nature-based solutions, adaptation planning, and sustainable, resilient infrastructure development in WWF priority conservation landscapes around the world. He holds a master's degree in environmental management from Duke's Nicholas School of the Environment.

Elena Calcerrada Romero, Infrastructure Sustainability Manager at IDOM, is a Civil Engineer (MEng) with over 8 years of experience and currently works as Infrastructure Sustainability Manager at IDOM. She specializes in Sustainable Urban Drainage Systems and the development of sustainability strategies for infrastructure projects, with a strong focus on climate adaptation, carbon footprint reduction, and circular economy principles. Her professional background spans key infrastructure sectors—including rail, airport, water, port, and urban—with international project experience across Europe, Latin America, the Middle East, and Central Asia.

Daniel Hornett is a professional civil engineer specializing in sustainable infrastructure and the integration of nature-based and nature-positive approaches into development planning and design. As Senior Program Officer for the WWF-US Sustainable Infrastructure initiative, he develops technical guidance to help engineers, planners, and developers integrate biodiversity and climate considerations across the infrastructure lifecycle, while supporting the conservation community to more effectively engage with and influence the infrastructure sector. Before joining WWF, Daniel worked for the Napa County Planning, Building, and Environmental Services Department in his home state of California as an assistant engineer, inspector, floodplain manager, and safety assessment evaluator.

Shail Joshi is a Senior Program Officer on WWF's Climate Adaptation team, driving climate adaptation initiatives across Asia and globally. His work includes climate vulnerability assessments, community-led planning, aligning conservation with climate priorities, and supporting large-scale NbS investments in low- and middle-income countries. He has experience in urban planning, sustainable infrastructure, and environmental conservation across the US, Latin America, Africa, and South Asia. Shail is co-author of "Along the Betwa," a book on urbanization's impact on river systems—featured by UNESCO and National Geographic. He holds a Master in City Planning from Massachusetts Institute of Technology (MIT).

Anthony Kane is President & CEO of the Institute for Sustainable Infrastructure (ISI), a Washington, DC-based nonprofit advancing sustainable, resilient, and equitable infrastructure through the Envision Sustainable Infrastructure Framework. Envision is a holistic system that guides and assesses the sustainability and resilience of all types of infrastructure. Under his leadership, ISI supports thousands of professionals and oversees billions of dollars in verified infrastructure projects worldwide. Anthony directs ISI's strategic vision, guides Envision development, and represents ISI on key national and global advisory groups. Previously, Anthony directed research at the Zofnass Program for Sustainable Infrastructure at Harvard University, where he co-led development of the initial version of Envision (v2). He later oversaw the release of Envision v3 as ISI's VP of Research and Development and Managing Director. He holds a Master in Design Studies from Harvard University's Graduate School of Design and a Bachelor of Architecture from Virginia Tech. Anthony serves as a commissioner on the DC Commission on Climate Change and Resiliency and participates in international coalitions focused on sustainable infrastructure.

Dr. Cris B. Liban is Chief Sustainability Officer at LA Metro and Board Chair of the Philippine Development Foundation (PhilDev). A licensed civil engineer and geologist, and an environmental systems thinker by practice, he works at the intersection of people, infrastructure, and nature—translating ambitious goals into durable public value.

Roberto Mezzalama has a B. Sc. in Natural Sciences from Pavia University and an M. Sc. in Environmental Engineering from Torino Polytechnic. He started his professional career as a civil servant in local governments in Italy, before joining WSP (formerly Golder Associates) as an environmental and social impact assessment expert specializing in biodiversity. In this capacity he has led dozens of environmental and social impact assessment studies for infrastructures in the energy, transportation, mining and industrial sectors in over 25 countries in Europe, North America, Africa and Central Asia. For over 10 years he has been a member of the Industry Advisory Group of the Zofnass Program at Harvard and has contributed to events and papers prepared by this research team. He served as a member of the Board of the Torino Polytechnic and is currently a contract professor of the Master course in Environmental Economics at Torino University.

Daniel Rubio Blanco, Climate Change Director at IDOM. PhD in Environmental Planning with over 25 years of experience in environment, risk assessments and climate change, developing more than 50 projects for public, private and international organizations, such as AFD, EIB, IDB, WB, CABEI, GIZ, KfW, EC, and UNDP, among others, and has led more than 50 projects in climate change adaptation, carbon footprint reduction and nature-based solutions (NbS) across Latam, Europe, the Middle East and Africa. He is the author of books and publications and has been university professor in several countries.

INDEX

Page numbers in **bold** indicate text within tables; page numbers in *italics* indicate text within figures.

0.2% annual chance flood *177*, *235*

1% annual chance flood *177*, *235*

100-year storm event 7, 236, 250; *see also* FEMA 100-year floodplain

100-year design flow (or flood) 177, 179, 188, **200**, **208**

500-year design flow 177

A

absorb 44, *63*, 74-75, **88**, 93, 95, *99*, **141**, **147**, 169, **195**, **240**; *see also* aspects of resilience; mitigation hierarchy

action gap 16, 20

action step 55-56, 58-59, 76, *78*, 88, 92-93, 111, 118, *124*, 138, **146**, 148, 162-163

acute shock 44, 74-75, 88-90, 93, 95-96, **141**, **147**, **176**, **226**

adaptability 36, 75, **88**, 96, *99*, 101-104, 128-130, *142*, 144, **147**, **150**, 155-157, **207**, **209**, **211**, 231, **242**, 250, **252**, **253**, **255**, **257**, **259**; *see also* core resilience principles

Adaptation Gap Report 21, *23*

afforestation 33, 83, 127

AGR *see* Adaptation Gap Report

Aichi Biodiversity Targets 12, 21, 23-25, 35

Aichi Targets *see* Aichi Biodiversity Targets

air filtration 113-114, *123*, *130*, *143*, **150**, **180**, **203**, **225**, **249**; *see also* ecosystem services

air pollutant 121, 129, 175, 182

air pollution **118**, 121-122, *124*, 129-130, 138-139, *142*, **146**, 151-157, 175, **179**, **196**, **199**, **201**, **207**, **209**, **211**, 226, 239, **241**, **243**, **245-247**, **253**, **255**, **258-259**; *see also* biodiversity

air quality 18, 40, 176, 181; standards 129, 175

Alameda County 167-169

Alliant Energy 162, 216-217, 219, 224, 227-228, 230, 234, 236-237, 250

anticipate 74-75, 91, 93-96; *see also* aspects of resilience

Apex Clean Energy Holdings 216-217, 221

Apex *see* Apex Clean Energy Holdings

aquaculture provisioning **180**, **202**, **225**, **248**; *see also* ecosystem services

AR^3T framework 61-62, 126

Asian Development Bank 8

aspects of resilience 74-75, 93

Assess 55-59, 76-78, 88-92, 118-121, 123-124, 138-139, **146**, 157, 162-163, 175-177, 223-225; *see also* action step

assessment criteria 55, 65-66, 72, 90, 139, 148-149, **150**-**153**, 181, 233; *see also* Assess

atmosphere 4, 36, 77, 79, 83, **114**

atmospheric extreme 234

automatic monitoring 239, **246**, **258**

avian species *see* species

avoid 11, 41, 61-63, 75-76, 81, *86*, **88**, 93-94, *99*, 116-118, 121, 126, 128-129, 141, 144, **147**, 188, 192, 195-196, **207**, **209**, **211**, 227-228, 236, 238, 240-241, **244**, **246**, **253**, **255**-**257**, **259**; *see also* mitigation hierarchy

avoided emissions 76, 78, **139**, **146**, 162, 169, 176

avoided GHG emissions *see* avoided emissions

B

Bald and Golden Eagle Protection Act 223

BART *see* Bay Area Rapid Transit

BART Silicon Valley Berryessa extension (SVBX) project 172

baseline 11, 56, 111, 119, 239; conditions *55*, *58-59*, 175; flow maintenance **180**, 203, **225**, 249; & peak flow 113-114, *123*, *130*, *143*, **150**

Bay Area 167-169, 171, 173, 175-176

Bay Area Rapid Transit (BART) 162, 166-169, 170-173, 175-176, 182-184, 186, 188, **198**, **202**, **206**

Bay Area Ridge Trail 171

Berryessa BART station *see* Berryessa Transit Center

Berryessa BART Urban Village 169; *see also* transit-oriented urban village

Berryessa station *see* Berryessa Transit Center

Berryessa Transit Center 162, 167-169, 171-178, 184, 186-187, 205

biodiversity 4-5, 8, 11-12, 17-25, 32-36, 40, 43-44, 50-53, 56, 58, 60, 62-64, 111-112, 115-121, 123-127, 129-131, 137, 141-142, 147-148, 162-163, 183, 227-229, 238; conservation 16; crisis 7, 11-12, 16, 25, 32, 34, 137; decline 16, 33, 111-112, 120; -driven risks *57*; goals 50, 52, 56, 62-63, 124, 156; net gain 115-116, 125-127, 156; offset 126; protection 11; strategy *see* strategy; target *see* target; pressures on *54*, *58-59*, 62, *66*, 116-120, *122*, 124-128, 130-131, 137-139, *142*, 144, 146-149, **151**-**153**, 155, *179*, 181, 196-197, **199**, **201**, **207**, **209**, **211**, 224, 226, **240**-**247**, **253**, **255**, **257**, **259**; state of *58-59*, 116-119, **139**, **146**, 175, **178**, **224**

biodiversity action 6, 20-21, 32-33, 49-52, 58, 60-62, 64-66, 117, 131, 137, 141, 145, 148, 154-157, 163, 181, 205, **207**, **209**, **211**, 251, **253**, **255**, **257**, **259**

biodiversity action criteria 52, *54-55*, *59*, 62, 65, 111-112, 116, **118**, 148, 162, 229 biodiversity loss 6-7, 9-10, 16-19, 24, 32, 34-35, 40, 42, 49-52, 56, 60, 64, 71, 120, 126, 137, 139, 141, 156, *163*, 176, 195, action against 50; drivers of 18-19, 21, 24, 56, 119, 137, 155, 197, **198-201**, 205, 224, **240-247**, 251; halt and reverse **20**, 117-118, 123, 125, **140**, **146**, 223

bioenergy with carbon capture and storage (BECCS) 83

bioengineering 127, 173, 190-191, **200**, **202**, **208**

biological control 113-114, *123*, *130*, *143*, **152**, **180**, **203**, **225**, **249**

biological diversity *see* biodiversity

biomass provisioning *113*, **180**, **202**, **225**, **248**

bioretention area *see* bioretention basin

bioretention basin 127, 171, 188-190, 194, **198**, **202**, 204, **208**, **210**

biosphere 4-5

Black & Veatch *216*, *227*

BREEAM Infrastructure 41

buffer 36, 43, 94, 104, 111, 127, 129, 192-193, 197, **200**, **202**, 205, **208**, 228, 237-238, 241, **246**, 251, **256**

byproduct 64, 84, 95, 121, 156, 186, **198**, **206**, 231-233, **242**, **254**

C

California 162, 166-167, 172, 175-176, **178**, 181-182, 185-186, 204

California Environmental Quality Act 175-176

California's Fourth Climate Change Assessment 176

California Global Warming Solutions Act 182

California red-legged frog 181

California tiger salamander 181

Caltrain *166*, 169

capacity factor 219, 230

carbon 4, 10-11, 16, 21, 24, 32-33, 36, 41, 44, 72-73, 79, 81, 83, 85, 91, 115, 137, 182, 184; accounting 10; capture 44, **76**, 82-83, *86*, 103-105, *142*, 144, **147**, **150**, **206**, **208**, **210**, **252**, **254**, **256**, **258**; credit 83; dioxide 4, 17, 21, 36, 77, 83, 120, 169, 185, 227; dioxide emissions 6, 21, 40; dioxide equivalent 72, 182, 224; emissions 219, 229, 233, 239, **246**, **258**; footprint 40-41, 230; intensity 7, 84, 155; intensive 56; offset 81, 83; pool 36, 83; removal 81, 83; sequestration 33, 36, 83, 114-115, 120, 123, **198**, **206**; sink 36, 80, 120, 139, 144, 156; storage 36, 83, *86*, 103-105, 112, **114**, 139-140, *142*, 144, **147**, **150**, 185, **206**, **208**, **210**, **252**, **254**, **256**, **258**; store 3, 111, 120-121; neutrality 81; embodied 44-45, 72-74, 78, 81-82, 84; end-of-life 72-74; operational 72-74; use stage embodied 72, *74*; upfront 72-74; whole-life 72-74

carbon-neutral project 81

Cartagena Protocol *22*

CBD *see* Convention on Biological Diversity

Central California Coast steelhead 172-173, **178**, 181, 192-193, **202**, 204

CEQA *see* California Environmental Quality Act

chronic stressor 74-75, 88-91, 93-96, 120, 127, **141**, **147**, **176**, 204, **226**

circular economy 41, 64, 84; *see also* resource circularity

City of San José 166-169, 172, 175, 179, 182-183, 192

City of Santa Clara Climate Action Plan (CAP) 182

City Resilience Framework *54*, 75

Clay *see* Clay County

Clay County 217, 221, 228

clean energy 24, 82, 184, 217, 219, 227-228

clean power *see* clean energy

climate 7, 10-12, 16-18, 20-21, 24-25, 32-36, 41-42, 45, 50, 52, 56, 72, 78-79, 81, 89, 98, 100, 115-116, 123, 126, 137, 140-141, **147**, 156, 162, 176, 182, 185, 190, 205, 251; ambition 91, 93, 141, 154; finance 24; goal 21, 36, 43, 50, 52, 56, 62-64, 87, 111; -related risk *see* risk; -related threat *see* threat; risk *see* risk; risk reduction strategy *see* strategy; scenario 41, 89-90; extreme 6-7, 36, 89, 94, 100, 111; neutrality 79, 87; resilience 7-8, 40, 42, 49; target *see* target; threat *see* threat

climate action 16-17, 21, 24, 32, 35, 40, 42, 49-51, 60-62, 64-65, 81, 89, 102, 117, 148, 182, 251; criteria 52, 65, 71, 111, 137-138, *142*, 145, 148, 156, 162, 205, 229

climate-biodiversity action *see* integrated climate-biodiversity action

climate change 4-6, 10, 16-25, 32-36, 40-43, 49-53, 56, 62, 64, 71-72, 79, 87, 89-92, 95, 98, 100, 102, 111, 115, 117-122, *124*, 127-128, *130*, 137-142, **146**, **151**, 153-157, 176, 182, **196**, 224, 226, **241**, 250; adaptation 6, 10, 20-21, 33, 36, 42, 50, 52-53, 58, 62, *64*, 71, 74, 89, 92-93, 102-103, 111, 115, 117, 137, 140-143, 148, *154-157*, 251; crisis 10-12, 32, 137-138, 162-163, 195, 205, 240; mitigation 10, 20, 33, 36, 42, 50, 52-53, 58, 62, *64*, 71, 76-77, 87, 92, 102-103, 111, 115, 126-127, 137-138, 140-143, 148, *154-157*, 162-163, 182-184, 195, 205, 240, 251; -related shock *see* acute shock; vulnerability assessment 121; drivers of 16, 56, 60, 64, *66*, 71-72, 77-78, 82, 86-87, 102-103, 137-138, 141, *143*, 148-150, **152**, 155, 163, 175-176, 197, **198-201**, 205, 224, 241-248, 251

climate change adaptation criteria 52, *54-55*, *58-59*, *65*, *71*, **88**, 102, 111, **114**, 128-129, 137-138, 145, 147-148, **207**, **209**, **211**, **253**, **255**, **257**, **259**

climate change mitigation criteria 52, *54-55*, *58-59*, *65*, 71-72, 76, 81, 102, 111, 128-129, 137-138, 147-148, **206**, **208**, **210**, **252**, **254**, **256**, **258**

Climate Disclosure Standards Board (CDSB) 53

CO_2 *see* carbon dioxide

CO_2e *see* carbon dioxide equivalent

Commit 55-59, **76**, 79, **88**, 92-93, **118**, 125, 138, **140**, **146**-**147**, 163, 182-183, 227-228; *see also* action step

commitment criteria 55, 140, 148 *see also* Commit

Common International Classification of Ecosystem Services (CICES) 53

Condition Monitoring System 230, **242**, 250, **252**

Conference of the Parties (COP) 17, 20-24, 33-35, 92

conservation 16, 19, 21, 24, 32-33, 35-36, 40, 62, 92, 95, 111, 115-117, 119, 126-128, 140, 183, 190, 227-228, 236-237, **244**, **256**

constructed wetland 10, 127

construction 10-11, 40, 45, 72, 77, 81-82, 96, 126, 129, 172-173, 179, 181, 183-185, 188, 190-191, 193-194, **198**, **200**, **202**, **206**, **208**, **210**, **221**, 224, **226**, 233-234, 238-239, **242**, **246**, **252**, **256**, **258**; activity 121, 126, 128, 167, **178**, 181, 183, 193-194, 239, **246**, **258**; contractor 80, 87, 233; cost 7; equipment 83, 121, 229; job 228; impact 176, 179; industry 81; material 231, **242**, **252**; quality 96; phase *see* stage; policy 182; site *73*, 80, 194, 239, **258**; stage 41, 60-61, 72-74, 77, 80, 84, *86*; subcontractor 80; waste *see* waste; offsite 96; post- 194, **200**, **210**, 234, 236

Construction Technical Specifications 185

Convention on Biological Diversity 16-17, 20-22, 24, 34-35, 111-112, 125-132

COP *see* Conference of the Parties

core resilience principles *54*, *75*, 94, 96-98, 101-103, *130*, 154

corporate action 24-25, 50, 63-64

corporate reporting 52

Coyote Creek *169-170*, 172, 179, 181, 188, 193

cradle-to-grave LCA 224

criteria 49-53, 55 56, 58-62, 64-65

crop **114**, **150**, **180**, **202**, **224-225**, 238, **248**; fields 221; production 112, **114**, 221; crop provisioning 113-114, *123*, *130*, *143*, **150** *see also* ecosystem services

cross-cutting criteria 138, 141, 145, 148, **207**, **211**, **253**, **255**, **257**, **259**

cross-cutting principle 60-61, 64-65, 81, 98, 111, 148, 156

cross-cutting strategies *see* cross-cutting criteria

cultural services 112-113, **180**, **203**, **225**, **249**; *see also* ecosystem services

D

decarbonization 7, 10, 12, 40, **76**, 82-83, *86*, 103-105, *142*, **144**, **147**, **150**, **206**, **208**, **210**, **252**, **254**, **256**, **258**

decommissioning 49, 60, 221

deconstruction 41, 60, 72-73, 183

Deep Decarbonization Pathways Project 33

degradation 7, 32, 35-36, 89, 94-96, 119-121

dependency 7, 25, *54*, *58-59*, 64, 66, 90-91, 104, 111-112, 116, **118**, 123-124, 127, 139, **146**, 157, 163, 175, **180**, 197, 225-226, 241

Desktop Analysis 223

developing countries 24

Dickinson *see* Dickinson County

Dickinson County 217, 221, 228

digital replica 231, **242**, 250, **252**

direct air carbon capture and storage (DACCS) 83

direct driver 21, 24, 33, 117, 119-120, 141

disassembly 84

disaster risk *see* risk

disclosure 25

disease control *113*, **180**, **203**, **225**, **249**; *see also* ecosystem services

downstream emissions *73-74*, 76-78, 80, 84-86, 102-103, 138-139, *143*, **146**, **150**, **152**, *154-157*, **195**, **198**, **200**, **240**, **242**, **244**, **246**; *see also* Scope 3 GHG emissions

drivers of biodiversity loss *see* biodiversity loss

drivers of climate change *see* climate change

durability 75, 84-85, **88**, 95-96, *99*, 101-104, 128-130, *142*, 144, **147**, **150**, 155-156, **207**, **209**, **211**, 219, **253**, **255**, **257**, **259**; *see also* core resilience principles

E

ecological value 178-179, 192, 224, 237, **244**, **256**

ecosystem 7-8, 10-12, 16, 18, 20-21, 25, 32-36, 40, 89, 111-112, 115-117, 119-121, 123, 125-126, 128, 140, 144, 157, 176, 205, 237-238, 250, 251; assessment and accounting system 53, 112; -based approach 32, 35; -based adaptation 35; condition 36, 112, 119; state of **118**, **139**, **146**, **178**, **224**

ecosystem services 7-8, 21, 32, 36, 44, 49, 51, 53-54, 94, 111-121, 123-127, 130-131, **139**, *143*, **146**, 157, *163*, **180**, 197, **202**, **203**, **225**, 241, **248**, **249**; classification system 53-54

EGR *see* Emissions Gap Report

EIR *see* Environmental Impact Report

EIS *see* Environmental Impact Statement

electricity 40, 43, 72, 74, 77, 82-83, 119, 184-185, 219, 227, 229-230, 236

electrification 40, 44, **76**, 82-83, *86*, 103-105, *142*, **144**, **147**, **150**, 184, **198**, **206**, **208**, **210**, **252**, **254**, **256**, **258**

eliminate *see* avoid

eliminate threat *63*, *75*, **88**, 93-94, **141**, **147**, **190**, **195**, **240**; *see also* mitigation hierarchy

embodied carbon *see* carbon

Emissions Gap Report 21, *23*

endangered species 119, 223

Endangered Species Act 223, 227, 237

end-of-life 224; activities 77; facilities 72; project components 103; carbon *see* carbon; stage *60-61*, *73*, 77, 82, *86*, 103, **179**, **226**; repurposing **186**, **197-198**, **205-206**, **233**, **244**, **254**

energy 12, 24, 33, 40, 42-45, 49-50, 56, 72-73, 77-83, 85, 90-91, 95, 113-114, 123-124, *130*, 143-144, **150**, 156, 162, 176, **180**, 182-185, **198**, **202**, 204, **206**, **210**, 219, 221, 229, 251; efficiency 5, 24, 41, **76**, 82, *86*, 103-105, *142*, **144**, **147**, **150**, 183-185, **198**, **206**, **208**, **210**, 220-221, 229-231, **242**, **252**, **254**, **256**, **258**; -efficient see energy efficiency; generation 78, 82, 162, 204, **210**, 219, 221, 227, 236, **244**, 251, **256**; production 5, 77, **114**, 219; use 10, 33, 79, 81-82, 182, 185, **198**, **206**, **210**, 229, **242**, **252**; clean 8, 24, 82, 184, 217, 227-230, 236; operational 10, 72, 82, 185, **198**, **206**; renewable 40, 78-80, 184, **198**, **206**, 217, 227-229, **242**, 250, **252**; wind 215, 219, 227-229, 236, **244**, **256**

environmental impact assessment 5, 61, 112

Environmental Impact Report 176

Environmental Impact Statement 176

Environmental Product Declaration (EPD) 83, 155

Environmental Protection Agency (EPA) 53

Environmental, Social, and Governance (ESG) 25, 50, 52-54

Envision *see* Envision Rating System

Envision Rating System 5, 10-12, 41, 43, 52, *54*, 75, 128, 162, 173, 229, 232

erosion control 7-8, 113-114, *123*, *130*, *143*, **150**, **180**, 184, 190, **200**, **203**, **208**, **225**, 239, **249**; *see also* ecosystem services

equity 11, 16, 44-45, 64, 115, 125, 137, 156, 175

ESG *see* Environmental, Social, and Governance

European Environmental Agency 53

Everly 216-218, 233, 237

exposure 43, 89, 94-95; to climate-related hazards/threats 88-90, **139**, **146**, 175-176

extended lifespan 96, 230-231, **242**, 250, **252**

extended operating life *see* extended lifespan

F

fail safely *see* safe-to-fail

farming activity 217

farmland 221, 237-238, **248**; prime 221, 237

Federal Aviation Administration (FAA) 226, 240

Federal Emergency Management Agency 177, 226, 234-235, **245**

feeder bay 230, **242**, **252**

FEMA *see* Federal Emergency Management Agency

FEMA 100-year floodplain 226, 234

final service 112, **114**; *see also* ecosystem services

finance gap 24-25

financial risk 25

Flood Insurance Rate Maps (FIRMs) 177, 235

flood mitigation 113-114, *123*, *130*, *143*, **152**, **180**, **203**, **225**, 249; *see also* ecosystem services

floodplain 94, 172-173, 176, 178-179, 181, 188, 190, 192-194, **198**, **202**, 204, **208**, 234, 236, 239, **244**, **254**; *see also* FEMA 100-year floodplain

freshwater body *see* water body

food crops *see* crop

fossil fuel 5, *73*, 78, 82-83, 224, 227, 229

G

General Electric (GE) 216-217, 219, 224, 227, 230, 234, 236, **242**, 250, **252**

GHG emissions *see* greenhouse gas emissions

GHG emissions reduction strategy *see* strategy

GHG inventory 72, 76-78, **139**, **146**, 175

GHG Management Hierarchy *58-59*, 62-63, *65-66*, **76**, 81-82, *86*, **141**, **195**, **240**

GHG reduction target *see* target

GHG removal 83

Glasgow Climate Pact *23*, 34-35

Glasgow–Sharm el-Sheikh work program 92; *see also* Glasgow Climate Pact

Global Biodiversity Framework (GBF) *see* Kunming-Montreal Global Biodiversity Framework (GBF)

Global Biodiversity Outlook 17, 23-24

global climate regulation 113-114, *123*, *130*, *143*, **150**, **180**, **203**, **225**, **249**; *see also* ecosystem services

Global Environment Facility (GEF) 8

Global Goal on Adaptation 92, 115; *see also* Glasgow–Sharm el-Sheikh work program

Global Reporting Initiative (GRI) 53

global temperature rise 17, 21, 79; *see also* global warming

global warming 16-17, 20-21, 35, 79, 102, 120, 224

grazed biomass provisioning *113*, **180**, **202**, **225**, **248**; *see also* ecosystem services

greenhouse gas 4, 16-17, 32, 71-72, 77, 79, 182

greenhouse gas emissions 4, 10, 17-18, 20-21, 24, 33, 36, 40-41, *54*, 56, 58-60, 62-63, *65-66*, 71-72, 74, 76-79, 81, 83, 87, 101, 103, 139, **141**, 147-148, *163*, 169, 175-176, 182, 184, **195**, 224, 229, 231, **240**

Greenhouse Gas Protocol *22*, *54*, 72-74, 77-78

Greening Transportation Infrastructure Development (GRID) 8

Grinnell town **224**

H

habitat 35, 43-44, 49, **114**, 119-121, 125-127, 131, 173, 178-179, 181, 183, 185, 188, 190, 193, **198**, 204-205, **208**, 219, 221, 223-224, 226-228, 237-238, **244**, **246**, **248**, **256**; connectivity 11, 179, 193; conversion 117; degradation 119-120, 162; disturbance 10, 126; enhancement 184, 193, 197, **200**, **208**; fragmentation 6, 10, 120; loss 36, 120; preservation 11, 228; protection 125, 193, **202**; restoration 172-173, 192, **200**, **202**, **208**; biodiversity-rich 33; prime 228

hardscape 173, 230, **242**, **252**

hazard 6, 32, 41, 56, 71, 89-92, 94-96, 98, 123, 139-140, 176; -prone area 93; scenario planning 97; climate-related *58-59*, 88-90, 94, 96, **139**, **146**

hierarchy *see* mitigation hierarchy

high-performance component 82, 219, 230, **242**, **252**; equipment 82; material 95

Hydrologic Engineering Center River Analysis System (HEC-RAS) 177, 192

I

impact 4-7, 11, 16-17, 20, 24-25, 32-36, 40-41, 50-51, 53, 56-57, 60-64, 71-72, 74-75, *78*, 83, 88-91, 93-96, 104, 111-112, 114-121, 123-124, 126-129, 139-141, 146-148, 155, 157, 162-163, 169, 175-176, 181-182, 194-195, 197, 205, **207**, **209**, **211**, 217, 219, 221, 223-224, 226-230, 234, 237-242, **246**, 251-253, **255-259**; driver 51, 55-62, 65-66, 72, *78-79*, 81-87, 137, 148, 154, 163, 205; cumulative 5, 11, 18, 64; net positive 61-62; permanent 18, 126, 172, 181, 192, 228; positive 25, 50, 84, 176; short-term *see* temporary; temporary 126, 172, 176, **179**, 181, 192, 228, 237, 239, **246**, **258**; whole-life 60, *78*, *124*

impermeable surface 94, 188, **202**, 234, 238-239, **246**, **248**, **256**

impervious surface *see* impermeable surface

inclusivity 64, *75*, **88**, 98, 101, 115, 125, *138*, **144**, 147-148, 156, **207**, **209**, **211**, 217, **253**, **255**, **257**, **259**; *see also* core resilience principles *54*, *75*, 94, 97-98, 101, *103*, *130*, 154

indicator 10, 16-17, 21, 52-54, 131; linked 116; performance *55-59*, 87, 131, 148

indirect driver 33, 64, 117, 141

Indonesia 7-8

infiltration 121, 127, 129, 188, 190, 194, 204, 234, 238, **244**, **254**

infrastructure 4-7, 9-11, 35, 40-45, 49-50, 72, 75, 89, 91, 93, 95-97, 100, 112, 141, 172, 177, 221, 224, 234, 237-238, **246**, **248**, **256**; development 4-7, 9-11, 40, 42, 49, 64, 92, 126, 163, 274; footprint 231; investment 10, 42; owner 11, 64, 81, 97, 162; planning 6-8, 11, 41, 64, 98, 221; project 5, 7, 10, 40, 42, 49-52, *54*, 56, 58, 60, 62, 64, 72, 75-76, 78, 88, 90, 92, 94, 98, 111-112, **114**, 119-121, 123, 155, 205, 251; resilience 97; sector 7-8, 81; service 42, 74, 90, 93; system 42, 49, 60, 64, 74-75, 89, 90, 92-94, 97; blue 115; energy 230, 250; green 44, 115; grey 5, 7, 44, 94, 127; natural 42; nature positive 44, 117, 126; regenerative 12; road 8, 78; resilient 40, 42, 49, 74; social 45; sustainable 5-6, 11-12, 40-42, 162-163, 229; transportation 8; solar 229, **242**, **252**

innovation 11-12, 32, 40-41, 44, 56, 64, 81, 95, 100, 102, 156, 186, 188, **206**, 230

Institute for Sustainable Infrastructure (ISI) 5, 10-11, 52, *54*, 162

Institute of Environmental Management and Assessment (IEMA) 62-63, 81

integrated action *see* integrated climate-biodiversity action

integrated action criteria *see* integrated climate-biodiversity action criteria

integrated approach 32-34, 49

integrated climate-biodiversity action 32-33, 49-52, 64-65, 115, 117, 126, 137, 141, 148, 154-156, 163, 197, 205, 241, 251

integrated climate-biodiversity action criteria *52*, *54*, *66*, 115, 138, **146**-**147**, *163*, 197, 205-211, 241, 251-259

integrated criteria *see* integrated climate-biodiversity action criteria

integration 75, **88**, 97, *99*, 101, *103-104*, 128-130, *142*, 144, **147**, **150**, **207**, **209**, **211**, **253**, **255**, **257**, **259**; *see also* core resilience principles

Intergovernmental Panel on Climate change (IPCC) 16, **20**, 22-24, 33, 71, 77, 79, 89-90, 100

Intergovernmental Science-Policy Platform on Biodiversity and Ecosystem Services (IPBES) 17-18, 20-21, *23*, 33-34, 53, 112, 119, 121

intermediate service 112, **114**; *see also* ecosystem services

International Energy Agency (IEA) *22*

International Union for Conservation of Nature (IUCN) *22*, 24, *54*, 61-62, 115-116, 119, 126-127, 131, 156

Interstate Power and Light (IPL) 216-217, 221, 224, 226-228, 230, 233, 238

introduction of invasive species 118-119, 121-122, *124*, 126, *130*, 139, *142*, **146**, **151**, **153**, **179**, 181, 194, **196**, **199**-**201**, **207**, **209**-**211**, **226**, **241**, **243**, **245**, **247**, **253**, **255**, **257**, **259**; *see also* biodiversity

invasive species 21, 117, 120-121, 129, **139**, **152**, *154-157*, 194, **200**, 204, **210**

investment gap 25, 42

investor 25, 53, 217

Iowa 162, 216-217, 219, 227-228, 233

Iowa Department of Natural Resources (IDNR) 221, 228, 236

IPBES-IPCC Joint Report *23*, 33-34, 115, 117, 121, 137

irrigation **114**, *123*, *130*, *143*, **150**, 173, **180**, 183, 190, **200**, **202**, 204, **208**, **225**, 236, **248**

IUCN Red List of Threatened Species 119

J

joint action *see* integrated climate-biodiversity action

Jurisdiction Delineation Report 223

K

Kunming-Montreal Global Biodiversity Framework (GBF) 20-21, 23-24, 32, 34-35, 112, 121, 125, 140, 156

Kyoto Protocol *22*, 77

L

Land-Based Wind Energy Guidelines (WEG) 224, 227-228

land/freshwater/sea change **118**, 120, *122*, *124*, *130*, 138-139, *142*, **151**, 153, **179**, **199**, **201**, **207**, **209**, **211**, **243**, **245**, **247**, **253**, **255**, **257**, **259**; *see also* biodiversity

landscape 7, 10-11, 44, 119, 186, 190, 223, 239

landslide mitigation 113-114, *123*, *130*, *143*, **150**, **180**, **203**, **225**, **249**; *see also* ecosystem services

LCA *see* lifecycle assessment

Leadership in Energy and Environmental Design *see* LEED

LEED 219, 229

Life Cycle Analysis *see* lifecycle assessment

lifecycle 6, 60, 72, 181, 183, 224; assessment (LCA) 84, 224, 272; considerations 60, 65; cost 43; savings 44; stage *61*, 72, 80, *86*; infrastructure project 40-41, 49, 60, 72, 74, 77, 82, 120, 125, 228, 274

light pollution **118**, 121-122, *124*, 126, 129-130, 138-139, *142*, **146**, 151-157, **179**, 181, **196**, **199**, **201**, **207**, **209**, **211**, 226, **226**, **240**-**241**, **243**, **245**-**247**, **253**, **255**, **257**-**259**; *see also* biodiversity

light spill *see* light pollution

Little Sioux River 222-223

livestock provisioning *113*, **180**, **202**, **225**, **248**; *see also* ecosystem services

local climate regulation 113-114, *123*, *130*, *143*, **150**, **180**, **203**, **225**, **249**; *see also* ecosystem services

local tax benefits 219

M

Major Investment Study 175

maintenance 8, 41, 44-45, 60, 72-73, 77, 80, 82, 84, 91, 95, 100-101, 121, 179, 183, 194, 204, 221, 224, **226**, 230-231, 236, 250; long-term 11, 43, 230, **242**, **252**; predictive 230, 239, 250; preventive 41, 230, **242**, **252**; proactive 96; routine 60, 96

material 10, 25, 40-41, 43, 45, 49, 56, *60*, 72-73, 77, 80-85, **88**, 90-91, 95-97, 101-103, **114**, 121, 127, 129, 138-139, 144, **146**, **151**, 153-157, 183, 185-186, 194-195, 197-199, **201**, 205-207, **209**, **211**, **224**, 231-233, **240**-**245**, **247**, 250-255, **257**, **259**; efficiency 156; extraction 49, *60*, 72, 77, 83, 120, 155; passport 84; production 72-73; sourcing 60, **76**, 83, *86*, 103-105, *142*, 144, **147**, **150**, 154-155, 157, 205-206, **208**, **210**, 227, 251-252, **254**, **256**, **258**; excavated 233, **244**, **254**; locally-sourced 80, 85, 185, **198**, 232, 241, 251, **254**; recovered 84; recyclable 186, **198**, **206**; recycled 41, 95, 231, **254**; recycled-content 95, 183, 186, **198**, **206**; regional 85, 185, **198**, 232, 241, 251; substitute 84, 95

MET weather stations 230

meteorological evaluation towers 219

metric 11, 24-25, 43-44, 53, 72, 182

Midcontinent Independent System Operator (MISO) 221

Migratory Bird Treaty Act 223

Millennium Ecosystem Assessment 17, 18, 22, 53, 112

minimize 32, 41, 56, 61-63, 81, 84, **88**, 93, *99*, 116-118, 120-121, 126, **141**, **147**, 155, 183, 188, 190, 192-193, **195**-**196**, 202, 217, 219, 221, 224, 228, 231, 236-242, **246**, 251-252, **256**; *see also* mitigation hierarchy

mitigation hierarchy 41, 55-56, *58-63*, 65, 75-76, 79, 81-82, *86*,

88, 92-94, *99*, 116-118, 125-126, 128, 131, 141, 147-148, 181, 195-196, 240-241

monoculture 33

monitoring 21, 55, **76**, 87, 96, 100, **118**, 128-129, 131, **145**, *163*, 173, 181, 204, **210**, 228, 230-231, 234, 237, 239, **242**, **246**, 250, 252, **256**, **258**; *see also* Track

monitoring data 56, 100, 230, 250

Mori Prairie 222-224, 237

Municipal Separate Storm Sewer System (MS4) 183, 188

Muwekma Ohlona Indian Tribe 171

N

National Biodiversity Strategies and Action Plans (NBSAPs) **20**, 32

National Determined Contributions (NDCs) 20-21, 32

National Ecosystem Services Classification System (NESCS) 53

National Pollutant Discharge Elimination System (NPDES) 183

National Wetlands Inventory (NWI) 223

Natural Resource Conservation Service (NRCS) 223

Nature Conservancy 223, 237

nature's contributions to people (NCPs) 18, 32, 53, 112, 125

nature-based solutions (NbS) 6-9, 33, 36, 43-44, 50-51, *54*, 115-116, **118**, 126-128, *130*, 143-144, **147**, **150**, **152**, 156-157, 197, **207**, **209**, **211**, 241, **253**, **255**, **257**, **259**

nature-positive 7, 9-11, 44, 60, 62, 84, 117, 126, 140-141, 155, 162-163

nature-related financial risk 25

NEPA *see* US National Environmental Protection Act

net gain 56, 61, 115, *117*, 125-127, 137

net negative emissions **76**, 81, **141**, 224, 229

net positive 60-62, 82, 115, 126

net resilience gain 75, **88**, 92-93, **140**, **146**

net-zero 7, 10-11, 21, 42-45, 49, 60-62, 71, 79, 82-84, 93, 141, 155, 162-163

nitrogen 4, 32, 77, 121, 137

no net loss 56, 61-63, *117*, 125-127

no resilience loss 75, **88**, 93, 137, **140**

noise attenuation 113-114, *123*, *130*, *143*, **152**, **180**, **203**, **225**, **249**; *see also* ecosystem services

noise pollution **118**, 121-122, *124*, 126, 129-130, 138-139, *142-143*, **146**, **151**, 153-157, 176, **179**, **196**, **199**, **201**, **207**, **209**, **211**, 226, **241**, **243**, **245**, **247**, **253**, **255**, **257**, **259**; *see also* biodiversity

nursery population and habitat maintenance 113-114, *123*, *130*, *143*, **180**, **203**, **225**, **249**; *see also* ecosystem services

O

Ocheyedan River 222-223

offset 11, 61-63, 72, **76**, 78, 81, 116-118, 126, **141**, **147**, 172, **195-196**, **240-241**; *see also* mitigation hierarchy

operating life 229, 230, **242**, 250, **252**

operation 8, 11, 25, 41, 63, 72-73, 77, 80-82, 91, 93, 95, 97-98, 100, 121, **179**, 183-184, 186, 221, 224, 231, 233, 236, 250; stage *60-61*, *73-74*, *86*

operational carbon *see* carbon

operational energy *see* energy

operational waste *see* waste

Operations Center *see* Spencer Operations Center

opportunity 7-8, 100, *103-105*, 128, *142-143*, 145, *149*, 154, 156-157, 162-163, 181-182, 186, 197-198, 205-206, 241, 251

Opportunity Matrix 65, 148-154, 197, 205, 241, 251

Organisation for Economic Co-operation and Development (OECD) 8, 93

P

Paris Agreement 17, 20-21, 23-25, 32, 34-36, 42, 50, 64, 71, **76**, 79, 82, 92, 102, 111, 140, **146**, 156

particulate matter (PM) 121, 175

pathway 64, 79; 1.5°C 21, 83; 2°C 21; business-as-usual 32; deep decarbonization 82; Paris-aligned 24; policy 33

peak flow mitigation **180**, **203**, **225**, **249**; *see also* water flow regulation; ecosystem services

permeable pavement *see* permeable surface

permeable surface 44, 185, 188, 194, **198**, **202**, **208**, 238

persistent organic pollutants (POPs)

pest control 36, 113-114, *123*, *130*, *143*, **152**, **180**, 194, **203**, **225**, **249**; *see also* ecosystem services

Philippines 7-8, 43-44

phosphorus 121, 238

physical asset risk *see* risk

plan-do-check-act 11, 56, 100

planetary boundaries 4, 10

Playbook for Nature-Positive Infrastructure Development 117, 127

pollination 36, 113-114, *123*, *130*, *143*, **180**, **203**, **225**, **249**; *see also* ecosystem services

pollution control 194, 239

Poweshiek County 162, *216*

prairie 223-224, 237-238, **246**, **248**, **256**

predictive maintenance *see* maintenance

pressures on biodiversity *see* biodiversity

Pressure-State-Response framework 56

prevention, abatement, and control measure 117-118, 127-130, 144, **147**, **152**, 205, **207**, **209**, **211**, 251, **253**, **255**, **257**, **259**

preventive maintenance *see* maintenance

prime farmland *see* farmland

prime habitat 228

Principles for Resilient Infrastructure 75, 92

procurement 7-9, 11-12, 43-44, 82, 183-184, 233; sustainable 233, 241, **244**, 251, **254**

Product Category Rules 155

Product Environmental Footprint (PEF) study 155

product stage *60-61*, 72-74, *86*, 144

provisioning services 112-114, 124, **180**, **202**, **225**, **248**; *see also* ecosystem services public engagement 64, 217; *see also* stakeholder engagement; inclusivity

Q

qualities of resilient systems 75, 94

R

rainfall pattern regulation 113-114, *123*, *130*, *143*, **150**, **180**, **203**, **225**, **249**; *see also* ecosystem services

recover 63, 74-75, **88**, 93, *99*, 141, **147**, **195**, **240**; *see also* aspects of resilience

recyclable *see* material

recycled *see* material

recycled-content *see* material

recycling 41, 78, 84, 96, 186, **198**, **206**, 231, 233, **244**, **254**

reduce *63*, **76**, 81, *86*, 126, **141**, **147**, **195**, **240**; *see also* mitigation hierarchy

reduce exposure/vulnerability to threat *63*, *75*, **88**, 93, *99*, **141**, **147**, **195**, **240**; *see also* mitigation hierarchy

redundancy 74-75, **88**, 96-97, *99*, 101, 103-105, *142*, **144**, **147**, **150**, **207**, **209**, **211**, 236, **244**, **253**, **255**-**257**, **259**; *see also* core resilience principles

reflective capacity *75*, **88**, 95-96, 98, 100-101, *138*, 141, 145, 147-148, 156, **207**, **209**, **211**, 250, **253**, **255**, **257**, **259**; *see also* core resilience principles

reforestation 33, 83, 127

refurbishment 72, 84, 96

Regional Transmission Organization 221

regulating and maintenance services 19, 112-114, 124, **180**, **203**, **225**, **249**; *see also* ecosystem services

rehabilitation 60, *73*, 77, 80, 91, 115

remanufacturing 84

renewable energy *see* energy

renewable energy capacity *see* energy

Renewable Energy Certificate (REC) 82

Renewable Portfolio Standard (RPS) 227

reparability 84, 95, *102*, 155

replacement 58, 60, 72-73, 90, 95, 101, 172

reputational risk 91, 104

research and development (R&D) 63, 81, 141

resilience 7-8, 10-12, 32, 35-36, 40-41, 43-44, 62, 74-75, 89, 91-94, 96-98, 100, 102, 111, 155, 234; hierarchy *58-59*, 75, **88**, 92-94, *99*, **141**, **195**, **240**; management steps 63; of ecosystems 112, 115, 125; strategy *see* strategy; target *see* target

resilient infrastructure 40, 42, 45, 49, 74-75

resilient system 75, 92, 94

resist 36, 74-75, 93, 95; *see also* aspects of resilience

resource 11, 97, 101, 104, **114**, 119, 155-156, 181; availability risk *see* risk; circularity 64, **76**, 84, *86*, 95, 102-105, 128, *130*, *142*, 144, **147**, **150**, 154-157, 205, **206**, **208**, **210**, 251-252, **254**, **256**, **258**; consumption 101; efficiency 75, 102, 155, 156, 183; exploitation 118-120, *122*, *124*, *126*, 128-130, 138-139, *142*, 144, **146**, **151**, 153-157, **196**, **199**, **201**, **207**, **209**, **211**, **241**, **243**, **245**, **247**, **253**, **255**, **257**, **259** *see also* biodiversity; diversification 102; natural 19, 25, 40-42, 60, 64, 84, 89, 91, 95-96, 120, 156, 182

resourcefulness 75, **88**, 95, *99*, 101-104, 128-130, *142*, 144, **147**, **150**, 154, 156-157, 205, **207**, **209**, **211**, 251, **253**, **255**, **257**, **259**; *see also* core resilience principles

Respond 55-59, **76**, 81-82, **88**, 93-95, 97, **118**, 126-127, 138, **141**, **144**, **147**, 162-163, 184, 229, 234; *see also* action step

response criteria 55, 61-62, 65-66, 72, 75, 82, 94, 115, 127, 137-138, 141, 144, 148-150, **152**, 154-155, 163, 205-211, 251-259; *see also* Respond

responsible material sourcing *see* material sourcing

restoration 6, 8, 21, 36, 83, **114**, 116, 126, 140, 232

ecosystem restoration *23*, 35, 40, 64, 115, 127, 140; of disturbed soils 129, 238, **246**, **256**; river 127; creek 162, *167*, *170*, 172-173, 178-179, 184, 186, 188, 193, 195, **202**, 204; habitat 173, 192, **200**, **208**

restore 61-63, *75*, **88**, 93, 95, 98-99, 112, 115-118, 126-127, **141**, **147**, **195**-**196**, **240**-**241**; *see also* mitigation hierarchy

retention and breakdown of nutrients **180**, **203**, **225**, **249**; *see also* ecosystem services

reusability *see* reuse

reuse 41, 78, 80, 84-85, 95-96, 103, 186, 188, 190, 194, **198**, **206**, 233, **244**, **254**

revegetation 173, 185, 190, *192*, **198**, **200**, **202**, **206**, **208**

Rio Earth Summit 16, *22*, *34*

risk 6, 16, 18, 24-25, 33, 41-43, 50, 55-62, 64-65, 83, 89-93, 97-98, 100-102, 104, 111-112, 120, 123-124, 128, 137, 139, 145, 155, 157, 163, 176-177, 190, 205, 223-224, 226, 234, 237-238, 251; assessment 41, 176, 228, 234; control 44; exposure 16, 25, 91, 175, 195, 240; identification 64, 90; management 25, 40, 53, 89, 96, 98, 100; mitigation 42, 62, 194, 234; profile 94; reduction 7-8, 32, 36, 40, 43-44, 94, 98, 111, 145, 227; climate-related 25, 33, 41-42, 53-54, *56-59*, 62, *65*, 88-100, 102-104, 137, 139, *143*, 145-146, 148-149, **151**, 153-155, 163, 175-176, **195**, 197-201, 205, 226, 240-248, 251; disaster 17, 33, 89, 115; nature-related 25; physical asset **88**, 90, *92*, 94-96, *99*, *104*, 138-139, *143*, **146**, *154-157*, **195**, **240**; resource availability **88**, 90-92, 95-99, 102, *104*, 120, 138-139, *143*, **146**, **151**, 153-157, **195**, **199**, **201**, **240**, **243**, **245**, **247**; service continuity **88**, 90, *92*, 96-99, *104*, 138-139, *143*, **146**, **151**, 153-157, **176**, **195**, **199**, **201**, **240**, **243**, **245**, **247**; supply chain continuity **88**, 90-92, 96, *99*, 102, 104, 138-139, *143*, **146**, **195**, **240**

Rockefeller Foundation *54*, 75, 94

rotor 219

RPS *see* Renewable Portfolio Standard

runoff 40, 120, 128, **179**, 188, 194, 202, 226, 234, 236, 238-239, **246**, 250, **258**; control 128, 183, 234, **244**, **254**; reduction and treatment 194, **200**, **210**

S

safe-to-fail 95, 234, **244**, **254**

San Francisco Bay 167-169, 171-172, 175-177, 182

San Francisco Estuary Institute 173, 179

San José *see* City of San José

Santa Clara County 162, 166-169, 179, 181-182

Santa Clara Valley Transportation Authority (VTA) 162, *166*, 182-183, 186, 188, 204

Santa Clara Valley Water District (SCVWD) 177-179

Science Based Targets Network (SBTN) 53-54, 56, 61-62, 126

science-based target 11, 53, 79

Science-Based Targets (SBTi) *54*

Science-Based Targets Network (SBTN) 53-54, 56

Scope 1 GHG emissions 72-74, 76-79, 82-83, *86*, *103*, *105*, 138-139, *143*, **146**, **150**, **152**, *154-157*, 185, **195**, **198**, **200**, **206**, 229, **240, 242, 244, 246, 252**

Scope 2 GHG emissions 72-74, 76-78, 80, 82-83, *86*, *103*, *105*, 138-139, *143*, **146**, **150**, **152**, *154-157*, **195**, **198**, **200**, **206**, **240**, **242, 244, 246**

Scope 3 GHG emissions 72-74, 76-78, 80, 82-86, *103*, *105*, 138-139, *143*, **146**, **150**, **152**, *154-157*, **195**, **198**, **200**, **240**, **242**, **244**, **246**

Scope 1 emissions *see* Scope 1 GHG emissions

Scope 2 emissions *see* Scope 2 GHG emissions

Scope 3 emissions *see* Scope 3 GHG emissions

SDG *see* Sustainable Development Goal

sediment control 194, 239, **246**, **258**

Sendai Framework for Disaster Risk Reduction *23*, 35, 89

setback 217, 221, 224, 226-228, 234, 237-238, **244**, **246**, **248**, **256**; riparian 183, *187*, 192, **202**

shock *see* acute shock

Silicon Valley 169

Silicon Valley BART extension project *see* BART Silicon Valley Berryessa extension (SVBX) project

Silicon Valley extension project *see* BART Silicon Valley Berryessa extension (SVBX) project

Silicon Valley Rapid Transit Corridor 168-169, 175

site characterization 223

site evaluation 223

siting 10, **88**, 93-94, *99*, *104*, 117, 128, *130*, *142*, 144, **147**, **150**, **152**, 162, 183, 192, 197, **200**, 205, **207**-**209**, **211**, 221, 223, 226-228, 234, 237, 239, 241, **244**, **246**, 251, **253**, **255-259**

social safeguards 60, 64, 156

soil and sediment retention 8, **180**, **203**, **225**, **249**; *see also* ecosystem services

soil erosion control 8, 113-114, *123*, *130*, *143*, **150**, **180**, **203**, **225**, **249**; *see also* soil and sediment retention; ecosystem services

soil pollution **118**, 121-122, *124*, 129-130, 138-139, *142*, **146**, 151-157, **179**, 194, **196**, **199-201**, **207**, **209-211**, **226**, **241**, **243**, **245**, **247**, **253**, **255**, **257**, **259**; *see also* biodiversity

soil quality regulation 113-114, *123*, *130*, *143*, **150**, **180**, **203**, **225**, **249**; *see also* ecosystem services

solid waste remediation 113-114, *123*, *130*, *143*, **150**, **180**, **203**, **225**, **249**; *see also* ecosystem services

species 18, 20, 33, 111, 116-117, 119-121, 125-126, 128, 131, 190, 193, **202**, 204, 223, 237; abundance 112, 120; alien *see* invasive species; avian 237, **246**, **256**; behavior 121; distribution 120; disturbance 224; diversity **114**, 120; endangered 223; extinction 18, 32, 89, 94, 120; population 6, **114**, 119-120, 126; -richness 10, 115; migration 11, 44; avian 237, **246**, **256**; native 115, 121, 125, 129, 193, 236; non-native 179, **202**; protected 227, 228; sensitive-status 129, **178-179**; special status 176, 179; state of 118-119, **139**, **146**, **178**, **224**; threatened 18, 21, 125, 180-181, 223

special flood hazard area *177*, *235*

Spencer *216*, *218*, 239

Spencer Operations Center 218-219, 221-222, 229-231, 233, 235-236, 238, 240, **242**, **246**, 250, **252**, **258**

Spicer Substation 218-219, 221-222, **224**, 230

SRI value 230, **242**, **252**

stakeholder engagement 44, 60, 64, 98, 148, 162, 221, 228, 232; *see also* inclusivity

state of biodiversity *see* biodiversity

Stony Creek 222-223, 236, **244**, **256**

storm mitigation 113-114, *123*, *130*, *143*, **152**, **180**, **203**, **225**, **249**; *see also* ecosystem services

stormwater 44, 173, **176**, 188, 194, 226, 239, **246**, **258**; basin *see* bioretention basin; control 184, 188, 197-198, **200**, 205, **208**, 234, 241, **244**, 251, **254**; detention pond 234, 236, **244**, 250, **254**; discharge 183; management 183, 234, 239, 250; overflow 89; runoff 128, 188, 194, **202**, 234; treatment 194

strategy 11, 36, 40-41, 50-51, *54-57*, 59-61, 64-66, 72, 77, 82-83, 85, 87, 89, 100-104, 112, 115-116, **118**, 120, 127-131, 137, *142*, 144-145, 147-149, 154-157, 162-163, 169, 173, 175, 181-182, 184-185, 190, 195, 197-202, 205-211, 226, 229-231, 240-247, 250-259; -specific target 56-57, 79, 87; biodiversity *58-59*, *66*, **118**, 127, *130*, 137, 144, 148-150, **152**, 162; climate risk reduction *58-59*, *65-66*, **88**, 94, *99*, *103-104*, **144**, 147-150; GHG emissions reduction *58-59*, **76**, *86*, *103-105*, 127; impact-specific *61*, 79; resilience 75, 94, 98-100, 102-104

stream corridor 173, 223, **224**

stress *see* chronic stressor

subset of criteria 55, 58, 76-77, 79, 81-82, **88**, 90, 93, 118-119, 123, 125, 127, **146**, 148

substitute material *see* material

supply chain 83-84, 91, 98, 126, 141, *155*; continuity risk *see* risk; disruption 90, 91

sustainability 6, 11, 25, 40-44, 51-52, 60, 112, 155-156, 162-163, 169, 183, 227-228, 230, 233; strategy *see* sustainable strategy

Sustainable Development Goal 16, 18-19, 21, *23*, 25, 34-35, 40-42, 50, 53, 92, 115

sustainable infrastructure *see* infrastructure

Sustainable Infrastructure Program in Asia (SIPA) 6-7

sustainable procurement *see* procurement

sustainable strategy 41, 83, 117, 155, 163, 184, 205, 229, 251

Sustainable Urban Drainage Systems (SuDS) 40, 127

sustainable use 16, 19-20, 36, 111

synergy 32-33, 35, 72, 135, 221

system recovery **88**, 94, 97, *99*, *103-104*, *142*, **144**, **147**, **150**, **207**, **209**, **211**, **253**, **255**, **257**, **259**

T

target 17-18, 21, 24, 32, 35-36, 50, 55-62, 64, 76-77, 79-80, 87-88, 92-93, 100, 112, 117-118, 121, 125, 128, 131, 140, 145-148, 156, 182, 194, 204, 227-228; -setting process *54*, 56, 60; adaptation 35, 92, 140; biodiversity 11, 24, 33, *58-59*, **118**, 124-125, **140**, **146**-**147**; carbon 11; climate 33, 35, 42, 52, 56, 100, **146**, 156; GHG reduction *58-59*, **76**, 79, **140**; interim 79, 125; long-term 79, 87, 182; impact-specific *57*, *61*, 87; measurable 56, 79, 87; mitigation

64, 156, 183; nature 52; nature-positive 61; near-term 79, 87; net gain 127; net positive 61, 115; net-zero 41, 61, 75, 79, 93, 141; no net loss 127; Paris 17, 35; performance 56, 80; resilience 11, *58-59*, **88**, 92-93, 98, **140**; science-based 11, 53, 79; strategy-specific 56-57, 87; sustainability 227; temperature 21; thematic 56, 140; whole-life *61*

Taskforce on Climate-related Financial Disclosures (TCFD) **20**, *23*, 25, 53-54, 58, 89-91, 95

Taskforce on Nature-related Financial Disclosures (TNFD) **20**, *23*, 25, 53-54, 61, 111-112

Terry Kouba 219

The Economics of Ecosystems & Biodiversity (TEEB) 53

threat 16-18, 20, 32, 88-90, 92-94, 98, 120-121, 139, **144**, **147**, 177, **207**, **209**, **211**, **253**, **255**, **257**, **259**; climate-related 56, *58-59*, 88-90, *92*, 120, 139, **146**, 175-177, **195**, 226, **240**

threatened species *see* species

Tire-Derived Aggregate (TDA) 186, 188

Track 55-59, 76, 87-88, 91, 100, **118**, 131, 138, **145**, 147-148, 162-163, 183, 204, 250; *see also* action step

tracking criteria 55, 148; *see also* Track

tracking mechanism *58-59*

training 8, 43-45, 98, 221, 236, **244**, **256**

transform 11, 33, 44, 50, 61-64, 74-75, 79, 81, 84, *86*, **88**, 93, 99-100, *102*, 116-118, 126, 138, 141, 147-148, 154-156, **195-196**, **240-241**; *see also* mitigation hierarchy

transformative action *see* transform

transit 43-44, 78, 162, 167, 169, 171-172, 175-176, 182, 184, 195, **198**, 205-206; center 167, 169-171; network 12, 167, 176, 182; node 169; -oriented communities 182; -oriented development 169; station 85; urban village 169; *see also* Berryessa Transit Center

transition 12, 25, 33, 40, 44, 49, 58, 82, 87, 182, 219, 227; scenario 82; clean energy 82; low-carbon 24, 91

triple-bottom-line supplier vetting strategy 233, **244**, **254**

U

UAE Framework for Global Climate Resilience *23*, *34*, 140, 156

UK National Infrastructure Commission 74, 93

UN 2030 Agenda for Sustainable Development 18, *23*, *34*

UN Millennium Development Goals (MDGs) *22*, *34*

UN SEEA-EA classification 51, 53-54, 112-114, 123-124

UN System of Environmental Economic Accounting—Ecosystem Accounting (SEEA-EA) *see* UN SEEA-EA classification

UNEP World Conservation Monitoring Centre (WCMC) 32

Union Pacific Railroad (UPRR) *168*, 175, *178*, 179, 188

United Nations 16, 18, *22*

United Nations Environment Programme (UNEP) 8, 16, 21-24, 53, 231

United Nations Framework Convention on Climate Change (UNFCCC) 16-17, **20**, *22-23*, 32, 34, 71, 77, 79, 115

United Nations Global Compact (UNGC) 62, 126

United Nations Office for Disaster Risk Reduction (UNDRR) *22*, 75, 92

United Nations Office for Project Services (UNOPS) 40, 50

United States 43, 223, 227, 233

United States Geological Survey (USGS) 223

upfront carbon *see* carbon

upgradability 84, 95

Upland Prairie wind farm 162, 216-220, 223-224, 226-229, 231-234, 237, 239, 241, 250-251

Upper Penitencia Creek 169-172, 179, 184, 187-190, 192-193, **202**, 204; improvement project 170-173, 178-179, 181, 185, *187*, *191*, 193, 204, **210**; restoration project *see* improvement project

upstream emissions *73-74*, 76-78, 80, 83-86, 102-103, *105*, 138-139, *143*, **146**, **150**, **152**, *154-157*, **195**, **198**, **200**, **240**, **242**, **244**, **246**; *see also* Scope 3 GHG emissions

US Army Corps of Engineers (USACE) 7, 179

US Fish and Wildlife Service 181, 223-224, 227-228, 237

US National Environmental Protection Act 175-176

use stage embodied carbon *see* carbon

USFWS Eagle Conservation Plan Guidance 228

V

VTA *see* Santa Clara Valley Transportation Authority

VTA's Climate Action and Adaptation Plan 182

VTA's Sustainability Plan 182

VTA's Sustainability Program 182

volatile organic compounds (VOCs) 121, 129, 183

vulnerability 7, 40-41, 71, 75, 88-94, 98, 102-104, 141, **147**; assessment **88**, 121, **139**, **146**

W

waste 19, *49*, 56, *73*, 78-79, 84-85, 121, 155, 182, 186, 233, 239; disposal *60*, 72-73, 78, 186; diversion 80, 129, 186, **198**, **206**, 233, **254**; pollution **118**, 121-122, *124*, 126, *130*, 138-139, *142*, 144, **146**, **151**, 153-157, **179**, **196**, **199**, **201**, **207**, **209**, **211**, **226**, **241**, **243**, **245**, **247**, **253**, **255**, **257**, **259** *see also* biodiversity; processing *60*, 72-73, 78; reduction 129, 155, 183, 186, 231, 239; construction 186, **198**, **206**; operational *73*, 186, **198**, **206**, 233, **244**, **254**

wastewater 11, *73*, 79, 95, 121; treatment 78, 127-128, *130*, **152**

water 32, 35, 43, 85, 90-91, 95, 102, 112, **114**; availability 18, **88**, 90-91, 95-96, 120, 138-139, **146**, **151**, 153-157, **195**, **199**, **201**, **240**, **243**, **245**, **247**; water body **114**, 120, 127-128, **178**, 184, 188, 192, 197, **200**, 205, **208**, *222*, 226-228, 237-238, 241, 246, 248, 251, **256**; circularity 85, 95, 128, 156; conservation 95, 100, 190, 236, **244**, **256**; demand 91, 236, 238; efficiency 42, 102, 183, 190; flow regulation 113-114, *123*, *130*, *143*, **150**, **180**, 183, **203**, **225**, **249** *see also* ecosystem services; insecurity 71; management 10, 92, 239; pollution **118**, 121-122, 124, 128, *130*, 138-139, *142*, 144, **146**, **151**, 153-157, **179**, 194, **196**, **199-201**, **207**, **209-211**, 224, **226**, **241**, **243**, **245**, **247**, **253**, **255**, **257**, **259**; resource 18, 91, 176, 181-183, 236; purification 36, 113-114, *123*, *130*, *143*, **152**, **180**, **203**, **225**, **249** *see also* ecosystem services; quality 18, 40, 56, 91, 100, **114**, 176, 180-181, 184, 192-194, 204, **225**, 236, 238-239, 241, **246**, 250-251, **256**; -related conflict 18; security 115; supply 8, 95, 111, 113-114, *123*, *130*, *143*, **150**, **180**, **202**, **225**, **248** *see also* ecosystem services; scarcity 89, 91; shortage 91; stress 91, 93, **176**; treatment 12; use 56, 120, 190, **200**, 204, **208**, **210**, 219, 236, 238, **244**, **256**; non-potable 113-114, *123*, *130*, *143*, **150**, **180**,

202, 204, **225**, **248**; potable 113-114, *123*, *130*, *143*, **150**, **180**, 190, **200**, **202**, 204, 208, **225**, **236**, **244**, **248**, **256**

waterway 228, 238, **246**, **256**

weather extreme 6, 17, 36, 40, 42, 71, 89-90, 111, 231, 234, 236, **244**, **254**

Wetland Delineation Report 223

wetland 10, 18, 36, 43, 111, 127-128, 172-173, 176, 183-185, 188, 190, 192-193, 197-198, **200**, **202**, 204, 205, **208**, 219, 223, 226-228, 237-238, 241, **244**, **246**, 251, **256**

White Construction Inc 216, 232-233, 239

whole-life carbon *see* carbon

whole-life impact *see* impact

whole-lifecycle approach 60

wildlife 6, 18, 49, 125, 127, 183, 193, **200**, 204, **210**, 223, 227, 237, **244**, **248**, **256**; analysis 221; bridge 127; dispersal *see* movement; exclusion fencing 193, **200**, **210**; movement 179, 181, 223, **224**; refuge 128

wind energy *see* energy

wind farm *see* Upland Prairie wind farm

wind power *see* energy

wind turbine 216-219, 221, 223-224, 226-231, 233-234, 236-240, **242**, **244**, **246**, 250, **252**, **254**, **256**, **258**

wood provisioning 113-114, *123*, *130*, *143*, **150**, **180**, **202**, **225**, **248**; *see also* ecosystem services

workforce 221; availability **88**, 90-91, 98, **139**, **146**, *154-157*, **195**, **240**; commuting *60*, 85; skills development 43, **88**, 91, 94, 98, 138, **144**, 147-148, **150**, **207**, **209**, **211**, 221, **253**, **255**, **257**, **259**; training programs *see* training

World Economic Forum (WEF) 18, 24, 53, 89

World Green Building Council (WGBC) 72-74, 78

World Meteorological Organization (WMO) 16-17, *22*

World Wildlife Fund (WWF) 7-8, 117, 127

Z

zero-emission 43, 78, 80, 184-185, **198**, **206**

zero-ready project 80

zero-waste *102*, 155

For Product Safety Concerns and Information please contact our EU representative GPSR@taylorandfrancis.com Taylor & Francis Verlag GmbH, Kaufingerstraße 24, 80331 München, Germany

Batch number: 10733645

Printed by Printforce, the Netherlands